Emergency Handling of the “5.12” Tangjiashan Barrier Lake

Qigui Yang · Wenjun Yang

Emergency Handling of the "5.12" Tangjiashan Barrier Lake

长江出版社
CHANGJIANG PRESS

Springer

Qigui Yang
Changjiang Institute of Survey, Planning, Design and Research
Wuhan, China

Wenjun Yang
Changjiang River Scientific Research Institute
Wuhan, China

ISBN 978-981-99-1705-1 ISBN 978-981-99-1706-8 (eBook)
https://doi.org/10.1007/978-981-99-1706-8

Jointly published with Changjiang Press (Wuhan) Co., Ltd.
The print edition is not for sale in China (Mainland). Customers from China (Mainland) please order the print book from: Changjiang Press (Wuhan) Co., Ltd.

The translation was done with the help of an artificial intelligence machine translation tool. A subsequent human revision was done primarily in terms of content.

This Springer imprint is published by the registered company Springer Nature Singapore Pte Ltd.
The registered company address is: 152 Beach Road, #21-01/04 Gateway East, Singapore 189721, Singapore

Preface

Barrier lake used to be an unfamiliar term for ordinary people which was only known by professionals before Wenchuan earthquake on May 12, 2008. In fact, barrier lake is a common geological disaster. It is the natural damming of a river by collapse or slide of the bank slope of river valley induced by earthquake or precipitation. The water plugged by barrier dam creates a lake with hazards in two aspects: the loss caused by the rise of water level and flooding of the upstream area, and the damage to the downstream area due to the overflow of the lake or even dam-break flood. Historical records on the earthquake-induced barrier lake-break show that the number of casualties caused by the dam-break flood was far greater than that by the earthquakes themselves.

A total of 136 barrier lakes were formed after a devastating earthquake at magnitude 8.0 struck Wenchuan, Sichuan Province on May 12, 2008. Among them Tangjiashan barrier lake, which was in a state of extremely high hazard, threatening the lives of residents and important infrastructures downstream. In disaster relief of Tangjiashan barrier lake, unprecedented challenges and hardships were faced in terms of the hazardous circumstances, environmental extremes, tight rescue time and technical difficulties, which made it one of the hot spots among Chinese people during the rescue and relief of Wenchuan earthquake. The hazard of Tangjiashan barrier lake was quickly mitigated in a short period of time as a successful practice of overcoming natural disasters, which was praised by the Earthquake Relief Headquarters under the State Council as "a world wonder in handling large-scale barrier lakes". It will be of great significance for similar disaster relief practice in the future through recording the study and implementation of the emergency handling scheme, sharing the engineering experience and summarizing the advanced philosophy of disaster relief of barrier lake.

In June 2008, MWR and CWRC made timely arrangements, assigned CISPDR as the leading party jointly with other units involved in the emergency handling and disaster relief to prepare the *Summary Report on Techniques for Emergency Handling and Disaster Relief of Tangjiashan Barrier Lake*. At the same time, CISPDR and GIWP, as the editors-in-chief, initiated the formulation of *Standard for Classification of Risk Grade of Landslide Lake* and *Technique Guideline for Emergency Disposal of Landslide Lake*. Barrier lake-related research programs funded by the Ministry of Science and Technology of China were also launched. In May 2009, the abovementioned two technical codes were officially published; In December, 2011, *Report of Researches on Barrier Lake Emergency Handling Technology and Experiences* was finished as a component of the National Science and Technology Support Program during the 11th Five-Year Plan Period.

This book is compiled on the basis of *Summary Report on Techniques for Emergency Handling and Disaster Relief of Tangjiashan Barrier Lake* and *Report of Researches on Barrier Lake Emergency Handling Technology and Experiences.* Chapter 1 introduces the formation and break of barrier lake, and the current status of theoretical and experimental studies with this regard worldwide with an assessment on hazard level of Tangjiashan barrier lake; in Chaps. 2 and 3, the hydrological and geological characteristics of Tangjiashan barrier lake area are analyzed; Chaps. 4 and 5 focus on the analysis of the routing of Tangjiashan barrier lake and the stability study on barrier dam with discontinuous wide graded mixture; Chaps. 6 and 7 describe the study, formulation and implementation of the emergency handling and disaster relief scheme of Tangjiashan barrier lake; Chap. 8 describes the emergency monitoring system to guarantee the successful implementation of the disaster relief scheme; Chap. 9 describes the discharge process of lake water and the effect assessment of disaster relief.

To present an authentic record of the emergency handling and disaster relief process of Tangjiashan barrier lake, this book fully retains the technique solutions and problem-solving methods proposed and adopted in accordance with the professional competence at that time, so as to keep complete historical data as better reference for future in-depth study. Meanwhile, I wish to express my gratitude to all the participating units for their immense support and intensive cooperation during the preparation of *the Summary Report on Techniques for Emergency Handling and Disaster Relief of Tangjiashan Barrier Lake*, including POWERCHINA Chengdu, CAPF Hydropower Force, Hydrology Bureau of MWR, Hydrology Bureau of CWRC, IWHR, SIMIT, Changjiang Engineering Construction Bureau of CWRC and Sichuan Hydrology and Water Resources Survey Bureau.

Wuhan, China Qigui Yang
March 2020 Wenjun Yang

Acknowledgements

This book has been translated by Yan Fugen, Dai Wanli, Zou Zhongtian, Luo Xue, Yang Lin, Xu Kun, Zhou Zhao, etc. About the Authors and Preface have been translated by Yan Fugen. Chapters 1 and 2 have been translated by Yan Fugen and Zou Zhongtian; Chap. 3 has been translated by Luo Xue, Yang Lin and Zhou Zhao; Chap. 4 has been translated by Yang Lin and Xu Kun; Chap. 5 has been translated by Yan Fugen, Dai Wanli, Zou Zhongtian, Xu Kun and Zhou Zhao; Chap. 6 has been translated by Luo Xue, Yang Lin and Zhou Zhao; Chaps. 7 and 8 have been translated by Zou Zhongtian, Dai Wanli, and Zhou Zhao. Chapter 9 and the Appendix: Timeline of Key Events have been translated by Dai Wanli, Zou Zhongtian, and Xu Kun. The entire book was compiled by Yan Fugen.

Contents

About the Authors

Qigui Yang has been awarded the title of "National Engineering Survey and Design Master" by Chinese government. He is a professor of engineering as well as a doctoral supervisor. He is currently chairman of the board of CISPDR Corporation. Mr. Yang has won state-level titles of honor in China such as "Outstanding Mid-aged Experts", and he is also granted National Award for Excellence in Innovation. He has long been dedicated to the design of Earth-Rock dam and geotechnical engineering, risk assessment and research on emergency rescue technology of barrier lakes. He proposed hazard assessment methods of barrier lakes and built a technical system of emergency response, which fills the gap in barrier lake emergency response technology worldwide. He has led his team to deliver or participated in more than 30 research programs, Chinese government-funded or under specific engineering projects. He led in proposing technical solution to disaster relief of the Tangjiashan barrier lake, which was praised by Earthquake Relief Headquarters under State Council as "a world wonder in handling large barrier lakes".

He has been honored twice the State Science and Technology Advancement Award (ranking 1st and 2nd), provincial or ministerial Science and Technology Progress Special Awards twice (all ranking 1st) and first Awards five times; National Excellent Engineering Survey and Design Gold Award once, Silver Award twice. He is the holder of 25 Chinese Invention patents (12 items ranking 1st), 14 International Invention patents (6 items ranking 1st) and wrote 7 technical standards (2 items ranking 1st). He is also the author/coauthor of more than 80 published research papers and the first author of 5 books.

Wenjun Yang is a professor of engineering and a doctoral supervisor. He is currently the vice president of the Changjiang River Scientific Research Institute and an expert enjoying special government subsidies from the State Council. He is selected to "The New Century Talents Project" and the first batch of the "5151 Talents Project" of the Ministry of Water Resources. He won the Outstanding Youth Award of the Chinese Hydraulic Engineering Society and the First Zhang Guangdou Outstanding Youth Award. He has been long engaged in technical researches on hydraulics and river dynamics, water environment and water ecology, and water and sediment measurement technology.

He led two the National Science and Technology Support Program during the 11th Five-Year Plan Period and was responsible for one the National Key R&D Program during the 13th Five-Year Plan Period. Three the Research Program of National Major Scientific Research Instrument of National Natural Science Foundation of China and the General Project of National Natural Science Foundation of China.

He has been honored the State Science and Technology Advancement Award for three times, the National Engineering Design Gold Prize, and seven Provincial and Ministerial Science and Technology Progress Prizes (first prize and outstanding prize). He is the first author of four books and the coauthor of three technical codes. He is also the holder of more than 20 invention patents and the author/coauthor of more than 100 published research papers.

Abbreviations and Acronyms

CAS	Chinese Academy of Sciences
CISPDR	Changjiang Institute of Survey Planning Design and Research
CRSRI	Changjiang River Scientific Research Institute
CWRC	Changjiang Water Resources Commission
DEM	Digital Elevation Model
GIWP	General Institute of Water Conservancy and Hydropower Planning and Design, Ministry of Water Resources
IWHR	China Institute of Water Resources and Hydropower Research
MOFA	Ministry of Foreign Affairs of the People's Republic of China
MOST	Ministry of Science and Technology of the People's Republic of China
MWR	Ministry of Water Resources of China
POWERCHINA Chengdu	POWERCHINA Chengdu Engineering Corporation
SIMIT	Shanghai Institute of Microsystems and Information Technology
TSDI	Tianjin Railway Survey Design Institute Group Corporation

Chapter 1
Overview

1.1 Formation of Barrier Lake

Barrier lake is formed by both internal causes that the geology and geomorphology of river valley and the hydrodynamics of riverbed create the conditions of landslide and collapse and external causes that the sudden earthquake, precipitation, volcanic eruption and large-scale excavation trigger the landslide and collapse to plug the river. Barrier lake mostly occurs in the alpine canyon areas where strong tectonic activities, such as frequent earthquakes and volcanic movements and developed rock joints and fissures, are conducive to the occurrence of landslide and bank collapse. Overall, precipitation and earthquake are the two main factors inducing the formation of barrier lake. According to the data statistics of a large number of barrier lakes worldwide, 90% of barrier lakes are caused by earthquake or precipitation [1–9].

Statistical analysis shows that the formation of barrier dam can be divided into four modes (Table 1.1) [6]. The landslide or collapse at a relatively high sliding speed, after a large horizontal displacement, stops due to obstruction or throws some materials to plug the river, shown as modes I and II in the table. Mode III is a special way in the formation of barrier dam. For example, in 1933, the natural rock-fill dam in a big lake caused by Diexi earthquake in the upper reaches of the Minjiang River was formed by the simultaneous landslides of Guanyinyan and Yinpingyan, which confronted both banks of the Minjiang River and combined in the river [10]. Mode IV, such as the one occurred during Diexi earthquake in 1933, two tall barrier dams were formed on the Minjiang River due to the Diexi-Jiaochang terrace landslide.

Q. Yang and W. Yang, *Emergency Handling of the "5.12" Tangjiashan Barrier Lake*,
https://doi.org/10.1007/978-981-99-1706-8_1

Table 1.1 Main modes of formation of barrier dam [6]

Mode	Features	Plane schematic diagram	Section schematic diagram	Typical examples
I	Landslide, collapse and mudslide cross the riverbed and rush to the opposite bank slope at a relatively high speed, with a certain ascent height, which is the maximum height of rock-fill dam	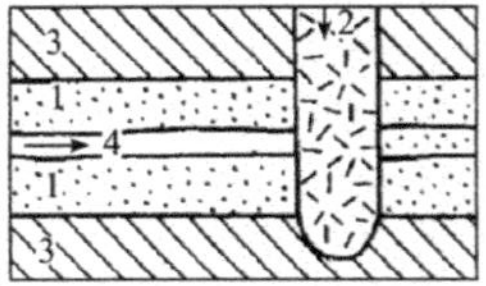	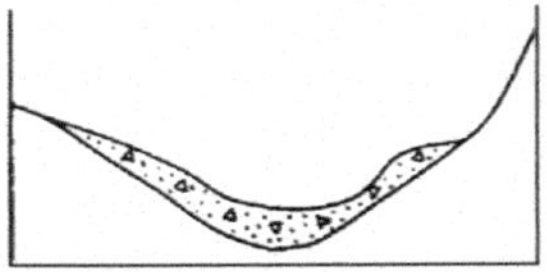	Tanggudong landslide, Maidipo landslide, Gongpeng Lake landslide, Shijiapo landslide and Zaoyang landslide
II	Landslide, in the form of whole or debris flow, rushes into the riverbed at a certain speed and moves along the river valley upstream and downstream for a distance, forming a broad rock-fill dam	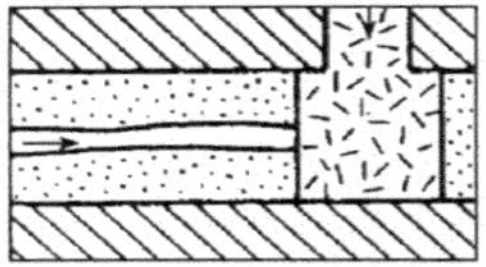	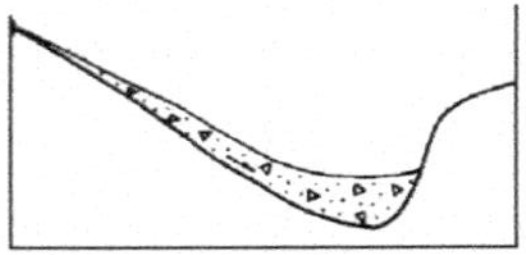	Koushan landslide, Wickre Mona landslide in New Zealand, Luquan landslide and Yigong Lake mudslide
III	The opposite slopes on both banks of the river are damaged and destabilized at the same time, moving to the river valley, and plug the riverbed with the combination of the heads of the damaged slopes	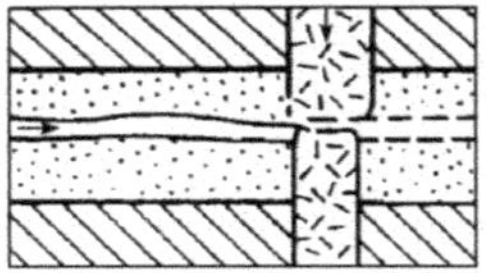	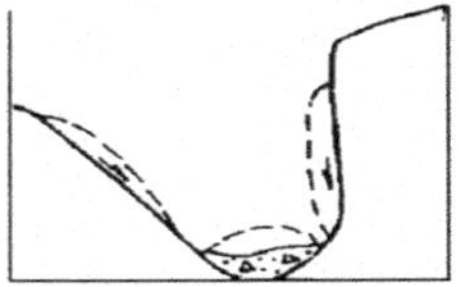	Guanyinyan-Yinpingyan landslide
IV	Landslides enter the riverbed in strands, forming two or more dams, with at least one dam completely plugging the river	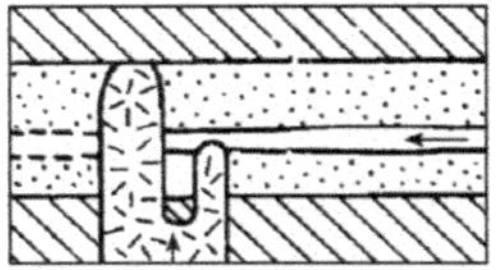	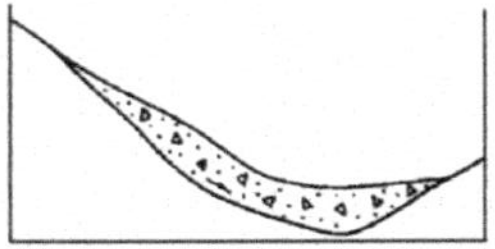	Jiguanling landslide, Diexi-Jiaochang terrace landslide

Note 1-riverbed; 2-damaged slope and its movement direction; 3-slopes of both banks; 4-river

1.2 Barrier Lake-Break

1.2.1 Study on the Forms of Barrier Lake-Break

The main reasons leading to thc barrier lake-break are overflow on dam crest, leakage and piping at dam body or foundation, landslide of dam body, etc. According to the statistical analysis of barrier lakes in China and abroad [6, 11–13], the forms of barrier lake-break can generally be divided into the following three modes.

(1) **Overflow Leading to Scouring and Break**

If the barrier dam has low water permeability, high strength and large upstream inflow, the water level of the lake will rise faster than the seepage velocity, then overflow is the main reason leading to the barrier lake-break. At the initial stage of overflow, due to the slow water velocity, there are little sediment incipient motion and slow scouring rate. As the water level of the barrier lake gradually rises, the velocity of breach is gradually accelerated, and there are lots of sediment incipient motion and increased scouring rate. The barrier lake finally breaks in the process of flow down-cutting and scouring deep and gravity collapsing and expanding.

(2) **Sudden Sliding Leading to Break**

If the barrier dam has high water permeability and low strength, although the water level of the lake rises rapidly, the seepage area will also expand downstream and upstream. When the water level reaches a certain critical value, the sliding surface will be created in the barrier dam, resulting in a sudden break. Since the earth-rock mass is not saturated at the time of the break, the collapsed earth mass will temporarily stay at the dam site, and then be taken downstream with the scouring of the flow, forming a large-scale mudslide.

(3) **Asymptotic Sliding Destruction**

If the barrier lake has high water permeability, the seepage water will appear on the downstream slope of the barrier dam before the water level of the lake rises significantly. If the strength of the earth-rock mass is insufficient, the local barrier dam will slide and lose stability when the seepage water appears. As time goes by, the sliding surface will expand upstream until the break of the dam crest.

1.2.2 Hazard Level of Barrier Lake-Break

Before 2008, there was no Chinese or international standard for hazard and disaster loss assessment of barrier lake. Based on the comprehensive analysis of literature worldwide on hazard assessment of barrier lake [14–16], the dam height, maximum storage capacity and dam structure of barrier lake were selected as classification indexes to divide the hazard of barrier lake-break into four levels: extremely high hazard, high hazard, medium hazard and low hazard (Table 1.2).

Table 1.2 Reference standard for hazard classification of barrier lake-break

Hazard level	Classification indexes		
	Dam height (m)	Maximum storage capacity (million m^3)	Dam structure
Extremely high hazard	> 80	100	Mainly earth
High hazard	50–80	10–100	Earth with boulders
Medium hazard	25–50	1–10	Boulders with earth
Low hazard	< 25	< 1	Mainly boulders

Note At least two of the three conditions are met

1.2.3 Disaster Loss Assessment of Barrier Lake-Break

Based on the analysis of Chinese and foreign literature [15, 17–25], population, important cities and facilities affected by the hazard were selected as the main contents of disaster loss assessment of break. The proposed classification standard of damage loss of barrier lake was shown in Table 1.3.

1.3 Theoretical and Experimental Study on Barrier Lake-Break

The barrier dam has many similarities with the artificial earth-rock dam in terms of dam structure and dam-break features and at the same time has its own features. Currently, there is little specialized study on barrier lake-break, which mostly is combined with the dam-break study of artificial earth-rock dam or refers to the study results of dam-break simulation of artificial earth-rock dam. In view of the fact that the material composition of the barrier dam is close to that of the earth-rock dam, the

Table 1.3 Classification standard of damage loss of barrier lake

Loss level	Classification indexes of degree of possible threat to downstream		
	Affected population (million)	Important cities	Important facilities
Extremely serious	> 1	Above county-level cities or urban districts	Very important facilities
Serious	0.1–1	Above town-level	Important facilities
Medium serious	0.001–1	Villages	Medium-important facilities
Slightly serious	< 0.001	Village-level communities or scattered households	General facilities

Note At least two of the three conditions are met

present dam-break study on earth-rock dam is often used for the analysis of barrier lake-break [26–30].

The dam-break of earth-rock dam is a gradual break by flood, indicating that it is a lasting process of interaction between water and dam materials. On the one hand, with the dam-building materials being scoured by flow, the breach continues to expand, its bottom elevation drops, and cross-sectional area increases gradually; on the other hand, the expanded breach results in the constant change of the dam-break flood discharge; meanwhile, the scoured dam materials entering flow will also cause the change of flow structure and sediment carrying capacity. The dam-break occurrence, expansion and degree of earth-rock dam are affected by various factors, such as the dam-break cause, the dam structure dimensions and materials, the reservoir storage capacity and the water level downstream, making the study very difficult. Theoretical analysis, hydraulic test study and numerical simulation are the main methodologies of study on barrier dam-break at present.

1.3.1 Theoretical Analysis

From the point of view of analysis of mathematical and physical problems, the study on barrier lake-break can be summed up as the Riemann problem of solving the Saint–Venant equations which control the flow movement. In 1871, Saint–Venant put forward the equations of unsteady flow control of open channels, namely Saint–Venant equations, laying a theoretical foundation for the hydraulic analysis of the dam-break problems [31].

In 1892, Riter, by assuming the downstream a dry riverbed and ignoring its fluctuation and shear friction, used the characteristic form of Saint–Venant equations to obtain the simplified solution of the instantaneous complete dam-break problem of rectangular section for the first time, namely Riter solution. However, this analytical solution was not in line with the reality. Due to the riverbed friction, the velocity and profile of the wave crest moving downstream are different from Riter solution. In 1949 and 1957, based on the study results of Courant and Friederich, Stoker [32] ignored the fluctuation and frictional action of riverbed and divided the flow patterns at the dam site into eight types, including continuous wave flow, critical flow, discontinuous flow, etc., extending Riter solution to the downstream riverbed with water.

In 1970, Su [33] extended the perturbation solution of Dresler [34] to open channels with different sections, raising the first-order perturbation solution of water surface curve and average velocity distribution of open channels in rectangular, triangular and parabolic sections with the consideration of friction. In 1980, Chen et al. [35] studied the dam-break problem of finite-length reservoirs with the consideration of friction. In 1980, Lin Bingnan [36] used the characteristic theory and Riemann method to obtain the analytical solution of instantaneous complete dam-break problem of the flat bottom of open channels of finite-length reservoirs in parabolic section. In 1982 and 1984, Hunt [37] approximately regarded the dam-break wave as a motion wave to study the dam-break problem of inclined riverbed. In 1982, Lu Jikang [38] et al. used the characteristic theory and Riemann method to study the analytical solution of instantaneous complete dam-break problem of inclined open channels in parabolic section with the consideration of friction.

In the study of the dam-break problem of inclined bottom slope, Su [33] used the disturbance series to obtain the approximate solution of the effective time for the retrograde peak to reach the upstream boundary. Hunt [37] offered a pure motion wave solution for the area where the effective distance covered by flood waves downstream of the steep riverbed is close to 4 times the length of the reservoir above the dam site. Smith [39] proposed a similar solution to the stable 3D flow with slow viscosity under the inclined plan, suitable for the flow deep in the ocean.

Due to the quasi-linearity of Saint–Venant equations and the discontinuity of flow, it is difficult to analyze and solve the dam-break problem. At present, the analytical solution of the dam-break problem is limited to a few simple examples, and even the study on the analytical solution of one-dimensional dam-break problem is not very deep. The study on the dam-break mechanism will be an important and difficult issue in the future.

1.3.2 Hydraulic Test Study

From the point of view of practical engineering application, it is necessary to study the physical model test of dam-break. Especially for large-scale projects, the parameters of model test are the main basis of engineering design and construction. At the same time, the results of model test can be used to verify the feasibility of numerical simulation calculation method. In the past few decades, a large number of dam-break tests [9, 11, 29, 40–46] on different scales (dam height 2~6 m) have been successively carried out in many countries. In recent years, numerous tests have been conducted in China to study the effects of dam height, inflow, dam material and other factors on barrier dam-break.

In the dam-break study of earth-rock dam, substantial laboratory tests were conducted in the USA and Austria in 1950 and 1960s. The analysis of the test results showed that the relation of dam-break time scale studied in the two countries was basically the same. In addition, regarding the same boulders for slope protection, when the slope becomes gentle, the critical water head breaking the dam slope will increase significantly. For example, when the slope of the downstream dam varies from 1:1.5 to 1:2, the critical water head will increase by 15%. Recently, Europe has invested heavily in the comprehensive comparative study of dam-break flood. By 1998, the CADAM project launched with the support of the EU had lasted for two years [40], and the subsequent IMPACT [41] project supported by the EU had conducted numerous model test studies on the dam-break mechanism of earth-rock dam. The break tests not only improved people's understanding of the dam-break process and

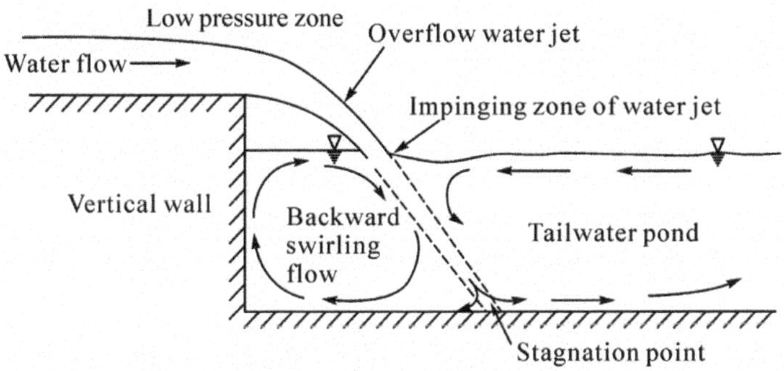

Fig. 1.1 Schematic diagram of headcut

mechanism of earth-rock dam, but also provided test data for the calibration and verification of the mathematical model of the dam-break process.

On the basis of the observation on numerous dam-break tests and prototype dam-break examples, researchers in the USA and Europe put forward a new overtopping dam-break scouring mechanism of earth-rock dam, namely headcut scouring [26], and believed that headcut scouring is the scouring mechanism that plays a leading role in the dam-break process for most earth-rock dams, especially the dams with cohesive earth materials. The headcut refers to the sudden drop in the elevation of riverbed, similar to the waterfall landform (Fig. 1.1). When the water flows through the headcut, the overflow jet impinges the riverbed surface downward and generates backward swirling flow, which exerts shear stress on the vertical or nearly vertical falling water surface and scours the riverbed surface and the foundation of the vertical falling water surface, resulting in the instability and collapse of the falling water surface and the continuous expansion upstream of the headcut.

In recent years, the Hydraulic Engineering Research Unit (HERU) under the U.S. Department of Agriculture has done a series of overtopping dam-break tests of large-scale earth-rock dams to simulate the overtopping scouring of earth-rock dams with different materials [26, 47]. Based on the observation of the test process and the analysis of the test data, the following dam-break process of earth-rock dam was raised: Firstly, the overtopping flow generates rill-like scouring on the downstream slope of the dam. The scouring continues to expand, forming a network of rills, and eventually larger gullies. Initially, the gullies contain a number of small stepped headcuts, which recede upstream and widen into large headcuts with time. Under the action of flow, the headcuts gradually expand upstream to the edge of the crest of the earth-rock dam. The continuous expansion upstream of the headcuts will reduce the dam crest at breach and increase the discharge of breach rapidly, eventually leading to complete dam-break. The widening of gullies and the expansion of headcuts upstream in the dam-break process are caused by the turbulence and shear stress of flow around the impinging zone of water jet of the headcuts downstream. The flow continuously scours the bottom of the headcuts, causing the instability and collapse of the vertical wall, which leads to the widening of the breach and the continuous expansion of the headcuts upstream.

By far, some preliminary results have been achieved in the study of headcut scouring. However, on the whole, this study is still in its infancy, with the formation, expansion and scouring mechanisms of headcut not very clear. Therefore, it is difficult to obtain the model of headcut scouring widely recognized.

1.3.3 Simulation of the Dam-Break Process of Earth-Rock Dam

At present, the developed mathematical models for simulating the dam-break process of earth-rock dam can be roughly divided into two categories.

(1) **Model Based on Parameters**

This model, by using some key parameters, such as the final width of breach and the dam-break duration, simulates the expansion of breach through simple time-varying process (e.g., the linear expansion theory of breach size), including the DAMBRK model and SMPDBK model of the National Weather Service of the USA [28, 29, 48, 49]. There are also some other models that simulate the dam-break process by establishing regression equations between key parameters such as the reservoir storage capacity and dam height, the expansion velocity of breach and the maximum dam-break flood discharge. In general, this parameter-based model is relatively simple, which does not involve the actual dam-break mechanism, requires less data input and is easy to use.

(2) **Model Based on Physical Process**

This model, by integrating the knowledge of hydraulics, sediment dynamics, soil mechanics, hydrology and other disciplines, builds a time-varying process to simulate the actual dam-break process and dam-break flood process, including the typical BEED model [50], BREACH model [48] and BRESZHU model [49]. These models with complicated structures can simulate the dam-break process accurately and in detail, but are limited by the current knowledge of the dam-break process and mechanism of earth-rock dam [51–54].

Since the dam-break process of earth-rock dam is rather complicated and contains various uncertain factors, there are many imperfections in the dam-break simulation technology developed on this basis. For example, the simulation objects of existing models are mostly earth-rock dams filled with non-cohesive earth materials, whereas the studies on earth-rock dams filled with cohesive earth materials in practical projects are insufficient; almost all the dam-break models are only confined to homogeneous dams, and it is difficult to simulate composite structure dams and slope protection on dam surface; most models fail to distinguish the difference between the scouring velocity of dam-break flow and the sediment carrying capacity of flow; the reasonable relation widely recognized has not been established between the dam scouring velocity and the features of dam-building materials and the characteristic parameters of dam-break flow.

1.3.4 Study on Simulation of Dam-Break Flood

The study on the simulation of dam-break flood in the early stage was mostly based on the analysis of historical dam-break data and some assumptions (such as the instantaneous complete barrier dam-break) to establish the empirical relation between the dam-break parameters, including peak discharge and final entrance width, and the reservoir storage capacity, front water head of the dam, dam height, etc. and the routing process of dam-break flood waves downstream. Later, it gradually developed into the study on the numerical simulation model of dam-break flood.

1.3.4.1 Empirical Formula for Calculation of Dam-Break Peak Discharge

At present, there are several empirical formulas commonly used in China for the calculation of dam-break peak discharge, including the Saint–Venant Formula, the empirical formula of Water-ways Experiment Station in USA, Stoke method, positive and negative wave intersection method, wave rated discharge method, Schoklitsch method, the empirical formula of TSDI, the empirical formula of Hydrological General Station of Liaoning Province and the "unified formula" raised by Professor Xie Renzhi in 1982 [55].

Each method has certain applicable conditions; for example, the Saint–Venant Formula requires rectangular riverbed and drying up under the dam; Stoke method requires a certain static water depth downstream; positive and negative wave intersection method is suitable for deep water downstream and slow flow under the dam; wave rated discharge method properly considers the effects of the water depth downstream and assumes that the upstream and downstream of the dam site are flat-bottomed channels without resistance; the empirical formula of Hydrological General Station of Liaoning Province requires that the reservoir velocity and discharge before dam-break be ignored. Table 1.4 lists some empirical formulas for the calculation of dam-break peak discharge.

1.3.4.2 Empirical Formula for Calculation of Downstream Routing of Break Flood

The calculation of downstream routing of dam-break flood is the basis for assessing the dam-break disaster loss and is related to such factors as the dam-break form, hydrograph of dam-break flood, dam-break peak discharge, process of inflow, downstream water level, section shape of downstream river and on-way distance to the

Table 1.4 Some empirical formulas for calculation of dam-break peak discharge

Serial number	Formula name	Expression
1	Empirical formula of Water-ways Experimental Station	$Q_m = \frac{8}{27}\sqrt{g}\left(\frac{\mathrm{BH_0}}{\mathrm{bh}}\right)^{0.28}\mathrm{bh}^{1.5}$
2	Xie Renzhi Formula	$Q_m = \lambda B_0\sqrt{g}H_0^{1.5}$
3	Empirical formula of Hydrological General Station of Liaoning Province	$Q_m = 0.206\sqrt{g}BH_0^{1.5}$
4	Empirical formula of TSDI	$Q_m = K_{\mathrm{nb}}b\sqrt{g}(H_0 - H_2)^{1.5}$

Note λ is the flow coefficient; B_0 is the width of river valley at the dam site; H_0 is the upstream water depth of the dam; b is the average width of the breach; B is the average width of the section of dam site; H_2 is the downstream water depth of dam site; and $K_{\mathrm{nb}} = 1.35b/B$

dam site. Generally, the mathematical model of unsteady flow is used to calculate the downstream routing of dam-break flood [14, 56–58].

If the downstream data of the dam is scarce, Lisztvan Formula is often used to estimate the on-way maximum discharge downstream:

$$Q_{xm} = \frac{\mathrm{W}Q_m}{(W + Q_m\mathrm{XK})}$$

where

- Q_{xm} Maximum discharge at X downstream of the dam.
- Q_m Maximum dam-break discharge at the dam site.
- W Total dam-break discharge.
- X Distance from the downstream section to the dam site.
- K Section coefficient of river valley.
- H_{mx} The peak height somewhere downstream from the dam site is estimated by the following formula:

$$H_{mx} = H_{10}\left[1 + \frac{4\alpha^2(2n+1)H_0^{2n+1}}{n(n+1)^2 i_0 V^2}x\right]^{\frac{1}{2n+1}}$$

where

- H_{mx} Peak height at x from the dam site.
- H_{10} Maximum dam-break water depth.
- α Section coefficient of river valley, $\alpha = A/H$ where: A is section area and H is section height.
- i_0 Riverbed gradient.
- V Reservoir storage capacity of barrier dam.
- X Distance from the downstream to the dam site.

1.3.4.3 Calculation Model of Routing of Dam-Break Flood

With the rapid development of computer performance and numerical calculation methods, numerical simulation of the routing of dam-break flood has increasingly become the main means to study the spread of dam-break flood. In the middle and late twentieth century, great progress was made in the study on the simulation of dam-break flood of earth-rock dam, with the establishment of a series of calculation models of the routing of dam-break flood, such as DAMBRK in the USA, DE.FT in the Netherlands and DHI in Denmark. At present, the widely used model abroad is the calculation model of dam-break and flood routing developed by the National Weather Service of the USA. This model includes three mathematical models and corresponding calculation programs, namely DAMBRK—the calculation model of the dynamic routing of dam-break and dam-break flood waves; BREACH—the model of breach erosion of earth dam; and SMPDBK—the simplified dam-break model, of which DAMBRK model is the most detailed and complicated, consisting of two parts: one is the calculation of dam-break flood, including the calculation of the hydrograph of dam-break flood discharge and the dam-break water level at the dam site; the other is the simulation of the spread and routing of dam-break flood downstream. China began the systematic study on dam-break numerical models in 1970s, and then conducted the study on the calculation models of the routing of dam-break flood according to the.

1.4 Assessment of Hazard Level and Handling of Tangjiashan Barrier Lake

At 14:28 on May 12, 2008, an earthquake with magnitude 8.0 struck Wenchuan, Sichuan. On May 14, the Institute of Remote Sensing and Digital Earth, Chinese Academy of Sciences, identified Tangjiashan barrier lake in the aerial image after the earthquake.

Tangjiashan barrier lake was located in the valley of Tongkou River, branch of Fujiang River, 3.2 km upstream of Beichuan County. The barrier dam was 803 m long along the river and 612 m long across the river, with the height of 82–124 m and the volume of 20.37 million m^3. Upstream of the lake, the catchment area was 3550 km^2, and the maximum water storage capacity was 316 million m^3.

Tangjiashan barrier dam was a secondary disaster caused by the strong earthquake. After the river was plugged, the barrier lake water flooded and caused damage to Yuli and Xuanping towns upstream. The reservoir formed by backwater and the water pollution also became prominent problems in the earthquake-stricken area.

Downstream of Tangjiashan barrier lake, there were important cities such as Mianyang, the second largest city in Sichuan Province. Moreover, there were important infrastructures including Baoji-Chengdu railway, the transportation artery and Lanzhou-Chengdu-Chongqing oil pipeline, the strategic energy channel. Once

Tangjiashan barrier dam-break occurred, the life safety of more than 1.3 million people in Mianyang and Suining and the safety of important infrastructures along the way would be threatened directly.

On May 18, with the support of the Air Force, water resources experts observed the barrier dam from the air for the first time. On May 20–21, the experts basically identified the data such as the scale of the dam, the relative height difference between the water level and the dam and the water volume in the lake via field survey and measurement. The data showed that the catchment area above Tangjiashan was $3550km^2$, and the water level of the lake rose at a velocity of more than 2 m per day. Even if there was no heavy precipitation, the daily water inflow of the lake reached 7.2 million m^3, and the height difference between the dam pass and the water surface was only 39 m.

On May 21, the water resources group of Earthquake Relief Headquarters under the State Council determined that Tangjiashan barrier lake was the only extremely hazard out of the 35 counted barrier lakes formed by Wenchuan earthquake, with the combination of the classification standard for assessment of break hazard and disaster loss and with the consideration of the dam height, the maximum potential water storage capacity, the dam structure and the possible threat to the residents and facilities downstream. Therefore, the group decided to make use of the limited time to simultaneously implement engineering measures to relieve the disaster and transfer personnel to avoid the hazard in upstream- and downstream-affected areas.

The dangerous situation, tight rescue time and great technical difficulty of Tangjiashan barrier lake were all unprecedented. It was the flood season, with continued aftershocks, the barrier lake area in high mountains and valleys was faced with complex geological environment, harsh meteorological conditions and frequent landslides, collapses and mudslides. If there were strong rainfall process in this area, the barrier lake would be filled up soon, inducing a great dam-break hazard. About 3 km upstream of the area, there was Malingyan ancient landslide mass with a total volume of about 16.5 million m^3. Under the action of aftershocks and rainfall, it would slide into the area quickly, being pushed by the surging waves to the front of the dam, with the consequences unpredictable. At the same time, due to the complete interruption of traffic, electricity and communication, rescue teams and materials and equipment were difficult to be sent to the site, further increasing the difficulty of rescue and disaster relief. The rescue and disaster relief situation of Tangjiashan barrier lake was very severe, with extremely arduous task and great responsibility.

On May 22, Premier Wen Jiabao presided over a dedicated meeting in Mianyang to study the emergency handling of barrier lakes after the earthquake, focusing on the emergency handling scheme of Tangjiashan barrier lake, and instructed that the emergency handling should be "proactive, early, with the combination of disaster relief and hazard avoidance". Then the Tangjiashan barrier lake emergency handling headquarters was established, with officer from Sichuan provincial government as the chief commander and officer from MWR as the executive deputy commander.

On May 23, the water resources group reported to Earthquake Relief Headquarters under the State Council on the *Technical Scheme for Design and Construction of Tangjiashan Emergency Handling Project*.

According to the construction conditions and capacity of emergency handling and disaster relief of Tangjiashan barrier lake, the water resources group judged the construction period in combination with hydrometeorological forecast. Meanwhile, based on the material structure features of the barrier dam as well as the conditions of extraordinary air transportation and material support from the Air Force and local people's governments for the emergency project, the water resources group recommended the project scheme of disaster relief, namely to excavate a 5–10 m deep drainage channel, also known as discharge channel, with machinery, and the alternative scheme, to form a drainage channel about 5 m deep by blasting. After the discharge from the barrier lake, the drainage channel would be widened by water power to improve the discharge capacity, gradually reducing the water level and water volume of the barrier lake to avoid sudden dam-break. At the same time, a new river course with a certain flood discharge capacity would be formed through the scouring of the discharged lake water, thus eliminating the threat of the barrier dam to the upstream and downstream.

On May 26, officers and soldiers of CAPF Hydropower Force began to excavate the drainage channel with machinery. In the early morning of June 1, the construction task of drainage channel was overfulfilled ahead of schedule. At 07:08 on June 7, the water of the barrier lake began to be discharged through the drainage channel. During the discharge, there were neither casualties nor losses of important infrastructures in the downstream.

On June 11, the drainage channel formed a new river course with the capacity for flood of 20 years return period. During the discharge, the flow washed away about 500,000 m^3 of dam deposits, and the remaining dam had strong anti-seepage and anti-scouring abilities and good anti-sliding and overall stability. Finally, the disaster of Tangjiashan barrier lake was relieved, completely preventing the sudden break which would bring significant damage to the downstream area.

References

1. Yan ZW, Wei YQ, Cai H. Formation mechanism and stability analysis of Barrier Dam [J]. Chinese J Geol Hazard Control. 2009;20(4):55–9.
2. Huang RQ. Large-scale landslides and their sliding mechanisms in china since the 20th century [J]. Chin J Rock Mech Eng. 2007;26(3):433–54.
3. Chai HJ, Liu HC, Zhang ZY, et al. The distribution causes and effects of damming landslides in China[J]. J Chengdu Univ Technol. 2000;27(3):302–7.
4. Hejun C, Liu H, Zhang Z (1995) The catalog of Chinese landslide dam events [J]. J Geol Hazards Environ Preservat10n 6(4):1–9
5. Echuan Y, Zhuoyuan Z, Hanchao L et al (2003) A study on the landslide event and its environmental effects in China [J]. Acta Geoscientica Sinica z1:205–09
6. Costa JE, Schuster RL. The formation and failure of natural dams[J]. Geol Soc Am Bull. 1988;100(7):1054–68.
7. Gongzhong N, Jianguo G, Yan D. Preliminary study on earthquake-induced dammed lake [J]. Q Sci. 2004;4(3):293–301.
8. Costa JE, Schuster RL (1991) Documented historical landslide dams from around the World[R]. Denver, United States GeologicalSurvey

9. Ponce VM. Documented cases of earth dam breaches[R]. San Diego: San Diego State University; 1982.
10. Yvchuan L. Diexi earthquake record [J]. World Literature History. 2002;6:39–42.
11. Zhu Y, Liao H, Wu Z. Summary of foreign earth dam break simulation [J]. J Yangtze River Sci Res Inst. 2003;4:26–9.
12. Hejun C, Hanchao L, Zhuoyuan Z, et al. Preliminarily stability analysis of natural rock-field dam resulting from damming landslide [J]. Geol Sci Technol Inf. 2001;20(1):77–81.
13. Dai F, Lee C, Deng J, et al. The 1786 earthquake-triggered landslidedam and Subsequent dam-break flood on the Dadu River, Southwestern China[J]. Geomorphology. 2005;73(3):277–8.
14. Apel H, Thieken AH, Merz B, et al. Flood risk assessment and associated uncertainty[J]. Nat Hazard. 2004;4(2):295–308.
15. Ermini L, Casagli N. Prediction of the behaviour of landslide dams using a geomorphological dimensionless index[J]. Earth Surf Proc Land. 2003;28(1):31–47.
16. Clague JJ, Evans SG. Formation and failure of natural dams in the Canadian Cordillera[J]. Bull Geol Surv Canada. 1994;464:1–35.
17. Wang B, Fu Q. A review on evaluation method of economic loss of flood in the World [J]. J Catastrophol. 2007;22(3):95–9.
18. Feng MQ, Zhou XD, Zhang GG. Review on development of assessment of flood damage loss [J]. Northwest Water Resour Water Eng. 2002;13(1):32–6.
19. Wang Y. A review of flood disaster loss assessment models at different scales [J]. Res Water Conservancy Develop. 2002;12(2):66–9.
20. Li X, et al. Study on Chinese statistical and estimating system of disaster economy and its index system [J]. J Natijral Disasters. 1993;1:5–15.
21. Shi G (1990) Flood loss rate and its determination method [J]. Water Conservancy Econ 2:37–42
22. Wanxiu C. Flood disaster loss assessment system-application of remote sensing and GIS technology [M]. Beijing: China Water Conservancy Hydroelectric Press; 1999.
23. Das S, Lee R. A nontraditional methology for flood stage damage calculation [J]. Water Resour Bull. 1998;24(6):1263–72.
24. Srikantha H, Dushmanta D. Flood inundation modelling and loss estimation using distributed hydrologic model, GIS and RS [C]. Proc Int Workshop Utilization Remote Sens Technol Natl Disaster Reduction. 1998;10(26):239–50.
25. Schuster RL, Costa JE. Effects of landslide damming on hydroelectric projects[C]. Proc Vth Int Congress Int Assoc Eng Geol Buenos Aires. 1986;4:1295–307.
26. Hanson SJ, Robinson KM, Cook KR. Prediction of headcut migration using a terministic approach[J]. Trans ASAE. 2001;44(3):521–31.
27. Macdonald TC, Langriidge-Monoplis J. Breaching characteristics of dam failures[J]. J Hydraul Eng. 1984;110(5):567–85.
28. Coleman SE, Andrews DP, Webby MG. Overtopping breaching of noncohesive homogenous embankments [J]. J Hydraulic Eng. 2002;128(9):829–38.
29. Schmocker L, Hager WH. Modelling dike breaching due to overtopping [J]. J Hydraulic Res. 2009;47(5):585–97.
30. Kuang S, Wang X, Huang J, Wei Y, et al. Risk analysis and impact assessment of dam -break in barrier lake [J]. China Water Res. 2008;16:17–21.
31. Wang LH, Hu SY. Study on dam failure-related problems[J]. Adv Sci Technol Water Res. 2007;27(1):80–5.
32. Stoker JJ. Water wave[J]. New York: Interscience Publisers; 1957. p. 334–41.
33. Su ST. Dry bed wave due to sudden water releases [D]. Fort Collins: Colorado State University; 1970.
34. Dressler RR. Unsteady non-linear waves in sloping channels[J]. Processdings. 1958;247:186–98.
35. Chen C, Ambruster JT. Dam-break wave model: formulation and verification[J]. J Hydraulic Eng. 1980;106(5):747–67.
36. Lin BN, Gong ZY, Wang LX. Dam-site hydrographs due to sudden release [J]. J Qinghua Univ. 1980;20(1):17–31.

37. Hunt B. Dam break solution[J]. J Hydraul Eng. 1984;110(6):675–86.
38. Lu JK, Lin BN (2001) Dam-breakwaveinprismaticchannels[G]. Lin Bingnan 's Essays. Beijing, China Water Resources and Hydropower Press 419–423
39. Smith P. A similarity solution for slow viscous flow down an inclined plane[J]. J Fluid Mech. 1973;58(2):275–88.
40. Morris MW, Galland JC (1998) CADAM: Dambreak modelling guidelines and best practice[R]. Munich: HR Wallingford Ltd
41. Morris MW, Fabbri K, Brelen H (2005) IMPACT: Investigation of extreme flood processes and uncertainty contract No. EVG-CT-2001–00037, Final technical report [R]. Wallingford: HR Wallingford Ltd
42. Zhang JY, Li Y, Xuan GX, et al. Overtopping breaching of cohesive homogeneous earth dam with different cohesive strength. Sci China Ser E-Tech Sci. 2009;39(11):1881–6.
43. Hunt SL, Hanson GJ, Cook KR. Breaching observations from earthen embankment tests [J]. Trans ASAE. 2005;48(3):1115–20.
44. Morries MW, Hassan MA, Vaskinn KA. Breach formation: Field test and laboratory experiments[J]. J Hydraul Res. 2007;45(1):9–17.
45. Yan J, Cao ZX, Liu HH, et al. Experimental study of landslide dam-break flood over erodible bed in open channels[J]. J Hydrodyn. 2009;2(1):124–30.
46. Niu ZP, Xu WL, Zhang JM, Li J, Yue C et al (2009) Experimental Investigation of Scour and Dam-break of Landslide Dam[J]. Ournal of Sichuan University (Engineering Science Dition), 41(3):90–95
47. Hanson GJ, Cook KR, Temple DM (2002) Research results of large-scale embankment, overtopping breach tests[R]. Washington, DC: 2002 ASDSO Annual Conference
48. Fread DL (1988) Breach: an erosion model for Earthen Dam failures[R]. Silver Spring: Hydrologic Research Laboratory, NOAA, NWS, U. S. Dept. of Commerce
49. Zhu Y, Visser PJ, Vrijling JK. Soil headcut erosion: process and mathematical modeling[C]. Proc Marine Sci. 2008;9:125–36.
50. Singh VP, Scarlatos PD, Collins JG et al (1988) Breach erosion of Earthfill dams (BEED) model[J]. Natural Hazards 1(2):161–180
51. EvansSG T. maximum discharge of outburst floods caused by the breaching of man-made and natural dams. Canadian Geotech J. 1986;23:385–7.
52. MacDonald TC, Langridge-Monopolis J. Breaching characteristics of dam failures[J]. J Hydraul Eng. 1984;110:567–86.
53. Simmler H, Samet L. Dam failure from overtopping studied in a hydraulic model[C]. In: Transactions of 14th International Congress on Large Dams 1:427–445
54. Singh VP. Dam breach modeling techology[M]. Norwell, Mass: Kluwer Acad Publ; 1996.
55. Renzhi X. Dam-break hydraulics [M]. Jinan: Shandong Science and Technology Press; 1993.
56. Sarmaa K, Dasm M. Analytical solution of a flood wave resulting from dike failure[J]. Proc Inst Civil Eng Water Maritime Eng. 2003;1:41–5.
57. Xu YS, Xu YX, Liao PM, Cheng GX, et al. Study on sedimentation in alluvial channel of landslide induced at Tsao-lin [J]. J Sediment Res. 2004;3:22–9.
58. Zech Y, Soares-Frazo S, Spinewine B, et al. Dam-break induced sediment movement: experimental approaches and numerical modelling[J]. J Hydraul Res. 2008;46(2):176–90.

Chapter 2
Hydrological Analysis of Tangjiashan Barrier Lake Area

The basic data of Tangjiashan barrier lake area are the most basis for formulating the emergency handling scheme of the barrier lake. Since Tangjiashan barrier was formed in a short time by the earthquake-induced landslides, and there were no surveying and mapping data on the reservoir area and dam area of the barrier lake that could meet the requirements of engineering accuracy, it was impossible to accurately draw up the basic data such as reservoir storage capacity and lake water volume. For this reason, many units involved in the hazard analysis and formulation of disaster relief collected the existing topographic data with various precision by using possible information systems, and moreover, some units used remote sensing technology to quickly acquire the river topography, so as to measure the storage capacity curve of the barrier lake and split the section of flood routing. In a very short time before May 22, the units provided the hydrological and topographic data of the barrier lake and its affected areas that meet the requirements of emergency scheme, laying a foundation for the formulation of emergency handling scheme.

2.1 Basic Situation of the Affected Area

2.1.1 Basic Situation of Fujiang River Basin

2.1.1.1 Overview of the Basin

Fujiang River is the largest tributary on the right bank of Jialing River, located at 29° 59′ 13″–32° 44′ 6″ N and 103° 44′ 4″–106° 16′ 10″ E. It originates from the north slope of Xuebaoding of Minshan Mountain in Songpan County, northwest Sichuan, flows from northwest to southeast, passing through Pingwu, Jiangyou, Mianyang, Santai, Shehong, Suining and Tongnan and other cities (counties, districts), and joins Jialing River in Hechuan, Chongqing. The total length of the main stream is 675 km, with an average gradient of 1.4% and a basin area of 36,272 km^2. The basin is a long

Q. Yang and W. Yang, *Emergency Handling of the "5.12" Tangjiashan Barrier Lake*,
https://doi.org/10.1007/978-981-99-1706-8_2

and narrow shape inclined from north to southeast. The upper reaches are separated by Motianling, Longmen Mountain and Bailong River in the north and Minshan Mountain and Minjiang River in the west; the middle and lower reaches are adjacent to Tuojiang River in the west and the main stream of Jialing River in the east.

The section above Wudu (Denglongqiao) is the upper reaches of Fujiang River, with a length of 238 km and an average gradient of 5.7%, which is a mountainous river; the section from Denglongqiao to Guojunba, Suining is the middle reaches, with a length of 284 km and an average gradient of 0.83%; the section after Guojunba, Suining is the lower reaches, with a length of 153 km and an average gradient of 0.47%.

The Fujiang River system is developed in pinnate-shape, covering 21 tributaries (including 11 primary tributaries), each with an area more than 500 km^2 and 10 tributaries (8 primary tributaries), each with an area over 1000 km^2. Among them, Huaxi River, Pingtang River and Tongkou River are mountainous rivers, with large stream gradient; Kaijiang River, Zitong River, Najiang River, Anju River and Xiaoanxi River flow through hilly areas, with gentle stream gradient. The water system and hydrological station network of Fujiang River Basin are shown in Fig. 2.1.

The topography of the basin inclines from northwest to southeast with an elevation of 188–5588 m and can be divided into three types: mountains, hills and plains, which account for 37.9%, 56.9% and 5.2% of the total area, respectively. The north of Wudu and Anzhou is the upper reaches, which include Minshan Mountain and Longmen Mountain, and belong to the middle and high mountain areas with an elevation of 600–5588 m and an area accounting for 35% of the total area of the basin. The area covers high mountains, deep valleys and slopes, with complex geological structure and metamorphic rocks as the main strata, followed by Paleozoic carbonate rocks and clastic rocks under cracking development, some cracking activities are moderate and strong. In Back Longmen Mountain and Beichuan area, the crust stability is poor, with the basic earthquake intensity of above VIII and relatively developed mountain disasters such as landslide, collapse and mudslide. The south of Wudu and Anzhou is the middle and lower reaches, which belong to hilly flat dam area, with an area accounting for 65% of the total area of the basin and an elevation of 190–800 m generally. Except for scattered mountains along the watershed and alluvial plains in the valley, most of the rest are hills, where cultivated land is concentrated, vegetation is poor, geological structure is simple, Jurassic and Cretaceous strata are distributed, and the basic earthquake intensity is equal to or less than 6°.

2.1.1.2 Water Conservancy Projects in the Basin

Fujiang River Basin enjoys superior geographical location, rapid economic and social development, and numerous water conservancy and hydropower projects. Among the built 131,000 water conservancy projects, there are 114,000 water storage projects such as Tangku reservoir, with a total reservoir storage capacity of 1.737 billion m^3; 135 levees with a length of 128.5 km; 10 hydropower plants on the main stream of Fujiang River, namely Shilongzui, Kaiyuan, Sanjiang, Luxi, Wenfeng, Jinhua,

Fig. 2.1 The water system and hydrological station network of Fujiang river basin

Luosichi, Baichansi, Anju and Weituo, all of which are daily regulation reservoirs, and developed for the purpose of power generation and shipping.

2.1.2 Basic Situation of Tongkou River Basin

2.1.2.1 Overview of the Basin

Tongkou River Basin, located in the mountainous area on the northwest edge of Sichuan Basin, is a primary tributary of Fujiang River, with geographical coordinates between 31° 39′–32° 26′ N and 103° 44′–104° 39′ E. The basin has two headstreams: the mainstream Baicao River and the tributary Qingpian River. Baicao River, originated in Huaziling on the east slope of Minshan Mountain, Songpan County, flows from northwest to southeast, with the joining of Siergou, Taibaigou and Houyuangou successively, and then to Zhicheng. Qingpian River, originated in Chaqishan Mountain, Qingpian Town, flows from west to east and gets its name after joining Zhenghe River. Qingpian River flows southeast, with the joining of Tumengou at Huatouzui, and then changes to flow northeast. After joining Baicao River at Zhicheng, it is called JianJiang River, flowing eastward through Xuanping and Qushan. After the joining of Fuxing River at Dengjia, it turns southeast to Tongkou Town and then called Tongkou River. Tongkou River flows southward to Huazui, suburb of Jiangyou City, and then into the right bank of Fujiang River. The catchment area is 4520 km^2, and the length of the main river is 173 km in this basin.

Tongkou River Basin is fan-shaped with terrain high in northwest and low in southeast. The northwest is typical medium–high and high mountain canyon landform. The section from Heyuan to Beichuan is situated at the fold belt of Longmen Mountain. Affected by long-term erosion and cutting of rivers, it has high mountains and deep valley, and large relative height difference, with the average elevation of 1000–4000 m and the highest peak, Chaqishan Mountain of 4769 m above sea level. The southeast is shallow-cut low hill and plain terrain, with the average elevation of 500–1000 m. Downstream of Qinglian, the river enters the basin hilly area, with the average elevation of 500 m or so.

Tongkou River Basin is divided into the upper reaches from its river head to Zhicheng, the middle reaches from Zhicheng to Tongkou Town, and the lower reaches after Tongkou Town. The reaches above Beichuan County is a typical "V"-shaped canyon. The river valley below the county is wider, with developed floodplain and I-class terrace. The upstream channel is steep, whereas the downstream channel is gentle. The reaches above Huilongqiao of Baicao River has the average gradient of 36.0%; the reaches from Huilongqiao to Zhicheng is 49 km long, with the natural fall of 473 m and the average gradient of 9.65%; the reaches from Zhicheng to Hekou is 68 km long, with the natural fall of 241 m and the average gradient of 3.54%.

In the basin, there are virgin forests in the upper reaches and abundant vegetation in the middle and lower reaches. The upper reaches of river head are mostly metamorphic strata, whereas the middle and lower reaches are sedimentary strata dominated by carbonate and clastic rocks. The bedrocks on both banks of the basin are exposed, mainly phyllite and sandstone. Hydropower plants, including Kuzhu, Qushan, Tongkou, Xiangshui and Qinglian, have been built successively along the middle and lower reaches. Among which, Kuzhu hydropower plant is located 4.0 km upstream of Beichuan County, with an installed capacity of 12 MW. Qushan hydropower plant is located in Beichuan County, with an installed capacity of 3.75 MW. Tongkou hydropower plant is located 1.5 km upstream of Tongkou Town, with an installed capacity of 45 MW. Xiangshui hydropower plant is located about 5 km downstream of Tongkou Town, with an installed capacity of 20 MW. Qinglian hydropower plant is located in Jiangyou City, with an installed capacity of 17.6 MW.

2.1.2.2 Climatic Characteristics of the Basin

Tongkou River Basin is located on the east of Longmen Mountain, belonging to subtropical monsoon climate zone. The terrain in the basin is complex, with wide height difference and large vertical variation of climate. Sunshine hours, frost days, snowfall and wind will increase with the elevation, whereas temperature will decrease with the elevation. The upper reaches of the basin have a dry and cold climate in winter, with the top of the mountain above 4200 m covered with snow for 8 months. The middle and lower reaches of the basin have a mountainous subtropical humid monsoon climate, with mild climate, abundant rainfall, less sunshine, low solar radiation, long frost-free period, plenty clouds, high humidity and four distinct seasons. The temperature rises rapidly in spring, with frequent cold air activities, low temperature and rainy weather, often cold in late spring. It is not extremely hot in summer, with concentrated rainfall, more floods and droughts. The temperature drops quickly in autumn, with continuous rainfall. It is dry in winter, with little rain.

Affected by climate and topography, the precipitation in Tongkou River Basin is uneven in regions and seasons. According to the statistics of measured data, the precipitation from May to October accounts for 91.1% of that of the whole year. The distribution of precipitation decreases from southeast to northwest. There are more rainstorm in the southeast, less rainfall in the central and sharp decline in the northwest, with the average annual precipitation of 600–900 mm.

According to the statistics of Beichuan meteorological station from 1961 to 1990, the average temperature was 15.6 °C, with the extreme highest temperature of 36.1 °C and the extreme lowest temperature of – 4.5 °C. The average relative humidity was 76%, with the minimum relative humidity of 15%. The average annual evaporation was 992.9 mm (20 cm evaporating dish), with the average annual precipitation of 1355.4 mm. Precipitation was mainly concentrated in May–September, accounting for 86.3% of the year's total, with the maximum precipitation of 323.4 mm on a single day. The average wind speed was 1.3 m/s, with the maximum wind speed of 12.0 m/s and the corresponding wind direction of southwest. According to the statistics of water temperature data of Jiangjunshi hydrological station from 1963 to 1975, the average water temperature was 13.2 °C.

According to the statistics of Piankou, Baishi and Zhicheng meteorological stations upstream of the barrier lake, the average annual precipitation of the basin in early June was 27.8 mm. Beichuan meteorological station, with the elevation of 638.9 m, recorded the meteorological characteristic values as shown in Table 2.1.

2.1.2.3 Main Hydrological Stations

There are Maoba, Jiangjunshi, Tongkou and Beichuan hydrological stations, Zhicheng water level station, Baishi and Piankou precipitation stations, and Tangjiashan temporary precipitation station, which have been set up successively along the main stream of Tongkou River Basin.

(1) Maoba Hydrological Station

Maoba hydrological station was set up in 1962, located about 1 km downstream of Beichuan County, with the catchment area of 3577 km^2. It collected the data of water level, discharge and precipitation for three years until it was revoked in 1964.

(2) Jiangjunshi Hydrological Station

Jiangjunshi hydrological station is located in Xiping Town, Jiangyou City, with the catchment area of 4162 km^2, used as the Tongkou River control station. The station was set up by the Hydrological General Station of Department of Water Resources of Sichuan Province in May, 1957, began the test of water level and discharge on June 1, the observation and test of precipitation in 1958, and the test of suspended load in 1959. The station was handed over to POWERCHINA Chengdu in 1963 and changed to water level station in February 1982. Since Qinglian hydropower plant completed for water storage in 1987 affected its discharge measurement, Jiangjunshi hydrological station was moved upstream 8.6 km to Tongkou Town. The measured maximum peak discharge was 4820 m^3/s (July 21, 1964), and the maximum peak discharge of historical flood was 8060 m^3/s (July 27, 1902).

(3) Tongkou Hydrological Station

Tongkou hydrological station is 1.0 km downstream of Tongkou Town, with the catchment area of 4102km^2. In 1987, a temporary water gauge was set under the chain bridge in Tongkou Town to observe the water level. In July of the same year, the observation post was moved down 2 km to set up Tongkou hydrological station. It began the observation and test of water level, precipitation and discharge in January 1988. In 1997, due to the construction of Xiangshui hydropower plant 5 km downstream, the test section of Tongkou hydrological station was located in the backwater zone of the plant, making the test conditions lost. It was revoked at the end of 1997. The measured maximum peak discharge was 5390 m^3/s (August 1995).

Table 2.1 Statistics of meteorological characteristic values of Beichuan meteorological station

Items		Jan.	Feb.	Mar.	Apr.	May	June	July	Aug.	Sept.	Oct.	Nov.	Dec.	The whole year	Notes
Precipitation (mm)	Average	6.1	9.5	23.5	56.5	100.9	141.9	370.2	337.8	219.4	65.0	20.5	4.1	1355.4	1961–1990
	Maximum precipitation on a single day	9.5	10.5	15.3	54.2	90.3	201.4	243.4	228.9	323.4	53.6	36.6	6.7	323.4	
	Average precipitation days (days)	6.1	8.9	12.1	14.1	17.1	15.5	18.2	17.2	18.7	17.5	10.3	5.1	160.8	
Evaporation (mm)		39.0	39.4	64.9	102.4	148.2	135.4	130.7	123.8	74.5	52.8	44.6	37.3	992.9	1961–1990
Temperature (°C)	Average	5.3	6.8	11.3	16.2	20.5	23.3	24.8	24.3	20.3	16.1	11.2	6.7	15.6	1961–1990
	Extreme highest	17.0	21.5	28.3	31.9	35.9	35.7	36.1	35.7	33.4	28.1	24.7	19.8	36.1	
	Extreme lowest	– 4.0	– 2.8	– 0.9	2.8	8.2	12.9	16.4	15.9	12.0	3.9	0	– 4.5	– 4.5	
Humidity (%)	Average	73	74	72	72	72	72	80	81	84	84	79	75	76	1961–1990
	Minimum	20	17	17	16	15	16	33	31	32	22	23	15	15	
Wind speed (m/s)	Average	1.0	1.1	1.4	1.6	1.7	1.7	1.4	1.3	1.0	0.9	1.0	1.0	1.3	1961–1990
	Maximum	7.0	7.3	8.0	8.7	12.0	10.0	9.0	8.7	8.3	8.3	7.3	6.0	12.0	
	Corresponding wind direction	E	E	E	ENE	SW	NNE	ENE	ENE	ENE	NE	ENE	2G	SW	
Ground temperature (°C)	Average	5.8	7.4	11.4	17.5	22.0	24.8	26.4	26.0	21.5	17.2	12.1	6.8	16.6	1961–1990

(continued)

Table 2.1 (continued)

Items		Jan.	Feb.	Mar.	Apr.	May	June	July	Aug.	Sept.	Oct.	Nov.	Dec.	The whole year	Notes
Sunshine (h)	Average	56.6	43.7	63.0	88.9	111.5	111.9	119.3	119.3	51.9	43.7	51.2	59.1	919.9	
Water temperature (°C)	Average	5.7	6.9	10.4	13.2	15.7	17.8	19.6	20.0	17.2	14.1	10.8	7.1	13.2	Jiangjunshi hydrological station 1963–1975
	Highest	9.0	11.6	15.7	18.9	22	24.8	26.5	24.5	21.5	17.7	16.1	11.5	26.5	
	Lowest	0.6	3.0	5.5	8.0	12	13.4	16	16.4	12.4	1.5	5.0	1.5	0.6	

Table 2.2 Main hydrological stations in Tongkou river basin

Name of station	Catchment area (km^2)	Start time	End time	Notes
Beichuan	3577	1998	Present	
Maoba	3577	1962	1964	Revoked
Tongkou power station	4102	January 1988	1997	Revoked
Jiangjunshi hydrological station	4162	May 1957	1987	Revoked
Zhicheng		1962	Present	

(4) Beichuan Hydrological Station

Beichuan hydrological station was set up in downtown of Beichuan County in 1998, with controlled basin area of 3577 km^2, and has been used for observation till now.

The main hydrological stations in Tongkou River Basin are shown in Table 2.2.

2.1.2.4 Runoff Characteristics of the Basin

Runoff in Tongkou River Basin mainly comes from precipitation, partially from snow melting and groundwater recharge. The change of runoff is consistent with the change of precipitation. The average discharge of the basin is 118 m^3/s, and the annual runoff is 3.71 billion m^3. The distribution of runoff is uneven during the year, mainly concentrated from May to October, when the average discharge is 187 m^3/s, and the corresponding runoff is 2.98 billion m^3, accounting for 80% of the year's total. Of which, from July to September, the average discharge is 249 m^3/s, and the corresponding runoff is 1.98 billion m^3, accounting for 53.5% of the year's total. The average discharge and runoff of Jiangjunshi hydrological station and Tongkou River Basin are shown in Table 2.3.

The runoff of Tongkou River Basin has little inter-annual change. The measured maximum annual runoff was 4.70 billion m^3 (1964), and the measured minimum annual runoff was 2.52 billion m^3 (1969), with the extreme value ratio of 1.87. The average water inflow of Beichuan hydrological station from mid-May to late June is shown in Table 2.4.

2.1.2.5 Characteristics of Rainstorm Flood

Tongkou River flood is formed by rainstorm. Since the basin is located in the rainstorm area of Longmen Mountain, the rainstorm centers are mostly in Zhicheng, Beichuan and Jiangyou area. Every May, rainstorm can occur in this area, but is not intense. In June, with the western Pacific subtropical high (WPSH) extending westward and northward, rainstorm occurs from time to time in this area. In July and August, the axis of WPSH is at the north of the 30° N line, and this area is at the northwest edge of WPSH. Therefore, frequently, there are cold and warm air

Table 2.3 Average discharge and runoff of Jiangjunshi hydrological station and Tongkou river basin

Items		Jan.	Feb.	Mar.	Apr	May	June	July	Aug.	Sept.	Oct.	Nov.	Dec.	Year
Jiangjunshi hydrological station	Average discharge (m^3/s)	28.8	25.3	32.6	65.6	108	112	227	251	211	125	67.1	40.6	108
	Runoff (million m^3)	77.2	61.7	87.2	170	289	289	609	672	546	334	174	109	3420
	Percentage (%)	2.26	1.81	2.55	4.98	8.46	8.46	17.8	19.7	16.0	9.78	5.09	3.19	100
Tongkou river basin	Average discharge (m^3/s)	31.3	27.5	35.4	71.3	117	121	247	273	229	135	72.9	44.1	118
	Runoff (million m^3)	83.9	67	94.7	185	313	314	661	730	593	363	189	118	3710
	Percentage (%)	2.26	1.81	2.55	4.99	8.44	8.47	17.8	19.7	16.0	9.79	5.10	3.18	100

Table 2.4 Statistics of average water inflow of Beichuan hydrological station from Mid-May to late June

Time	Mid-May	Late May	Early June	Mid-June	Late June
Average daily discharge (m^3/s)	75.2	80.1	82.4	86.8	77.0
Ten-day water inflow (million m^3)	65.0	69.2	71.2	75.0	66.5

masses intersecting with each other, precipitation system activities, and rainstorms. In September, the WPSH gradually weakens and retreats eastward, sometimes accompanied by rainstorm. In October, the WPSH obviously retreats eastward, and the precipitation system further weakens, mainly with continuous drizzle. Accordingly, the annual maximum flood in the basin mainly occurs from June to September, mostly in July and August, accounting for 75.6% of the total frequency. See Table 2.5.

The statistical analysis of the daily precipitation of Beichuan hydrological station from 1961 to 2007 shows that the maximum average monthly precipitation in Tangjiashan area occurred in July and August, with the data of 343.5 mm and 337.5 mm, respectively. The average days of heavy rain with daily precipitation more than (equal to) 25.0 mm were mainly concentrated in June–September, mostly in July–August, 4.3 days and 3.9 days, respectively, followed by 2.0 days in September and less than 2.0 days in June. The average days of rainstorm with daily precipitation more than (equal to) 50.0 mm were mainly concentrated in June–September, mostly in July–August, 2.4 days and 1.6 days, respectively, followed by 0.8 days in September and only 0.7 days in June.

The flood process is characterized by steep rise and fall. A flood usually lasts 5–7 days. The historical maximum peak discharge was 6720 m^3/s. According to the statistics of the annual maximum flood at Jiangjunshi hydrological station, the annual maximum flood first occurred on June 15 (1986, 530 m^3/s) and on September 30 at the latest (1960, 3410 m^3/s). The maximum value of the measured annual maximum discharge series was 5390 m^3/s (August 11, 1995), and the minimum value was 505 m^3/s (September 16, 1991), the ratio of which was 10.7 times, indicating the great inter-annual change of floods. The statistics of annual maximum discharge frequency of Jiangjunshi (Tongkou) hydrological station are shown in Table 2.5.

Table 2.5 Statistics of annual maximum discharge frequency of Jiangjunshi (Tongkou) hydrological station

Time	June	July	Aug.	Sept.	Total
Frequency of occurrence	5	15	16	5	41
Percentage of total (%)	12.2	36.6	39.0	12.2	100

2.1.3 Design Flood of Fujiang River Main Stream

Fujiang River main stream mainly includes Fujiangqiao, Shehong and Xiaoheba hydrological stations, all of which have a long series of measured data, and have conducted a large number of surveys of historical floods.

2.1.3.1 Determination of Flood Survey and Return Period

POWERCHINA Chengdu, Chengdu Technological University, CWRC, and other units have conducted a large number of surveys of historical floods in Fujiangqiao, Shehong and Xiaoheba sections. The main historical floods occurred in 1519, 1781, 1782, 1840, 1882, 1889, 1900, 1902, 1903, 1917, 1921, 1927, 1936, 1937, 1945, 1954, 1961, 1964 and 1966. Among them, the long-term floods in 1519, 1781, 1782 and 1882 were only found in the stone carvings with flood marks of the Great Buddhist Temple in Tongnan, as well as the flood marks of other sections. The survey results of historical floods of Fujiang River main stream are shown in Table 2.6.

2.1.3.2 Calculation of Design Flood of Main Hydrological Stations

The frequency calculation of the floods of Fujiangqiao, Shehong, Xiaoheba and other hydrological stations followed the principle of independent sampling according to the annual maximum value of measured flood series. For Fujiangqiao hydrological station, the historical floods in 1889, 1902, 1917 and 1945 were considered. For Shehong hydrological station, the historical floods in 1840, 1889 and 1945 were considered, and the flood in 1889 was only counted with its water level instead of water volume. The measured flood in 1981 was taken as extra-large value. For Xiaoheba hydrological station, the historical floods in 1840, 1945, 1937 and 1921 were considered, and the measured flood in 1981 was taken as extra-large value. The measured flood series and historical floods of the three stations constituted non-sequential flood series, which were used for the frequency analysis and calculation, respectively.

The empirical frequency calculation of historical floods used the formula $P_M = \frac{M}{N+1} \times 100\%$, whereas the empirical frequency calculation of measured sequential series used the formula $P_m = \left(\frac{a}{N+1} + \left(1 - \frac{a}{N+1}\right) \times \frac{m-l}{n-l+1}\right) \times 100\%$. With P-III curve as the type of frequency curve, the statistical parameters were determined by curving fitting method. The results of design peak discharges of each frequency are shown in Table 2.7.

Table 2.6 Survey results of historical floods of Fujiang river main stream

River section	Catchment area (km^2)	Survey results of measured floods (ranked by discharge)			
		Year	Peak discharge (m^3/s)	Return period (year)	Reliability level
Fujiangqiao	11,903	1902	14,800	120	Relatively reliable
		1889	14,000	60	For reference
		1917	12,900	40	For reference
		1945	12,600	30	Relatively reliable
Shehong	23,574	1840	30,400	226	For reference
		1889		113	
		1945	26,200	75	Relatively reliable
		1981	25,700	57	Measured with instruments
		1937	20,400		Relatively reliable
Xiaoheba	29,488	1870			
		1981	28,700	168	Measured with instruments
		1840	28,100	84	For reference
		1903			
		1945	25,500	56	Reliable
		1937	24,000		Relatively reliable
		1921	22,600		Relatively reliable

Table 2.7 Results of design peak discharge of Fujiangqiao, Shehong and Xiaoheba of Fujiang river main stream

Name of station	Statistic parameters			Design peak discharge of each frequency				
	Mean value	C_v	C_s	1%	2%	5%	10%	20%
Fujiangqiao	5680	0.5	3.0	15,100	13,500	11,200	9470	7640
Shehong	9780	0.58	3.0	29,400	25,800	21,000	17,300	13,500
Xiaoheba	8820	0.58	3.5	27,300	23,800	19,000	15,500	11,900

2.2 Hydrological Characteristics of Barrier Lake Area

2.2.1 Hydrological Overview of Barrier Lake Area

The average annual precipitation in Tongkou River Basin is 1355.4 mm, of which the precipitation from May to October accounts for 91.1% of the year's total. The distribution of precipitation decreases from southeast to northwest. The catchment area of Tangjiashan barrier lake is 3550 km^2, accounting for 85.3% of the drainage area controlled by Jiangjunshi hydrological station. According to the calculation of the runoff results of the station, the average discharge of Tangjiashan barrier lake area is 92.3 m^3/s, and the annual runoff is 2.91 billion m^3; the average discharge from May to October is 147 m^3/s, and the corresponding runoff is 2.34 billion m^3, accounting for 80% of the annual runoff; the average discharge from July to September is 196 m^3/s, and the corresponding runoff is 1.56 billion m^3, accounting for 53.5% of the annual runoff. It is calculated that the maximum annual runoff of Tangjiashan barrier lake area was 3.70 billion m^3 (1964), and the minimum annual runoff was 1.98 billion m^3 (1969), with the runoff extremum ratio of 1.87.

Since Tongkou River Basin is located in the rainstorm area of Longmen Mountain, the rainstorm centers are mostly in Zhicheng, Beichuan and Jiangyou area. Corresponding to the rainstorms, the annual maximum flood in the basin mainly occurs from June to September, mostly in July and August. The flood process is characterized by steep rise and fall, with sharp and thin peak pattern. Generally, a flood process lasts 5–7 days.

Tangjiashan barrier dam was located at the reservoir tail of Kuzhu hydropower plant in Tongkou River. It was 1.2 km away from Kuzhu hydropower plant and 4.6 km away from Beichuan hydrological station downstream. There were two tributaries joining the reservoir area of the barrier lake: one was the Baicao River joining near Zhicheng and the other was the Yuxi River joining near Xuanping. The capacity of the barrier lake was 316 million m^3 when the water level was 752 m, 238 million m^3 when the water level was 742 m, and 86 million m^3 when the water level was 714 m. When the water level of the barrier lake was 740 m, the length of backwater along the main stream of Tongkou River was about 45 km, and the width of the lake surface was generally narrow. The water surface from Xuanping to Zhicheng was slightly wider, with the width of 340 m in front of the barrier dam, 990 m in Xuanping, 690 m in Zhicheng and 160 m near the tail of the lake. The elevation of lake bottom was about 662 m in front of the dam, 660 m in Xuanping and about 710 m in Zhicheng. When the water level of the barrier lake was 740 m, the water depth was 78 m in front of the dam, 78 m in Xuanping and 30 m in Zhicheng.

Table 2.8 Design flood of Tangjiashan barrier lake

Items	0.5%	1%	2%	5%	10%	20%
Peak (m^3/s)	6970	6040	5120	3920	3040	2190
Discharge (million m^3)	343	299	256	200	158	117

2.2.2 *Design Flood*

The catchment area of Tangjiashan barrier lake was close to that of Xuanping hydropower plant, with a difference of only 1.1%. Xuanping hydropower plant is the controlling cascade plant of the "one reservoir seven cascades" scheme of Tongkou River, the primary tributary of Fujiang River, which was proposed in the *Planning Report of Fujiang River Basin in Sichuan Province.* It is a comprehensive large-scale water storage project with power generation as its main function, combined with irrigation, industrial and domestic water supply, and helping downstream cities against excessive floods. Xuanping hydropower plant has the catchment area of 3511 km^2, normal water storage level of 785 m, total reservoir storage capacity of 599 million m^3, regulating reservoir storage capacity of 392 million m^3, maximum dam height of 140 m, and total installed capacity of 150 MW.

After interpolation, prolongation and reduction, Maoba hydropower plant had the annual maximum discharge series of 47 years from 1957 to 2003, which formed non-sequential series together with the historical floods in 1902 and 1934. The empirical frequencies were calculated, respectively, with P-III curve as the fitting curve, the parameters estimated by moment method and the statistical parameters determined by curve fitting method.

Based on the results of design flood of Maoba hydropower plant, the design flood of Tangjiashan barrier dam site was calculated by hydrological analogy. The design peak and discharge of natural flood of Tangjiashan barrier lake are shown in Table 2.8, and the typical hydrographs of design flood are shown in Table 2.9.

2.3 Acquisition and Application of Basic Data

2.3.1 *Acquisition of Basic Data*

"Safe, scientific and fast" are the basic principles of emergency handling of barrier lake. During the calculation and analysis of the break-dam flood of barrier lake, attention should be paid to the rapid acquisition and processing of data, so as to provide decision-making basis for the emergency handling of barrier lake as soon as possible. In view of the lack of basic data of the barrier lake area, the units involved in the emergency handling shall work closely together, actively contact partner institutions at home and abroad, acquire the latest remote sensing data, collect

Table 2.9 Hydrographs of design flood of Tangjiashan barrier dam site

$P = 2\%$ flood hydrograph		$P = 5\%$ flood hydrograph		$P = 10\%$ flood hydrograph		$P = 20\%$ flood hydrograph	
T (h)	Q (m^3/s)	T (h)	Q (m^3/s)	T (h)	Q (m^3/s)	T (h)	Q (m^3/s)
1	1540	1	1200	1	947	1	704
2	1900	2	1480	2	1170	2	868
3	2260	3	1770	3	1390	3	1030
4	2620	4	2050	4	1610	4	1200
5	3000	5	2340	5	1850	5	1370
6	3470	6	2710	6	2130	6	1590
7	4100	7	3200	7	2520	7	1870
8	5120	8	3920	8	3040	8	2190
9	4240	9	3310	9	2610	9	1940
10	3760	10	2940	10	2310	10	1720
11	3490	11	2730	11	2150	11	1600
12	3400	12	2660	12	2090	12	1550
13	3280	13	2560	13	2020	13	1500
14	3160	14	2470	14	1940	14	1440
15	3040	15	2380	15	1870	15	1390
16	2910	16	2270	16	1790	16	1330
17	2810	17	2200	17	1730	17	1280
18	2710	18	2120	18	1670	18	1240
19	2610	19	2040	19	1610	19	1190
20	2510	20	1960	20	1540	20	1150
21	2410	21	1880	21	1480	21	1100
22	2310	22	1800	22	1420	22	1060
23	2210	23	1730	23	1360	23	1010
24	2130	24	1660	24	1310	24	970
25	2060	25	1610	25	1270	25	940
26	2000	26	1560	26	1230	26	910
27	1940	27	1520	27	1190	27	890
28	1880	28	1470	28	1160	28	860
29	1820	29	1420	29	1120	29	830
30	1750	30	1370	30	1080	30	800
31	1690	31	1320	31	1040	31	770
32	1630	32	1270	32	1000	32	740
33	1570	33	1230	33	966	33	720
34	1510	34	1180	34	929	34	690
35	1440	35	1130	35	886	35	660

and process basic data, and provide relevant basic data for the calculation of the dam-break flood in time.

After the formation of Tangjiashan barrier lake, in order to determine the capacity of the barrier lake and analyze the flood routing, many units involved in the hazard analysis and formulation of disaster relief collected the existing topographic data with various precision by using possible information systems, and moreover, some units used remote sensing technology to quickly acquire the river topography, so as to measure the storage capacity curve of the barrier lake and split the section of flood routing. In a very short time before May 22, the units provided the hydrological and topographic data of the barrier lake and its affected area that meet the requirements for the formulation of disaster relief scheme.

2.3.1.1 Full Utilization of Existing Measurement Results

There was a 1:50,000 topographic map of the affected area of Tangjiashan barrier lake, but lack of complete large-scale topographic and section data. In the past engineering design and construction, there were still some large-scale topographic data available.

POWERCHINA Chengdu, when designing Xuanping hydropower plant, measured the 1:2000 topographic maps of Xuanping reservoir area and the section from Xuanping to Tangjiashan, the 1:5000 topographic map of Beichuan County and its vicinity, as well as the 1:10,000 topographic map of Tongkou hydropower plant and part of its downstream range.

Hydrology and Water Resources Survey Bureau of Mianyang measured the large section data from Zhicheng to Tangjiashan before the earthquake. The barrier lake hydrological emergency monitoring group for earthquake relief of MWR also emergently measured the large section data from Jiangjunshi to Qinglian town.

2.3.1.2 Wide Application of Remote Sensing and Other New Technologies to Acquire Data

Remote sensing technology is an efficient means to acquire information in emergency rescue [1]. It has the features of quick acquisition and convenient application and can be applied to volume calculation, extraction of topographic section, etc.

(1) DEM Data

The 1:50,000 DEM data of Tangjiashan barrier lake area are shown in Fig. 2.2.

(2) TM Image

The pre-earthquake TM image data by TM color-synthetic remote sensing, with 15 m resolution, are shown in Fig. 2.3.

Fig. 2.2 DEM data diagram

Fig. 2.3 Pre-earthquake TM image data

(3) Multi-spectral Data

The image of post-earthquake Beichuan County on May 19 by Fuwei-2 satellite, with 8 m resolution, is shown in Fig. 2.4.

(4) Radar Remote Sensing Data

The radar image of Tangjiashan barrier lake area on May 20 (Fig. 2.5), with 3 m resolution, very sensitive to the edge of water body, suitable for extracting the edge of barrier lake and calculating the surface area.

(5) Aerial Digital Photography Data

The picture of downtown Beichuan County acquired by UCD sensor on May 18, with 0.3 m resolution and high accuracy, can be used to confirm the damage of houses, roads and bridges.

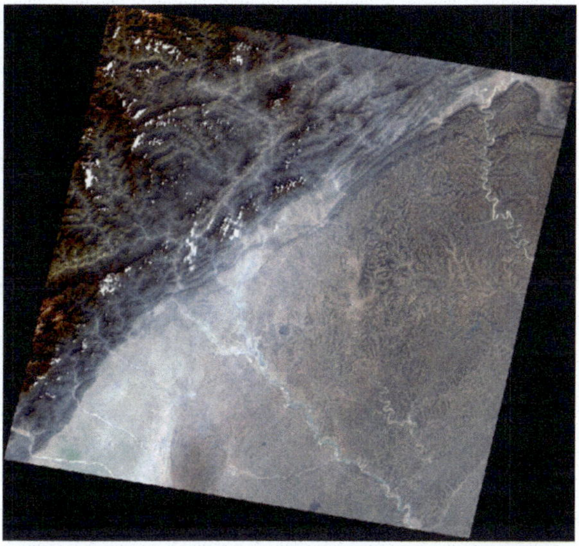

Fig. 2.4 Post-earthquake multi-spectral data on May 19

Fig. 2.5 Radar Image of Tangjiashan barrier lake on May 20

(6) Airborne LIDAR System

The collected LIDAR 3D topographic map of Tangjiashan barrier lake, with 0.15 m resolution, is shown in Fig. 2.6.

Fig. 2.6 LIDAR 3D topographic map

2.3.2 Application of Acquired Data

2.3.2.1 Digital Processing of Image Data

(1) Digitization of Paper Topographic Map

The topographic map of a large range of possibly flooded area above and downstream of Tangjiashan was only available with the aerial map of 1:50,000 in 1966, which adopted the Yellow Sea elevation system in 1956 as the elevation system and Beijing coordinate system as the plane control system. After scanning, the contour line was vectorized by CAD application software and converted into DEM.

(2) Digitization of Remote Sensing Image Data

By the use of ERDAS remote sensing image processing system, the DEM image was converted into COVERAGE file applicable to ARCVIEW, and then, the file was converted into DXF file with elevation by ARCVIEW 3D conversion module. Since the SRTM DEM image included cloud data, after the cloud data (with elevation greater than 20,000 m) was deleted, the DEM was reconstructed by interpolation.

2.3.2.2 Selection of Section Data in Flood Calculation

(1) Extraction of Topographic Section of DEM Data

During the calculation of the dam-break flood of one-dimensional unsteady flow, it was necessary to split the sections along the river channel according to the river

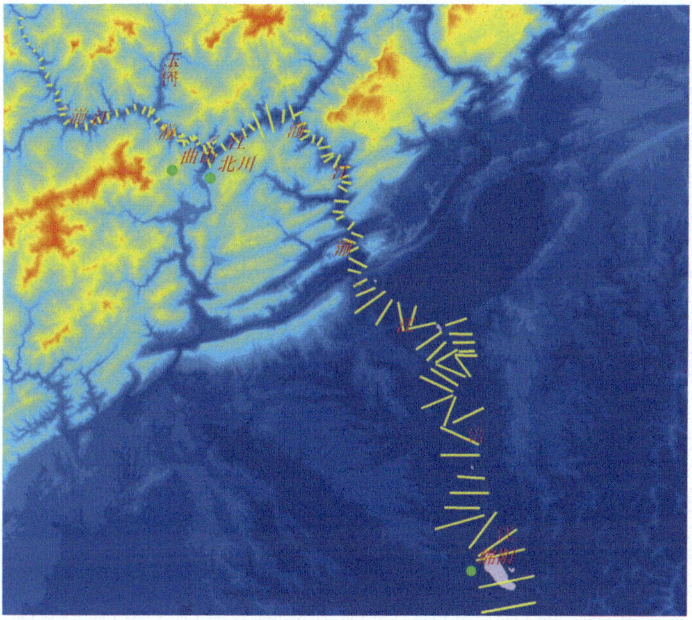

Fig. 2.7 Schematic diagram of section splitting based on DEM

regime, so as to acquire the elevation data of the sections with appropriate spacing (Fig. 2.7).

According to the river channel trend, multiple sections were determined and split on DEM, and the endpoint coordinate information of each section was saved. Then, the elevation information of each section was extracted, and based on the endpoint coordinate information of the sections and the spacing of sampling points, the coordinate position information of the sampling points was calculated. Next, According to the coordinate of the sampling point, its grid position was determined on the DEM, and its elevation data were finally acquired by bilinear interpolation combined with the elevation data of the grid center point; thus, the elevation data of each sampling point on the sections could be acquired by analogy.

(2) Selection of Large Section Data of Flood Routing

The collected topographic maps with various precision, measured large sections and other data should be selected according to the principle of "precision first". First of all, the measured large section data were directly adopted, and then, the required sections were split from the large-scale topographic maps as far as possible, and the topographic sections split with DEM data were used to make up the insufficient part.

2.3.2.3 Unification of Elevation System

The barrier dam was the new terrain formed after the earthquake. On May 21, surveyors set a relative elevation coordinate system on the Tangjiashan barrier dam, surveyed the topography of the barrier dam, and observed the front water level of the dam according to the relative elevation on the dam. However, the collected topographic data of the upstream and downstream areas of the barrier dam were all the data of national elevation system before the earthquake. After the earthquake, the land traffic near Tangjiashan barrier lake was completely interrupted, and it was impossible to conduct the lead point measurement of the elevation system. How to establish the relation between the relative elevation system of the barrier dam and the national elevation system was important for the determination of lake capacity, flood regulation calculation of barrier lake and deduction of downstream flood.

Establishing the relation between the sudden disaster body and the national elevation system by drawing on relative reference point is a method that can be adopted under extraordinary conditions, and its accuracy can meet the emergency needs depending on the experience of experts. The national elevation system adopted by Zhicheng water level station, upstream of Tangjiashan barrier in Tongkou River, was a relatively reliable elevation reference point after Wenchuan earthquake. After the Tangjiashan barrier lake was plugged, the flow velocity of the backwater formed in front of the dam was extremely slow. On May 19 when the backwater reached Zhicheng, the river channel entered a state of backwater. According to experts' experience, the difference of water level between Zhicheng and Tangjiashan dam was within 0.40 m at the same moment. Therefore, on May 22, based on the observation data of Zhicheng and in front of the dam at the same moment, the relation data between the front water level of the dam and the national elevation system were drawn up, and the elevation system of the measured topographic maps of the barrier dam was revised.

2.3.2.4 Determination of Storage Capacity Curve of Barrier Lake

The storage capacity curve of Tangjiashan barrier lake is one of the important parameters for the calculation of dam-break flood and flood regulation and the determination of disaster relief scheme. Hydrology Bureau of CWRC, CRSRI, Hydrology and Water Resources Survey Bureau of Mianyang, POWERCHINA Chengdu and other units, respectively, calculated the storage capacity curve of Tangjiashan barrier lake by different methods and data.

(1) POWERCHINA Chengdu calculated the stage-area curve and stage-storage capacity curve of Tangjiashan Lake area by using the measured topographic map of Xuanping hydropower plant upstream of Tangjiashan, ignoring the effects of earthquake on the river channel.

(2) Hydrology Bureau of CWRC and CRSRI, based on DEM data, applied the storage capacity calculation module developed on GIS platform to calculate the stage-area curve and stage-storage capacity curve of Tangjiashan Lake area.
(3) Hydrology and Water Resources Survey Bureau of Mianyang applied the measured large section data of river channel in Tangjiashan barrier lake area before the earthquake, and analyzed and calculated the storage capacity curve of Tangjiashan by using the section method. Of which, there was large spacing between the sections above Xuanping and dense sections from Xuanping to Tangjiashan.

Through comprehensive analysis and comparison, the calculation results of POWERCHINA Chengdu by using the measured large-scale topographic map was considered as having the highest accuracy among the various results above, although there existed errors between post-earthquake topographic map and pre-earthquake topographic map. Experts from the water resource group of the Earthquake Relief Headquarters under the State Council concluded that the accuracy was acceptable for lake capacity, and the basic application data with the main purpose of estimation of dam-break flood and deduction of discharged flood. The inversion of water inflow and the prediction of stage in front of the dam, with the principle of consistent inflow and outflow, could basically meet the application requirements. Therefore, the calculation results of storage capacity curve of POWERCHINA Chengdu were adopted, as shown in Table 2.10 and Fig. 2.8.

2.4 Relation Between Water Regime and Emergency Handling Project

2.4.1 Relation Between Rainstorm and Flood in Barrier Lake Area

Tongkou River Basin, where Tangjiashan barrier lake was located, is situated in Qingyi River-Lutoushan Mountain rainstorm area in Sichuan Province. A flood process is usually formed by 24 h rainfall, with the average maximum 24 h precipitation of 140 mm for years and the variation coefficient of 0.50. According to the precipitation-runoff relation in Qingyi River-Lutoushan Mountain rainstorm area analyzed in the *Calculation Manual of Rainstorm Flood in Small and Medium Basins of Sichuan Province*, the runoff coefficient in a precipitation process enlarges with the increase of precipitation, ranging from 0.85 to 0.91. The precipitation-runoff relation in Tangjiashan is shown in Fig. 2.9.

Since the duration of a flood in Tongkou River is 5–7 days, the design runoff depth was calculated back in terms of the 7-day design flood volume of each frequency of Tangjiashan barrier lake, and then the design areal precipitation of each frequency could be calculated based on the relation between precipitation and runoff, as shown in Table 2.11.

Table 2.10 Storage capacity curve of Tangjiashan barrier lake

Stage (yellow sea elevation, m)	667	670	680	690	700	710	720	730	740	750	752
Capacity (million m^3)	0	2.14	12.80	23.45	43.02	71.30	110.39	160.00	223.95	301.90	316.11

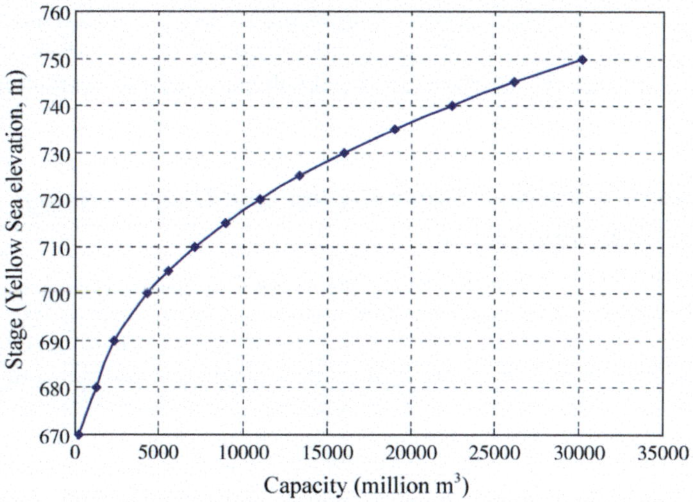

Fig. 2.8 Diagram of storage capacity curve of Tangjiashan barrier lake

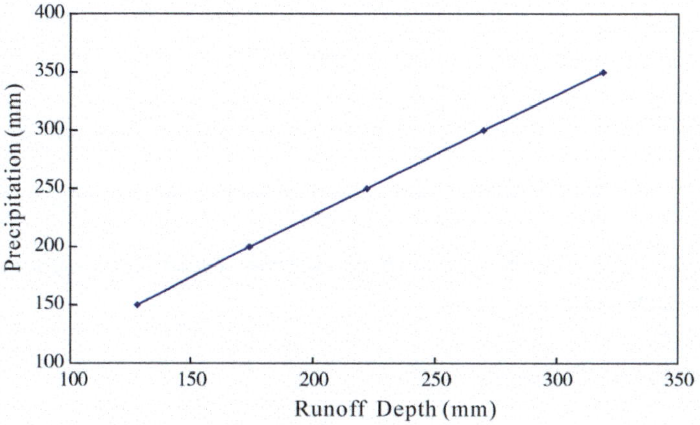

Fig. 2.9 Precipitation-runoff relation diagram of Tangjiashan

Table 2.11 Precipitation corresponding to each design flood in Tangjiashan

Frequency	0.5%	1%	2%	5%	10%	20%
W_{7d} (million m^3)	936	830	724	584	478	371
Runoff depth (mm)	264	234	204	164	135	105
Areal precipitation (mm)	296	263	232	189	158	123

If the flood of 5-year return period occurred, the average precipitation in the basin would be 123 mm, the peak discharge would be 2190 m^3/s, the 24-h flood volume would be 117 million m^3, and the 3-day flood volume would be 244 million m^3.

2.4.2 Relation Between Water Regime and Disaster Relief Construction Period

On May 23, the water resources group submitted the *Technical Scheme for Design and Construction of Tangjiashan Emergency Handling Project* to the front command of Earthquake Relief Headquarters under the State Council, recommending the scheme of excavating a drainage channel for water discharge and disaster relief. According to the comprehensive analysis of hydrometeorological conditions and reservoir storage capacity of the barrier dam, ignoring the effects of precipitation and landslide, the preliminary construction period was initially determined as 10 days. The project fully commenced on May 25 and was completed on June 3. This construction period was based on the average water inflow data for years. If there were rainstorm upstream during the construction period, the water level would rise sharply and the construction period be shortened by force. During the construction, the rising rate of water level of the barrier lake was predicted several times according to the actual water regime. The rise of actual water level of Zhicheng Station during construction was shown in Table 2.12.

2.4.3 Water Regime Prediction of Barrier Lake Area During Construction Period

2.4.3.1 Prediction Methods

The volume of water inflow to the lake was predicted by referring to the prediction scheme and storage capacity curve of Beichuan Hydrological Station downstream of Tangjiashan before the formation of Tangjiashan barrier lake. Then, the prediction scheme was revised according to the error between the predicted water level of Zhicheng and the actually observed value.

The flood of the basin above Beichuan was predicted by using the traditional precipitation-runoff prediction method. Through the setting of two precipitation stations in this section, the average areal precipitation of the basin was calculated by the arithmetic average method, and the average runoff depth in each period was the arithmetic mean value of runoff depth of each precipitation station, and the 6-h prediction scheme was used for the flood prediction scheme of the basin above Beichuan.

Table 2.12 Actual water level rise of Zhicheng station during construction

Time	Water level (m)	Daily rise (m)
8:00 am, May 19	709.84	–
8:00 am, May 20	712.49	2.65
7:40 am, May 21	714.78	2.29
8:00 am, May 22	717.48	2.70
8:00 am, May 23	719.81	2.33
8:00 am, May 24	721.65	1.84
8:00 am, May 25	723.58	1.93
8:00 am, May 26	725.20	1.62
8:00 am, May 27	727.02	1.82
8:00 am, May 28	728.66	1.64
8:00 am, May 29	730.13	1.47
8:00 am, May 30	731.92	1.79
8:00 am, May 31	733.44	1.52
8:00 am, June 1	734.71	1.27
8:00 am, June 2	735.85	1.14
8:00 am, June 3	736.93	1.08
8:00 am, June 4	738.22	1.29
8:00 am, June 5	739.16	0.94
8:00 am, June 6	740.06	0.90
8:00 am, June 7	741.09	1.03
20:00 pm, June 8	742.64	–
20:00 pm, June 9	743.74	1.10
23:15 pm, June 9	743.79	–

The rainfall in Tangjiashan barrier lake area was mainly predicted by two methods, one based on numerical weather prediction and the other based on predictor's experience, so as to achieve the daily rainfall prediction in Tangjiashan barrier lake area.

The water level of Tangjiashan barrier lake area was mainly predicted by the flood regulation calculation of static storage capacity, that is, using the known minimum regulated lake water level, and combining the storage capacity curve to calculate the future lake water level according to the predicted volume of water inflow to the lake.

From May 20 to 31, the volume of water inflow to the reservoir was estimated mainly based on precipitation, combined with the prediction of the water level process of Zhicheng Station and the daily increase record of water level in Tangjiashan barrier lake. From June 1, the prediction of specific water level of Tangjiashan (Zhicheng) was released, then the water inflow was predicted by the precipitation-runoff model, and the reservoir water level was predicted by the flood regulation calculation of static storage capacity.

2.4.3.2 Real-Time Prediction

Since May 21, the rainfall process had been predicted for three times. Among them, the two predictions on May 24–26 and June 5–6 were successful in the analysis and prediction on the possibility of occurrence, the time distribution and the total magnitude of rainfall process. Generally, the foreseeable period could be advanced by 3–4 h, and the rainfall information could be basically predicted. The rainfall process from May 24 to 26 was a critical prediction, which was of great significance to the discharge emergency response of the barrier lake.

Hydrology Bureau of CWRC started the rainfall prediction in Tangjiashan barrier lake area at 01:00 am on May 21 and conducted the first-stage rainfall prediction. Referring to the products of numerical weather model and the local topography, predictor released that the rainfall process from 24 to 26 was moderate rain, with the accumulated areal precipitation of about 30 mm. Eventually, upstream of Tangjiashan, there were local showers on 24, moderate rain and local heavy rain on 25 and sporadic light rain on 26, respectively.

(1) Prediction of Rainfall Process and Water Storage Capacity on May 25

At 08:00 am on May 25, it was predicted that there would be a moderate rainfall from 25 to 26, no rain from 27 to 28 and light rain from 29 to 31 in Beichuan area, with the accumulated rainfall of 38 mm and the total volume of water inflow to the lake of about 70 million m^3. When the water level (Zhicheng) was 723.58 m, the water level of the barrier lake would rise sharply if calculated as per the predicted volume of water inflow. According to the statistics afterward, from 08:00 am on May 25 to 08:00 am on May 31, the actual total volume of water inflow to the lake generated by the rainfall process was about 53.44 million m^3.

(2) Prediction of Discharge Time of Drainage Channel of Tangjiashan Barrier Dam

Due to the effects of the elevation change of the drainage channel and the damage of the water level observation equipment in Tangjiashan and Zhicheng in certain periods, the predicted objects varied from time to time.

At 08:00 am on May 30, according to the water regime and the foreseeable period of rainfall, it was predicted for the first time that the water level of barrier lake (Zhicheng Station) would reach 740 m on June 4.

At 20:00 pm on June 1, the prediction that "it is impossible that the water level of Tangjiashan barrier lake will reach or exceed 740 m before 08:00 am on June 5, and when it can reach 740 m after 8:00 am on June 5 will be further clarified, depending on the rainfall situation at 20:00 pm on June 4" was clearly released.

At 20:00 pm on June 3, according to the development of water and rain conditions, it was further clarified that Tangjiashan barrier lake would discharge at 02:00 am on the 7, with the foreseeable period up to 78 h. The prediction conclusion that the barrier lake would start to discharge at 02:00 am on the 7 was maintained in the following five consecutive predictions at 08:00 am on the 4, 20:00 pm on the 4, 08:00 am on the 5, 20:00 pm on the 5 and 08:00 am on the 6. Eventually, the water level of Tangjiashan barrier lake reached 740.00 m at 00:00 am on June 7.

Reference

1. Bichler A, Bobrowsky P, Best M, et al. Three-dimensional mapping of a landslide using a multi-geophysical approach: the Quesnel Forks landslide. Landslides. 2004;1:29–40.

Chapter 3
Geological Characteristics of the Tangjiashan Barrier Lake Area

3.1 Geological Conditions of the Barrier Lake Area

3.1.1 Original Topography

The Tangjiashan barrier lake was located on the Tongkou River, which was about 4.6 km north to the Beichuan County. This area was at the northwest edge of the Sichuan Basin, with a typical middle and high mountain valley landscape.

Tongkou River was the first-class tributary on the right bank of Fujiang River, which flowed through a narrow channel in this area from west to east, with a longitudinal average 3.57‰ slope drop of riverbed. The deep valley had a V-shaped cross section.

The Tangjiashan barrier dam was located at the end of the reservoir of Kuzhu Dam Hydropower Station, with a water level of 664.8–664.7 m, a water surface width of 100–130 m and a water depth of 0.5–2.0 m during the dry season.

The Tangjiashan Mountain was located on the right bank of barrier dam (Fig. 3.1), with a summit elevation of 1580 m and a slope height of 900 m approximately. The lower part had a steep slope of about 40° and exposed bedrock. The upper part had a gentle slope around 30°. A small shallow gully both appeared in the upper and lower region, respectively.

The Yuanhe Dam was located on the left bank of the barrier (Fig. 3.2), with a slope height of about 180 m. The lower part had a steep slope of about 45° and exposed bedrock. The upper part (gentle slope around 22°) was covered with residual slope deposits of gravel and soil. A small ridge with 20–50 m width sat on the top of the slope.

Q. Yang and W. Yang, *Emergency Handling of the "5.12" Tangjiashan Barrier Lake*,
https://doi.org/10.1007/978-981-99-1706-8_3

Fig. 3.1 Original view of the right bank river valley of the Tangjiashan barrier (from upstream to downstream)

Fig. 3.2 Original view of Yuanhe dam on the left bank of Tangjiashan barrier dam

3.1.2 Stratum Lithology

The upper part of the Lower Cambrian Qingping Formation, Quaternary residual slope deposit and alluvium were distributed in the river section.

(1) The upper part of the Lower Cambrian Qingping Formation ($\in_1 c$)

Gray and black thin to medium thick laminated feldspathic mica siltstone, siliceous slate, marl and mudstone form the rock layers alternating between soft and hard. They appeared in the lower part of the slope on both sides of the bank.

(2) Quaternary residual slope deposit (Q^{el-dl})

Yellow gravelly soil was composed of powdery loam, rock debris and boulders. The upper part was dominated by powdery loam which exceeded 60%, with scattering boulders. The content of rock debris and boulders increased in the lower part with thickness to be 5–20 m. It is distributed in the upper part of the slope on both banks.

(3) Quaternary alluvium (Q^{al})

Grey and black mud-bearing fine sand contains about 35% mud and a few small gravels. They laid in the bed of Tongkou River with about 20 m thickness.

3.1.3 Geological Structure

Tangjiashan was located at the Longmen Mountains thrust nappe structure which was the boundary fault of the Bayankara Massif and Southern China Massif from east in the Qinghai-Tibetan Plateau.

The Longmen Mountains thrust nappe structure started from Luding County in the southwestern direction and extended to the northeastern way through Guanxian County, Jiangyou County and Guangyuan County into the area of Mianxian County in Shaanxi Province. The total length was more than 500 km. This fault zone was the boundary between the Songpan-Ganzi geosynclinal fold system and the first geotectonic unit of Yangtze paraplatform. Through large numbers of tectonic changes, the rupture trace was relatively complex. The Longmen mountains front fault, main central fault and back fault composed the fault zone. The structure of Tongkou River region is shown in Fig. 3.3.

3.1.3.1 Longmen Mountains Front Fault

It started from Ningqiang and Mian County of Shaanxi Province in the north and ran southwest through Guangyuan, Jiangyou and Guanxian to Tianquan, with a total length of more than 500 km, and formed by several fracture feather-echelon. The overall occurrence of fracture was N35° with a northwestern dip of about 60°, which presented compressive features. The Dachuan-Tianquan section was a strong active fault since the Epipleistocene. The Longmen Mountains front fault was about 17 km away from Tangjiashan.

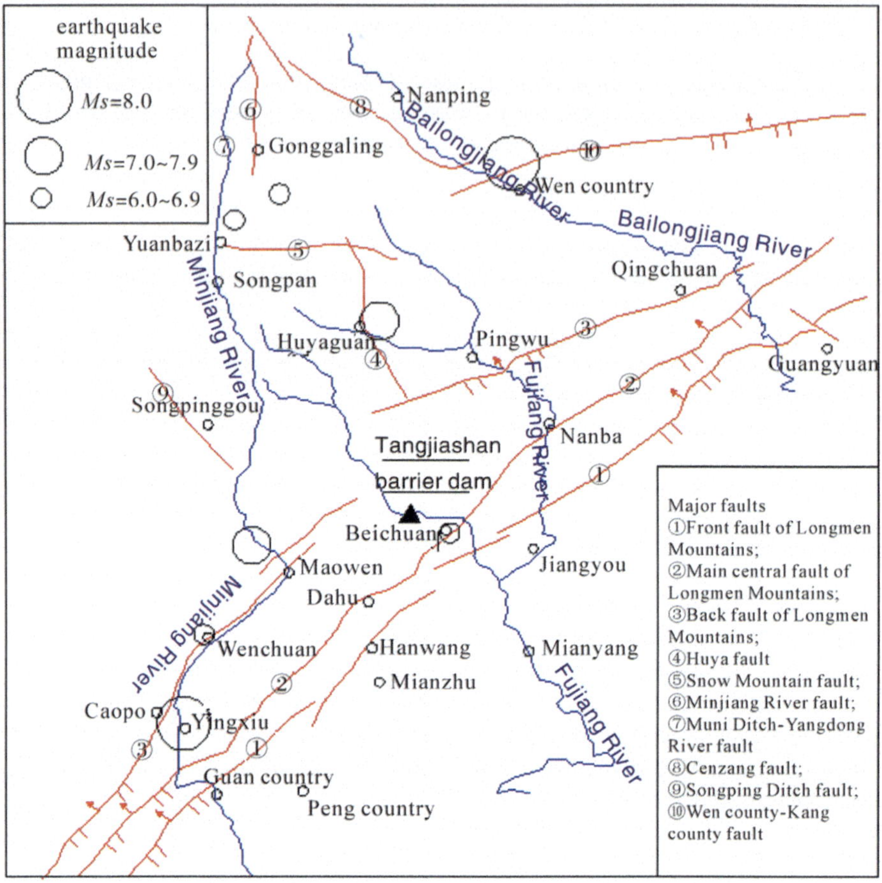

Fig. 3.3 Sketch of the structure of Tongkou river region

3.1.3.2 Longmen Mountains Main Central Fault

It began in the southwest near Luding and ran northeast through Yanjing, Yingxiu, Taiping, Beichuan, Nanba and Chaba into Shaanxi Province, which was diagonally through the entire Longmen Mountains for more than 500 km. It was called the Beichuan Fault near the project area. The overall occurrence of fracture was N45° E with a northwest dip of about 60°, and the width of shattered fault zone ranging from several meters to tens of meters. A series of parallel secondary faults on both sides of the main fault presented obvious compressive features. The Quaternary activity of the fault differed among the northern part of North Beichuan, the middle part of Beichuan-Taipingchang of An county and the southern part of South Taipingchang. It could be seen from the age data of each fault point that the latest activity age of the northern and southern sections of the fault is 110,000–1 million years, and the latest activity age of the middle section was 13,400 (± 1100) years. The activity of middle section was mostly present as stick–slip with a certain portion of right-lateral slip. The average vertical slip rate since the Epipleistocene was estimated to be 0.91 mm/a. The short leveling deformation data of the cross-fault showed that since 1985, the Chenjia dam site in the northern section of the fault was mostly present as creep-slip, and the northwestern side was relatively increasing at a rate of 0.1 mm/a, presenting compressive activity. In the southern section, the Yingxiu site was mostly present as creep-slip, and the northwestern side was relatively decreasing at a rate of 0.2 mm/a, presenting tensional activity. This indicated that the middle section of the Longmen Mountains main central fault was a strong active section since the Epipleistocene. The main central fault of Longmen Mountains was about 4 km away from Tangjiashan.

3.1.3.3 Longmen Mountains Back Fault

It started from the western side near the Lengji town of Luding and ran northeastern way through Longdong, Yuzixi, Gengda, Caopo, Wenchuan, Maowen, Pingwu and Qingchuan into Shaanxi Province, with total length of more than 500 km. The area near the project was called the Maowen Fault. The overall occurrence of fracture was N30°–50°E/NW∠50°–70°, with a fracture zone width of tens of meters, and several sections could reach more than 100 m, presenting compressive features, which consisted of fault mud, mylonite and structural lens. It could be seen from the age data of the fracture point thermoluminescence method that the age of the Maowen-Caopo section was 36,100 (± 2900) years old, while the age of the rest sections ranged from 150,000 to 480,000 years old. This indicated that the Maowen-Caopo section of the Longmen Mountains back fault was a relatively active section, and a 6.5 magnitude earthquake occurred, which was likely to be a full Pleistocene active fault. The Longmen Mountains Back Fault was 20–30 km away from Tangjiashan.

3.1.4 Earthquake

According to the seismogeological survey, the main seismic belts that had influence on the engineering area were Wudu, Pingwu, Songpan and Longmen Mountains.

Since the earthquake records in 186 BC, there have been 2 earthquakes of magnitude 8.0, 4 earthquakes of magnitude 7.0–7.9, 11 earthquakes of magnitude 6.0–6.9 and 41 earthquakes of magnitude 5.0–5.9. The earthquakes that had a large impact on the project area were:

On July 1, 1879, the strongest earthquake of magnitude 8.0 occurred in the Wen county-Wudu area of Gansu Province at about 04:00, with an epicenter intensity of 11°. The epicenter was 120 km north to the project area, and the earthquake response intensity on the project area was less than 6°.

On August 29, 1933, an earthquake of magnitude 7.5 occurred in Diexi of Maowen County, with an epicenter intensity of 10°. The epicenter was about 60 km northwest to the project area, and the earthquake response intensity on the project area was 6°.

On April 21, 1957, an earthquake of magnitude 6.5 occurred in Wenchuan County with an epicenter intensity of 8°. The epicenter was about 90 km southwest to the project area, and the earthquake response intensity on the project area is less than 6°.

On February 8, 1958, an earthquake of magnitude 6.2 occurred in the southwest of Beichuan County with an epicenter intensity of 10°. The epicenter was about 5 km southwest to the project area, and the earthquake response intensity on the project area was 6°.

On August 16, 22 and 23, 1976, three strong earthquakes with magnitudes of 7.2, 6.7 and 7.2 occurred continuously in Songpan-Pingwu, and the intensity of the epicenter reached 9° with single earthquake epicentral intensity of 8° and multiple earthquake superposition epicentral intensity up to 9°. The epicenter was 70–100 km northwest to the project area, and the earthquake response intensity on the project area was 6°.

On May 12, 2008, an 8.0 magnitude earthquake occurred in Wenchuan, right on the Yingxiu-Beichuan Fault, which was the result of southeastward extrusion of Longmen Mountains thrust nappe with clockwise shear. The southern section of the rupture plane was mainly present as thrusting with a certain right-lateral strike-slip component, while the northern section is dominated by right-lateral strike-slip with thrust component. The rupture plane cracked from the epicenter of Wenchuan County and spread at an average speed of 3.1 km/s to the northeastern direction of 49°. The rupture length was about 300 km, and the total rupture process lasted nearly 120 s. The major energy of the earthquake was released in the first 80 s, with a maximum fault rupture and dislocation of 9 m and depth of source was more than 10 km.

3.2 Formation Conditions and Inversion of the Tangjiashan Landslide

3.2.1 Formation Conditions of the Tangjiashan Landslide

3.2.1.1 Topographical and Morphological Conditions

Topography is the spatial factor for the development for allowing landslide blockage of a river, and the shape and the slope of the valley are two main control indicators. The depth of the river valley cut determines the critical volume for the stability and energy of the slope, and the slope of the landslide body has a direct impact on the stability. The critical slope of the landslide is 30°–45°, which allows a landslide blockage to occur. Slope less than is generally more stable, but when the slope is greater than 45°, the probability of landslide blockage to a river is even higher.

The Tangjiashan section of the Tongkou River was an asymmetric "V"-shaped valley with a steeper right bank slope of about 60°, which was much higher than the 30°–45° slope where landslide blockage to a river is likely to occur with a high probability. At the same time, the deep cut of the Tangjiashan dam site of Tongkou River section and the narrow valley topography made the Tangjiashan landslide body to have a large sliding volume and energy when it was destabilized. Therefore, the topographical conditions of the Tangjiashan dam site provided the basic conditions for landslide accumulation and river blockage when triggering factors were encountered.

3.2.1.2 Stratigraphic Conditions

Landslide damming is related to the existence of sliding-prone stratum in the region. Generally, strata composed of clay, mudstone, shale, marl and other metamorphic rocks such as schist, microphyllite, soft and hard inter-bedded rock or easily weathered rock are prone to trigger landslide blockage.

The survey showed that the Tangjiashan dam site was covered by a strong weathered zone with a thickness of 5–10 m, and the underlying weakly weathered bedrock was composed of thin siliceous rocks, sandstone, marl and mudstone of the Lower Cambrian Qingping Formation soft and hard inter-bedded, and the attitude was N60°E/NW∠60°. The right bank was a dip slope, and the bedrock fissures were developed so that the rock was more fragmented. Therefore, the weakly weathered bedrock consisting of thin siliceous rocks, sandstones, marls and mudstones of the Cambrian Lower Qingping Formation and the rock joint fissures developed along the slope provide the physical infrastructure for mass sliding when encountering inducing factors.

3.2.1.3 Triggering Conditions

Precipitation and earthquakes were the two main triggering factors for the formation of barrier lake. It would trigger a large number of barrier lakes where there topographical and geological conditions for landslide to block rivers, especially strong earthquakes.

According to the "Preliminary Determination of iso-intensity line of the Wenchuan Earthquake" compiled by the China Earthquake Administration on May 16, 2008, the seismic intensity at the Tangjiashan dam site was 10° or above. When an earthquake of magnitude up to 8.0 occurred, the rocky slope on the right bank of Tangjiashan was subjected to the instantaneous seismic forces from the horizontal and longitudinal waves of the earthquake, resulting in the sudden sliding of the rocky slope on the right bank of Tangjiashan into the river at high speed to form a barrier lake.

3.2.2 Formation Process Analysis of the Tangjiashan Barrier Dam

The Tangjiashan barrier dam was formed by the accumulation of bedding landslide, the slip cliff was about 634 m high, and the slide base was a bedrock layer.

The Tangjiashan landslide moved along the river to the left bank at high speed with 800 m long and climbed about 140 m up the slope. The fractured rock formed by the original bedrock in the lower part of the slope mainly accumulated on the left side, the upper part of the bedrock and the residual slope debris soil accumulated on the right side, forming a barrier which was high left and low right (Fig. 3.4). The huge sediment flow and air currents generated by the landslide washed up the left bank and wiped out the vegetation on the original slope (Fig. 3.5).

The lower reach of the Tangjiashan landslide did not rush directly to the left bank, but shifted nearly 90° toward to the lower reach. Undisintegrated fractured rock was visible on the downstream dam slope tending to lower reach (Fig. 3.6).

There were 2 landslides which did not completely slide down lying upstream and downstream of Tangjiashan landslide. The upstream landslide was 370 m long along the river, the elevation of the back edge of the slide bed was 1350 m, the elevation of the back edge of the accumulated mass was 1165 m, and the elevation of the front edge was 710 m. Downstream landslide was 450 m long along the river, the elevation of the back edge was 1100 m, and the elevation of the front edge was 670 m.

Therefore, the formation of the Tangjiashan barrier lake was determined by the topography, stratigraphic structure and hydrodynamics of the riverbed in this section, as well as the bedrock bedding slip of right bank triggered by seismic force.

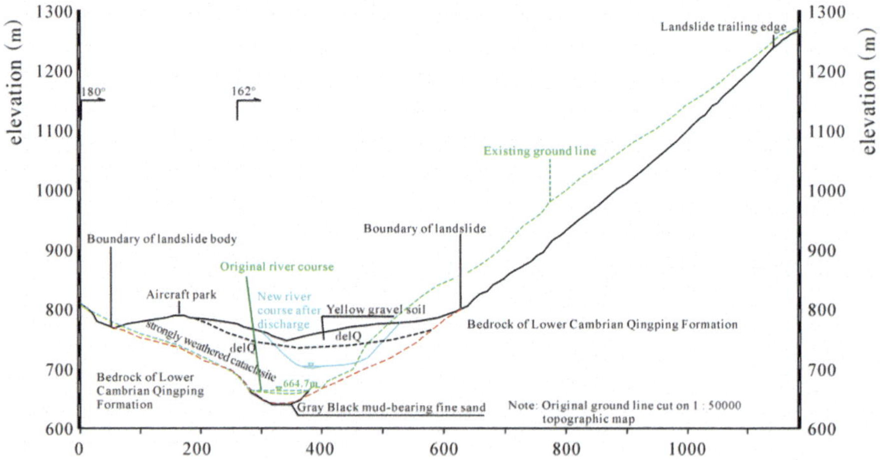

Fig. 3.4 Sketch of engineering geological cross section of Tangjiashan barrier dam

Fig. 3.5 Scouring phenomenon of vegetation on the left bank by high-speed sediment water flow

Fig. 3.6 Undisintegrated shifting fractured rock on the downstream dam

3.2.3 DDA Numerical Simulation of Barrier Dam Formation Process

In order to visually reproduce the formation process of the Tangjiashan barrier dam, and to verify the judgment of the experts on site, the CRSRI conducted a numerical simulation of the barrier dam formation process by the DDA method after the completion of the emergency treatment work.

3.2.3.1 Basic Principle

The DDA model is a general equilibrium equation which established by the displacement variational method of elasticity theory under the assumed displacement mode conditions. The non-embedded and tension-free contact criterion between the block unit interfaces is achieved by applying or removing the block interface stiffening springs.

The research shows that the DDA model can strictly satisfy three convergence conditions in the solution of the block system included the open-closed iteration convergence of the system, equilibrium condition of the force system and convergence of the dynamic solution.

3.2.3.2 Building Computational Models

The DDA Model units division of Tangjiashan Landslide was shown in Fig. 3.7. The maximum thickness of the landslide body reached more than 80 m, and the height was 634 m. In order to fully reflect the deformation and damage characteristics of the landslide body under the seismic load, the units of the landslide body should be divided into uniform size which was preferable not too large. The side length size of each unit here was 5–8 m. The representative block units were selected at the front, middle and back edges of the landslide and numbered as $438^{\#}$, $893^{\#}$ and $443^{\#}$, respectively. The location of each block was shown in Fig. 3.7.

In the DDA model, the landslide body was composed of four materials: the surface residual slope deposit, the strongly weathered rock layer, the weakly weathered rock layer and the lower bedrock. The thickness of the first two materials was 15–20 m. In addition, a 20 m thick mud-bearing fine sand at Kuzhu Dam site in the lower part of the landslide body was also considered.

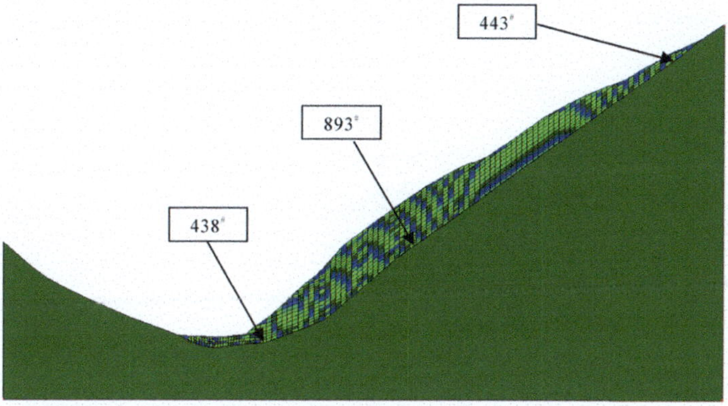

Fig. 3.7 DDA model of Tangjiashan landslide

3.2.3.3 Seismic Load Application and Material Parameters Selection of Landslide Body

(1) Selection of seismic load

According to the published Wenchuan seismic intensity zoning map, the seismic intensity at Tangjiashan landslide was 10°–11°, and the seismic intensity was considered as 10° in the calculation and analysis. According to the *Code for Slope Design of Water Conservancy and Hydropower Projects* (SL386-2007), the aseismic design of the slope based on the *Code for Aseismic Design of Hydraulic Buildings* (GB 51247–2018) specifying that the peak horizontal acceleration was 0.8 g, and the vertical acceleration was 2/3 of the horizontal acceleration. Since the actual seismic waves and related parameters of the Wenchuan earthquake were not collected, the acceleration seismic waves as shown in Fig. 3.8 were selected as the input.

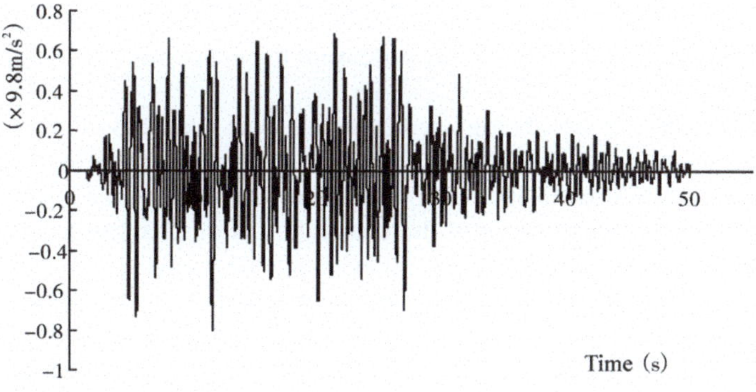

Fig. 3.8 Seismic acceleration input curve

Table 3.1 Selection of landslide body parameter

Parameters		Slope deposit	Strongly weathered rock mass	Weakly weathered rock mass	Sliding surface	Vertical interface
Strength	φ (°)	29–40	31–41	33–42	17–42	33–42
	C (kPa)	20–70	30–80	40–100	0–80	40–100
Deformation	E (GPa)	0.5	1.0	3.0	–	–
	μ	0.3	0.28	0.25	–	–

(2) Selection of Mechanical parameters

After the completion of the landslide, the location of the front edge and the shape of the barrier dam formed by the landslide were used as the inversion targets, and the landslide mechanical parameters that met the inversion targets were determined by inversion analysis within the range of possible values of each material area.

Based on the analysis of material composition and geological conditions of mountain mass on the right bank of Tangjiashan, values of various regional materials parameters were shown in Table 3.1. In order to evaluate the influence of the parameter values on the landslide sliding process and to ensure the appropriate parameters after the initiation of the landslide, the main intensity parameters were given in the table with a certain range of variation. The upper limit of the parameters was set by the stability calculation of DDA under the non-seismic condition; in other words, the Tangjiashan landslide should be in a stable state under the non-seismic condition.

3.2.3.4 Numerical Simulation of the Landslide Process in Tangjiashan

(1) Landslide state parameter value selection

First of all, the parameters of the landslide in the stable state under the non-seismic condition of self-weight were determined. After a variety of trial solutions, the values of the strength parameters in the stable state were the upper limits of the parameters in Table 3.1.

Based on the barrier morphology and landslide front edge elevation formed after the Tangjiashan landslide (Fig. 3.5), the inversion target function of the landslide DDA parameters was established. The results showed that under the seismic loading condition, the strength parameter of the landslide surface changed to the dynamic friction angle parameter, which was much lower than the strength parameter under the non-sliding condition. In the process of high-speed landslide, there were many factors influencing the parameters of slip surface, including sliding and rolling factors, as well as the influence of gas filling between the landslide body and bedrock surface,

Table 3.2 Selected parameters of land sliding state

Parameters		Slope deposit	Strongly weathered rock mass	Weakly weathered rock mass	Sliding surface	Vertical interface
Strength	φ (°)	28	30	32	17	31
	C (kPa)	20	30	40	0	40
Deformation	E (GPa)	0.5	1.0	3.0	–	–
	μ	0.3	0.28	0.25	–	–

and the parameters in the table (including slip surface parameters) should be regarded as comprehensive equivalent parameters (Table 3.2).

(2) Landslide process simulation

The three representative units 438# (foot of the landslide), 896# (middle of the landslide) and 443# (back edge of the landslide) were selected from the landslide block system as the object of analysis. Figures 3.9 and 3.10 showed the change process curves of slide velocity and slide distance of the abovementioned block unit with time, respectively. The results showed that the Tangjiashan landslide was a high-speed landslide, and its velocity change curve had significant nonlinear characteristics from the start to the stop of the landslide.

Under the strong seismic load, the middle and lower parts of the landslide on the right bank of Tangjiashan were activated first, and then, the upper part and the back edge of the landslide were activated. Then, the speed of the landslide accelerated, and the middle and lower parts slid across the bottom of the riverbed to the left bank. After the front edge of the landslide reached a certain elevation on the left bank, the landslide potential energy was gradually released due to the frictional energy dissipation effect of the left bank slope obstruction, and the internal disintegration

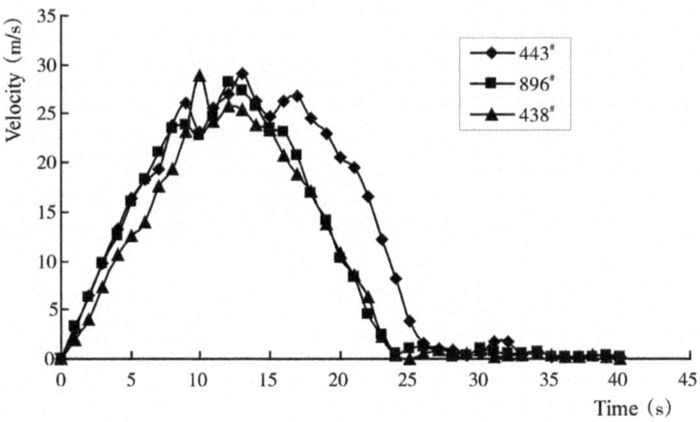

Fig. 3.9 Changing process curve of sliding velocity with time

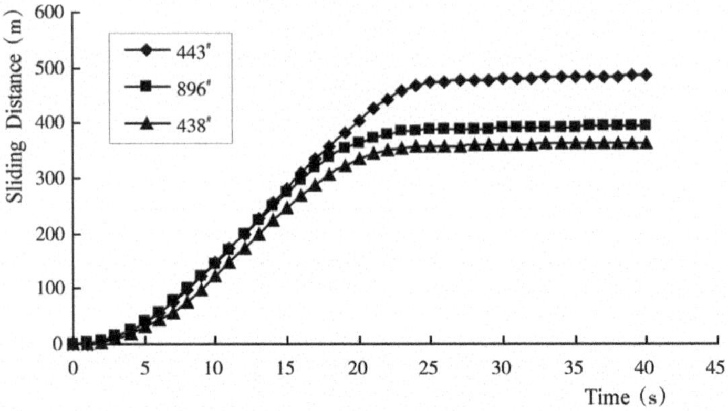

Fig. 3.10 Changing Process curve of sliding distance with time

of the bedrock surface and landslide body. In this way, the landslide sliding speed decreased and gradually stopped. Meanwhile, the rear part of the landslide, including the back edge, was incompatible with the front edge and the middle and lower part of the landslide at the later stage of the movement process. While the sliding speed of the middle and lower part of the landslide decreased at the left bank, the rear part of the landslide still had a large sliding velocity, which caused internal compression and uplift of the landslide body on the right bank of the riverbed, and the landslide body would be crushed and broken in this part, and the material at the rear part of the landslide body would be retained.

The results of the DDA method showed that the total duration of landslide sliding was about 40 s, in which the sliding speed was faster in 0–23 and 0–28 s for middle lower part of landslide and middle rear part of landslide, respectively, with the fastest sliding speed of 29 m/s, and the average sliding speed of 15–16.5 m/s.

Figure 3.11 showed the main stress distribution of the block within the barrier dam after the landslide stopped. The stress distribution pattern in the figure showed that the landslide body slid to the riverbed at high speed and then formed a barrier dam. The middle and lower part of the barrier was squeezed by the landslide movement process to produce the squeeze stress which was more consistent with the direction of the landslide movement, and the main stress directions were nearly parallel at the riverbed boundary and the bottom of the riverbed. The main stress value was generally 2–4 MPa, and the maximum main stress value could be 6–7 MPa. In a certain range between the front edge and the back edge of the landslide, the landslide body was disintegrated and the stress was small.

Figure 3.12 showed the displacement of the block system at each typical time step. Comparing the final shape of the landslide after 40 s, its shape and location were close to those in Fig. 3.8.

Fig. 3.11 Main stress distribution of barrier dam resulted from landslide

3.3 Geological Characteristics of the Tangjiashan Barrier Dam

3.3.1 Barrier Dam Morphology

According to the mapping and geological survey conducted by water experts at the barrier dam site from May 20–21, the barrier dam was 803 m long in the downstream direction and 612 m long in the horizontal direction, with a surface area of about 300,000 m^2, a dam height of 82–124 m and a volume of 20.37 million m^3. The shape of the barrier dam was shown in Fig. 3.13.

The top surface of the barrier was undulating, with the left side being high and the right side being low, the maximum elevation of the left side being 793.9 m and right side being 775 m. The upstream slope was about 200 m long with a gentle slope of about 20° (slope ratio 1:4). The downstream slope was about 300 m long with a foot elevation of 669.55 m. The upper steep slope was about 50 m long with a slope of about 55°. The central gentle slope was about 230 m long with a slope of about 32°, and the lower steep slope was about 20 m long with a slope of about 64°, with an average slope ratio of 1:2.4.

Three trenches were distributed in the barrier dam along the downstream direction. The right trench was right bow-shaped, running through upstream and downstream, with a width of 20–40 m at the bottom and a highest elevation of 752.2 m at the middle. The middle and left trench were located on the downstream slope of the dam, with a length of about 400 m and a width of 10–20 m at the bottom.

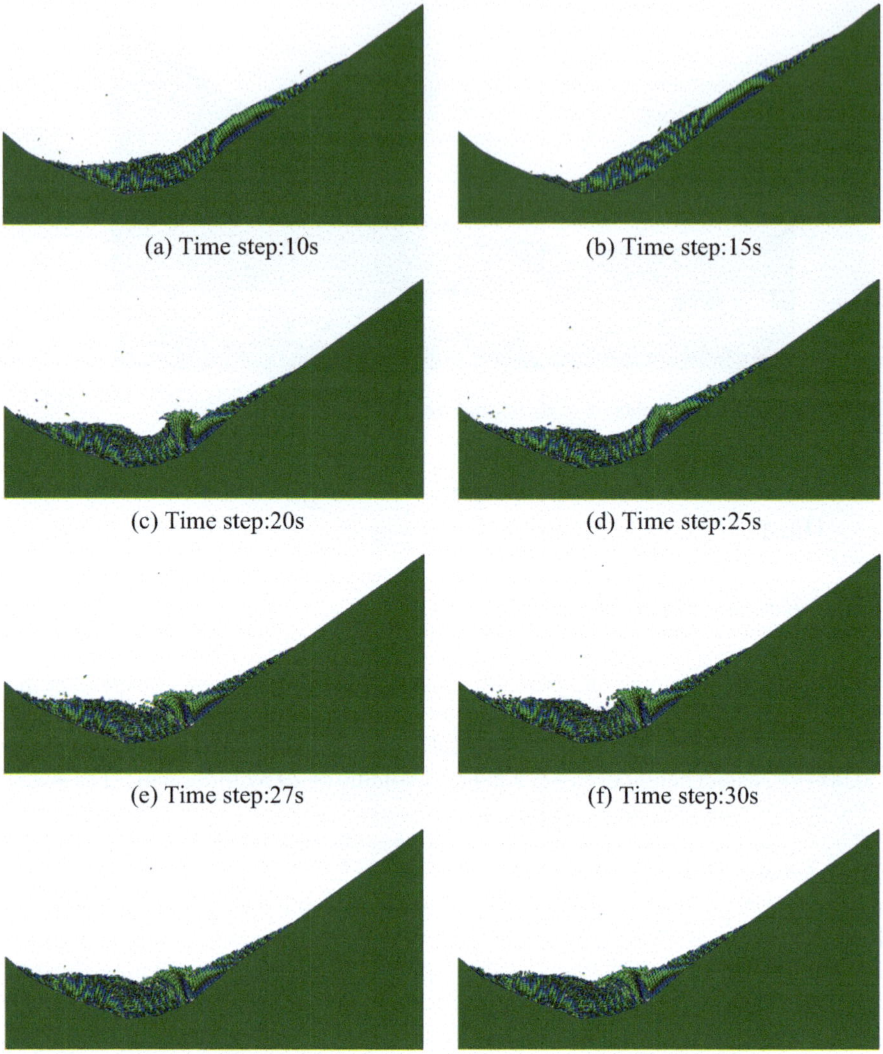

(a) Time step:10s (b) Time step:15s

(c) Time step:20s (d) Time step:25s

(e) Time step:27s (f) Time step:30s

(g) Time step:35s (h) Time step:40s

Fig. 3.12 Displacement diagram of the block system for each typical time step

3.3.2 Barrier Material Composition

The Tangjiashan barrier dam consisted of gravelly soil from the upper part of the original slope and fractured rock formed by sliding, compression and crushing of the bedrock in the upper part of the Cambrian Lower Qingping Formation. According to the geological survey of the barrier site, the gravelly soil accounted for about 14% and the fractured rock accounted for 86%.

Fig. 3.13 Original view of the Tangjiashan barrier dam

(1) Gravelly soil

The soil was earthy yellow in color and consisted of powdered loam and gravel (Fig. 3.14). Among them, powder loam accounted for about 60%, gravel accounted for 30–35%, and boulder (grain size 5–20 cm) accounted for 5–10%.

(2) Fractured rock

Fig. 3.14 Gravelly soil

Fig. 3.15 Fractured rocks formed by soft rocks (left: undisintegrated, right: disintegrated)

Fig. 3.16 Fractured rock formed by hard rock (left: disintegrated, right: undisintegrated)

The composition of the material was not homogeneous. The fractured rocks formed from soft rocks were mainly gravels with a grain size of less than 20 cm (Fig. 3.15). The fractured rocks formed from hard rocks were mainly boulders with a grain size of 1.0–3.0 m, and some boulders could reach several meters in size (Fig. 3.16). Some of the rocks were not disintegrated and retained the laminated structure of the original rocks.

According to the geological survey at the barrier site, the barrier dam was divided into three zones according to the different material composition.

Zone I: Located on the right side of the barrier dam, the river was low in the middle and high at both ends in the transverse direction, the lowest elevation was 747 m. The upper part of the barrier dam was mainly composed by gravel soil with a thickness of 15 m, which was shown as left side thin and right side thick. The lower part was fractured rock.

The right trench ran through the upstream to downstream above barrier dam. The original main riverbed was located in this area, and the elevation of riverbed was 664.7 m.

Due to the impact of the landslide sliding force, most of the 20 m thick mud-bearing fine sand deposited in the Kuzhu Dam reservoir area was washed away. The thickness was reduced by about 8.0 m, and the elevation of the underlying bedrock surface was 735–738 m.

Zone II: Located on the left side of the barrier dam, the topography was relatively high with the highest elevation of 793.9 m. It was composed of fractured rock with thickness of 35–107 m. There was gravel soil with thickness of 2–4 m on the upper part.

The original left bank slope was composed by residual slope deposit gravel soil and bedrock under the barrier dam.

Zone III: Located in the upstream of the barrier dam, it was formed by the mud-bearing fine sand deposited in the reservoir area of the Kuzhu Dam, which was squeezed and uplifted by landslide slide. It appeared below the upstream dam slope with elevation of 730–740 m.

3.3.3 Physical and Mechanical Properties and Permeability of the Barrier Dam

Physical and mechanical parameters of various types of soils based on the composition of the barrier dam material were proposed by field experts and experimental judgments, which could be seen in Table 3.3.

Table 3.3 Proposed values of physical and mechanical parameters for various types of soils in the Tangjiashan barrier dam

Layer	Density ρ(g/cm^3)		Angle of internal friction φ(°)		Cohesion c (kPa)		Permeability coefficient K (cm/s)	Allowable slope drop J
	Natural	Full water	Natural	Full water	Natural	Full water		
Grey-black mud-bearing fine sand	1.5–1.7	1.6–1.8	17	12	10	5	1×10^{-4}	0.10
Gravelly soil	1.7–1.8	1.8–1.9	25	20	30	10	1×10^{-4}	0.20
Strongly weathered fractured rock	2.0	2.1	33	30	30	10	1×10^{-2}	0.60
Weakly weathered fractured rock	2.1	2.2	37	34	60	30	1×10^{-3}	0.80

3.4 Evaluation and Analysis of the Barrier Dam Stability

3.4.1 Geological Macroscopic Evaluation of the Barrier Dam Stability

The barrier dam was mainly composed of fractured rock and part of the fractured rock was not disintegrated with high strength. The barrier dam was wide and thick with a dam bottom width of 803.4 m in the downstream direction. The geological experts made a macroscopic evaluation that overall stability of the dam slope was quite well without the possibility of full collapse.

The upper right side of the barrier dam was distributed with gravel soil, and the upstream slope was distributed with alluvial mud-bearing fine sand from the riverbed that was pushed up by the landslide slide, both of which had low permeability. And the barrier dam possesses good impermeability barrier dam. From the observation on site, it could be seen that more than a dozen leakage points were found in the area of 669 m elevation at the foot of the downstream dam. With the rise of the water level in the front of the barrier dam, the amount of leakage tended to gradually increase, but the water quality was clear and there was no turbidity, indicating that the barrier dam had good permeability stability. However, the phenomenon of partial overhanging of fractured rock could not be ruled out. Once the water level rose to a certain elevation, the concentrated leakage passage might appear and lead to the local collapse of the downstream slope.

When the water level of the barrier lake rose to overflow from the right side, the upper gravel soil and fractured rock formed by the original strong weathering would be easily washed away by the water flow, and the right trench might be rapidly undercut to result into the collapse of the left part of the barrier dam.

3.4.2 Analysis of the Possibility of the Tangjiashan Barrier Dam Collapse

Barrier lake was a kind of natural water storage facility with no special design, and the barrier dam material was not artificially selected and compacted. There were many engineering geological problems such as infiltration deformation, stability and settlement deformation. There presented great manifestations of overhead and material composition, which directly affected and determined the mode and process of barrier dam failure.

There are three common types of barrier dam collapse mode: barrier dam instability, seepage destruction and overtopping collapse. Among these three modes, the overtopping collapse is the most common one.

The dimensionless blockage index (DBI) method could be used to estimate the long-term safety of the accumulation:

$$\mathrm{DBI} = \log\left(\frac{A_b \times H_d}{V_d}\right)$$

in which:

V_d volume of the barrier dam
A_b catchment area
H_d barrier dam height.

The stability judgment criteria was that the barrier dam would stay stable for DBI < 2.75 and fall into the uncertain domain for 2.75 < DBI < 3.08, but falling into the unstable domain for DBI > 3.08.

The Tangjiashan barrier dam was 82–124 m high, with a volume of 20.37 million m^3 and a precipitation catchment area of 3550 km^2. Based on the average dam height of 103.45 m, DBI = 4.26, or based on the lowest dam height of 82.5 m, DBI = 4.16. According to the judgment criteria, the DBI for Tangjiashan barrier dam is more than 3.08, which would fall into an unstable domain with collapse risk.

The DBI method was an empirical method, which could quickly evaluate the barrier dam stability, but could not evaluate the potential barrier dam damage. Therefore, it was still necessary to evaluate the stability of the barrier dam by using geo-mechanics methods.

3.4.3 Analysis of the Collapse Process of the Tangjiashan Barrier Dam

Regardless the barrier dam collapse mode, the naturally formed barrier dam collapse could be divided into the instantaneous collapse (i.e., abrupt failure) or the progressive failure, which mainly depended on geometric and inducement factor. The occurrence, development and extent of the collapse were influenced by a variety of factors, such as the cause of the breach, the size and material of the barrier dam, the volume of the reservoir, and the downstream water level. It was quite difficult to make the simulation.

Combined with the material composition and collapse mode at for the Tangjiashan barrier dam, the upper part of the barrier dam was mainly composed of pulverized loam, rock debris and boulders, which made the overtopping flow gradually undercut the barrier dam. The discharge channel base would gradually decrease. The relatively thick weakly weathered rocks at the lower part of barrier dam had strong scouring resistant capacity. Therefore, it could be assumed that the process of barrier collapse belonged to the progressive collapse mode, and the elevation of the discharge channel base would maintain at about 720 m after the collapse (the roof elevation of the "fractured rock"). However, the progressive process was only relative to the abrupt collapse. The upper part of the barrier dam consisted of loose material with high head and enormous energy, and the progressive collapse process was still very fast and lasted only within a few hours.

3.4.4 Analysis of Barrier Dam Stability Calculation

Based on the morphological characteristics and material composition of the barrier dam, CWRC, POWERCHINA Chengdu and IWHR carried out different methods of analysis, calculation and evaluation of the stability of the barrier dam.

3.4.4.1 Results of CWRC

According to the morphological characteristics and material composition of the Tangjiashan barrier, the barrier top of the left side was the highest and the barrier dam was mainly composed of fractured rocks. But the right side of the barrier dam was composed of fractured rocks and gravelly soils with a low channel. Therefore, each cross section from the left and the right side of the barrier was selected for the stability analysis of dam slope.

Cross section of left side: the left side of Tangjiashan barrier dam was quite high. The elevation of dam top was 793.9 m, the elevation of dam bottom was 669.55 m, the barrier dam height was 124.35 m, and the width of the barrier dam bottom was 803.4 m. The barrier dam was mainly composed of fractured rock, and the upper part was covered with 4 m thick gravel soil. The base of the barrier was the original left bank slope and the slope body consisted of residual slope deposit gravel soil, which lied on the weathered bedrock. The upstream slope of the barrier dam was covered with silt-bearing fine sand with an elevation of 735 m and a slope of 20°. The downstream slope of the dam was a composite slope (Fig. 3.17).

Right trench section: emergency dredging works were used. The geological longitudinal section of the right trench was shown in Fig. 3.18.

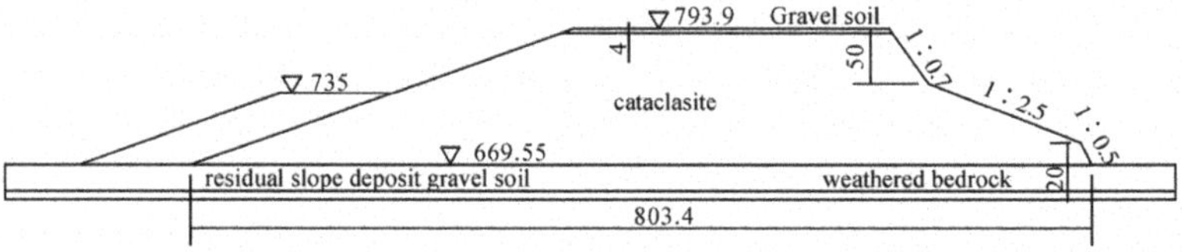

Fig. 3.17 Schematic diagram of the longitudinal section of the left side

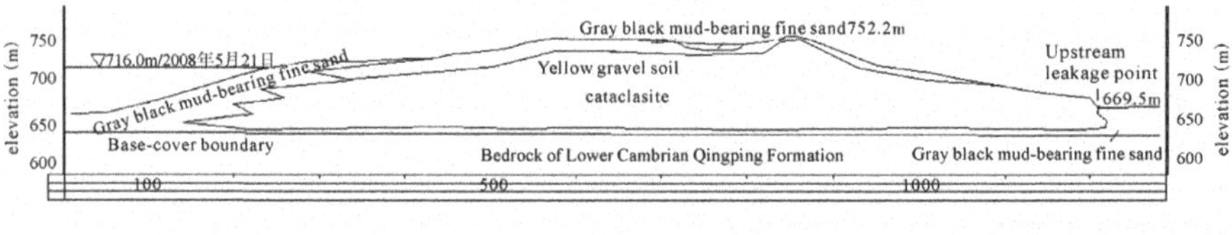

Fig. 3.18 Schematic diagram of the geological longitudinal section of the right trench

Table 3.4 Values of permeability parameters of various types of soils in the barrier proposed by the CWRC

Soil texture	Permeability coefficient *K* (cm/s)	Allowable slope drop *J*
Grey-black mud-bearing fine sand	1.0×10^{-4}	0.10
Strongly weathered Cataclasite	1.0×10^{-2}	0.60
Weakly weathered Cataclasite	1.0×10^{-3}	0.80
Gravelly soil	1.0×10^{-4}	0.20

(1) Barrier dam permeability stability analysis

The model adopted a two-dimensional in homogeneous isotropic saturated steady seepage finite element model, and the upstream water level boundary and downstream water level boundary were the given head boundary, i.e., the first kind boundary; the free surface and permeable surface belonged to the constant flow boundary, i.e., second kind boundary, the normal flow rate was 0. The upstream and downstream boundaries of the model were 1000 m away from the slope foot on the upstream and downstream sides of the barrier, respectively, and the lower boundary of the model was taken to be 100 m below the weakly weathered bedrock surface.

Based on the engineering geological evaluation data and the soil composition and empirical judgments, the values of various soil permeability parameters of the barrier dam were shown in Table 3.4.

The seepage analysis scheme was shown in Table 3.5, and the calculation scheme mainly considered the unfavorable working conditions of infiltration stability and sliding stability.

The SEEP2D finite element calculation program in ground water modeling system GMS5.0 was used to analyze the seepage pattern of the barrier dam, and the results of the four scenarios are shown in Table 3.6.

Near the exit of the dam slope foot, the distribution of isopotential lines was denser, and the head ratio was larger. When the upstream water level rose from

Table 3.5 Seepage analysis program of CWRC

Program number	Description
1	Upstream water level 752 m, downstream water level 669.55 m
2	Upstream water level 752 m, downstream water level 675 m
3	Upstream water level 745 m, downstream water level 675 m
4	Current upstream water level 717.48 m on May 22, downstream water level 669.55 m

Table 3.6 Seepage calculation results of CWRC

Programs	Upstream water level (m)	Downstream water level (m)	Single width flow rate (m^3/d)	Horizontal exit ratio drop	Vertical exit ratio drop
1	752.00	669.55	8.64	0.9	0.6
2	752.00	675.00		0.5	0.6
3	745.00	675.00		0.4	0.5
4	717.48	669.55		0.4	0.3

717.48 to 752 m, horizontal exit ratio drop at toe of cataclastic rock slope on left bank increased from 0.4 to 0.9 and the vertical exit ratio drop increased from 0.3 to 0.6. The single-width flow rate of the barrier was about 8.64 m^3/d.

Comparing the results of seepage calculations with the empirically developed values of various soil permeability parameters of the barrier dam, it could be seen that when the upstream water level raised to the elevation of 752 m, presenting a possibility of infiltration damage.

(2) Barrier dam stability analysis

A simplified Bishop method based on the limit equilibrium theory was used to analyze the stability of the barrier dam. Considering the large permeability coefficient of the barrier dam, the super pore pressure could be dissipated quickly and the effective stress method was used for the analysis. The simplified Bishop method used a circular slip fracture surface, assuming that there were only normal forces between the blocks, and no tangential forces.

The slope of the upstream dam was gentle, with a slope of about 20°. The mud-bearing fine sand deposited in the reservoir area of the Kuzhu Dam was extruded and raised by the landslide slide, which played a role of counter-pressure and impermeable cover to the upstream dam slope. So the upstream dam slope was relatively stable.

The slope of the downstream dam was steep in the upper and lower part and gentle in the middle part. The upper steep slope was about 50 m high with a slope ratio of 1:0.7, the middle gentle slope had a slope ratio of 1:2.5, and the lower steep slope was about 20 m high with a slope ratio of 1:0.5. Therefore, barrier dam stability analysis was conducted for the downstream dam slope.

During the stable seepage period, the unfavorable condition was that the upstream water level reached the overflow elevation of 752 m. The cross section of the barrier dam was quite large with long seepage path, so the water level in the downstream slope was slightly affected when the upstream water level dropped from the overflow elevation of 752–745 m. The downstream water level could be regarded as equal to the storage level of Kuzhu Dam, and the downstream water level in dry season was 669.55 m. The downstream calibration water level was set as 675 m with little influence on the stability of the downstream slope.

Table 3.7 Calculated working conditions for the stability analysis of the barrier dam of the CWRC

Working condition no.	Upstream water level (m)	Downstream water level (m)	Remark
1	717.48	669.5	The head difference between upstream storage water level on May 2 and downstream water level of Kuzhu Dam in dry season is 47.98 m
2	745	669.5	The head difference between upstream storage level 2 m higher than 743mscheduled in evacuation plan on May 22 and downstream water level of Kuzhu Dam in dry season is 75.5 m
3	752	669.5	The head difference between the upstream water level reaching the predetermined breach water level and the downstream water level of Kuzhu Dam in dry season is 82.5 m

Table 3.7 showed the water level working conditions for barrier dam stability analysis.

Based on the engineering geological evaluation data and the soil composition and empirical judgment, the proposed values of various soil physical and mechanical parameters of the barrier dam were presented in Table 3.8.

The safety coefficient of the downstream slope of the barrier dam for each working condition was shown in Table 3.9. The safety coefficient of the right side of barrier dam was smaller than that of the left section as the barrier base was covered by an 8 m thick layer of fine sand. Therefore, the stability of the barrier dam was controlled by the right section.

The sliding surfaces corresponding to each working condition are shown in Figs. 3.19, 3.20, 3.21, 3.22 and 3.23.

(3) Main acquaintance

(1) Seepage calculation showed that when the water level of the barrier lake raised to 752 m elevation, the barrier dam may present seepage collapse. There would present seepage collapse without drainage measures if the lake water was allowed to overflow freely.

Table 3.8 Suggested values of physical and mechanical parameters of various soil types of the barrier dam proposed by CWRC

Soil texture	Natural state		Saturated state		Saturated density (kg/m^3)	Wet density (kg/m^3)
	Cohesion (kPa)	Friction angle (°)	Cohesion (kPa)	Friction angle (°)		
Grey-black mud-bearing fine sand	10	17	5	12	1800	1700
Strongly weathered Cataclasite	30	35	10	30	2100	2000
Weakly weathered Cataclasite	60	40	30	35	2200	2100
Gravelly soil	30	25	10	20	1900	1800

Table 3.9 Safety factors of barrier body for each working condition as calculated by CWRC

Calculated cross section	Upstream water level 717.78 m	Upstream water level 745 m	Upstream water level 752 m
Left side	1.33	1.11	1.06
Trench	/	1.75	1.33

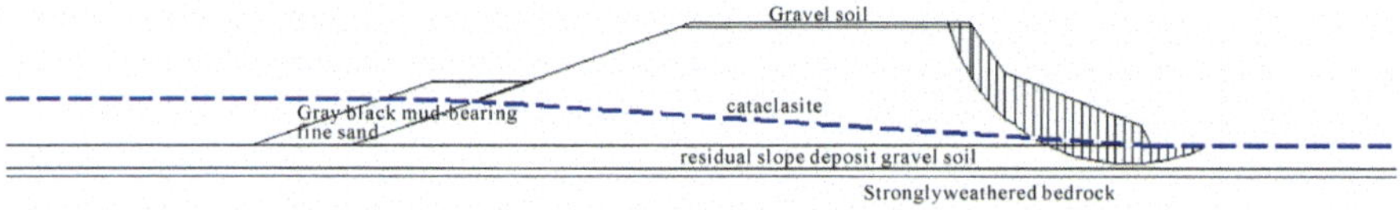

Fig. 3.19 Left cross section arc sliding search method minimum safety factor $F_s = 1.33$ with upstream water level of 717.48 m

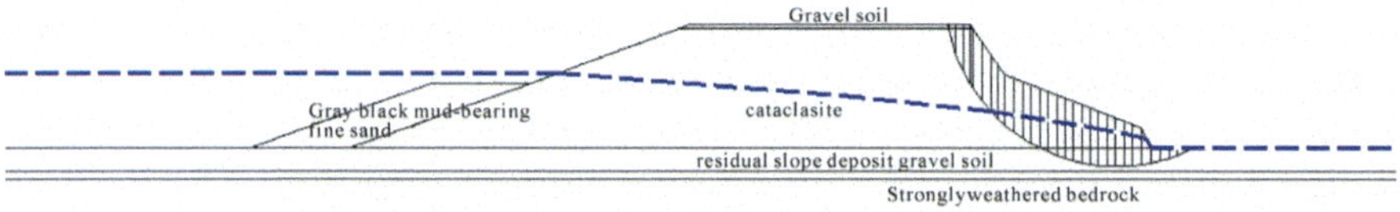

Fig. 3.20 Left bank section arc sliding search method minimum safety factor $F_s = 1.11$ with upstream water level of 745 m

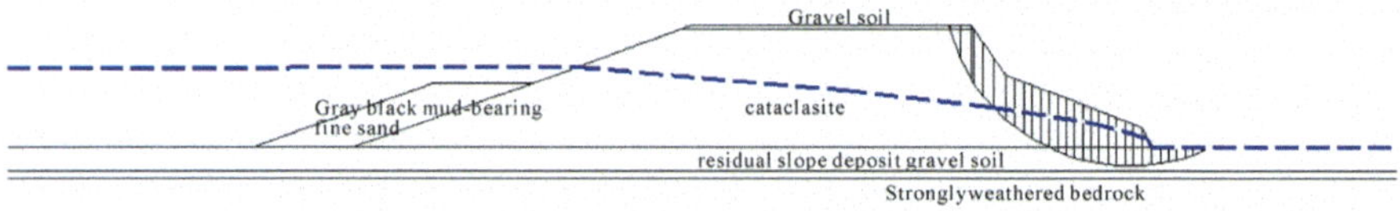

Fig. 3.21 Left bank section arc sliding search method minimum safety factor $F_s = 1.06$ with upstream water level of 752 m

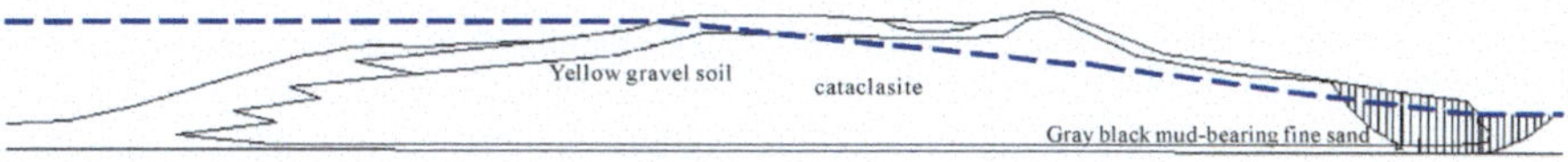

Fig. 3.22 Right trench arc sliding search method minimum safety factor $F_S = 1.75$ with upstream water level of 745 m

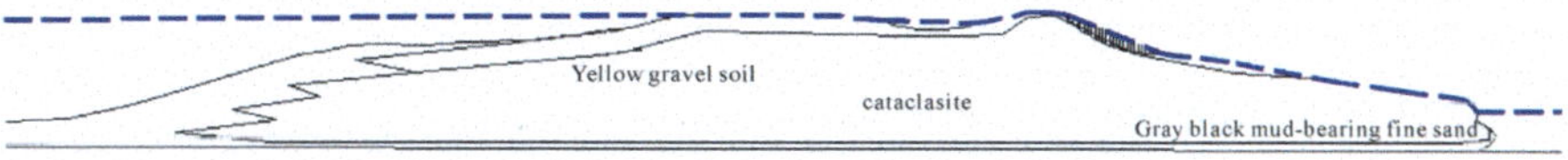

Fig. 3.23 Right trench arc sliding search method minimum safety factor $F_S = 1.3$ with upstream water level of 752 m

(2) The stability analysis of the barrier dam showed that the barrier dam was in a safe state under the water level of 717.48 m on May 22, but the stability of the barrier dam decreased as the water level rose. When the upstream water level reached 745 m (2 m higher than that of 743 m on May 22), the minimum safety factor of the barrier dam decreased down to 1.06, and the downstream slope was already in a critical state of sliding. The barrier dam reached the ultimate equilibrium state and the downstream dam slope was very likely to slide when the water level of upstream reached 752 m.
(3) Although the cross section of barrier dam was roughly described by geological information and the parameters were estimated empirically, a reasonable range of variation was considered in the calculation and a sensitivity analysis was performed. Therefore, the results were credible to reflect the current safety condition of the untreated barrier.
(4) The calculation results showed that if the discharge channel was ready for the discharge, the upstream water level should not exceed 745 m before the discharge, otherwise the downstream would be in a critical sliding state.

3.4.4.2 Results of POWERCHINA Chengdu Engineering Corporation

(1) Stability analysis of the side slopes of upper and lower barrier dam

In the stability analysis, the parameters determined by engineering analogy are shown in Table 3.10.

It was assumed that the upstream water level reached the maximum elevation of 750 m, no discharge channel construction was carried out, there was little water in downstream (water level of 669.50 m), and the barrier dam stayed in a saturated state. A typical longitudinal section of the barrier dam was used to analyze the stability of the upstream and downstream side slopes.

The calculation method was simplified Bishop method, the potential sliding surface was circular arc sliding surface, and the calculation program was STAB, a soil slope stability analysis program of IWHR.

Table 3.10 Values of physical and mechanical parameters of various soil types of the barrier dam proposed by POWERCHINA Chengdu Engineering Corporation

Layer	Density ρ (g/cm^3)		Angle of internal friction φ (°)		Cohesion c (kPa)		Permeability coefficient K (cm/s)	Allowable slope drop J
	Natural	Saturated water	Natural	Saturated water	Natural	Saturated water		
① Grayish black mud-bearing fine sand	1.5–1.7	1.6–1.8	17	12	10	5	1×10^{-4}	0.05
② Cataclasite	2.1	2.2	40	35	60	30	1×10^{-3}	0.25
③ Crushed stone	2.0	2.1	35	30	30	10	1×10^{-2}	0.10
④ Gravelly soil	1.7–1.8	1.8–1.9	25	20	30	10	1×10^{-4}	0.20

(1) Stability in the absence of earthquake

The side slope on the upstream barrier dam was mainly mud-bearing fine sand with low shear-resistant capacity. For its gentle slope, the overall side slope was stable but the local slope safety coefficient was low under the assumption that the upstream water level reached 750 m, little water in downstream, and the barrier dam was saturated. The safety factor of the most dangerous potential sliding surface was 1.009. The calculated profile and potential sliding surface were shown in Fig. 3.24.

The side slopes on the downstream barrier dam were mainly fractured rock with a small amount of gravel soil, which had a high overall and partial shear-resistant capacity. The side slope stability was high under the assumption that the upstream water level reached 750 m, little water in downstream, and the barrier dam was saturated. The safety factor of the most dangerous potential sliding surface was 1.345. The calculated profile and potential sliding surface were shown in Fig. 3.24.

(2) Stability in case of an earthquake.

Considering that the Tangjiashan barrier was located in the severe earthquake zone of Wenchuan, the continuous aftershocks and the sensitivity analysis was conducted for the barrier dam stability under the earthquake scenario.

If the assuming upstream water level reached 750 m with little water in downstream and saturate barrier dam, the stability of the barrier dam would be reduced if earthquake of VII and VIII intensity were encountered, respectively. In the case of superimposed VII-intensity earthquake, the minimum stability factor of the upstream side slope of the barrier was 0.751, which would be unstable and cause landslide, and the minimum stability factor of the downstream side slope of the barrier was 1.067, which would not be destabilized. In the case of superimposed VIII-intensity earthquake, the minimum stability factor was 0.605 for the upstream side slope of the barrier and 0.932 for the downstream side slope, both of which would be unstable

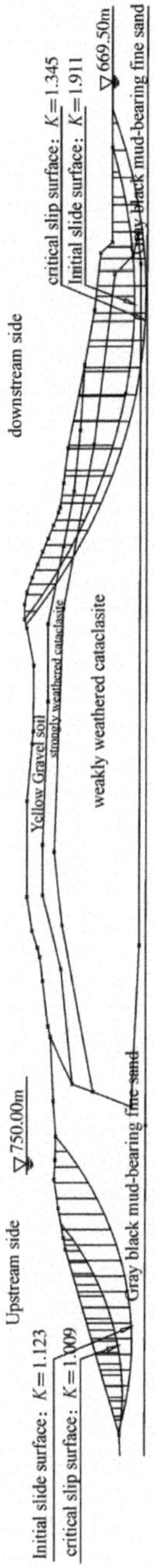

Fig. 3.24 Analysis diagram of upstream and downstream slope stability without earthquake

and cause landslides. In the case of VII and VIII-intensity earthquakes, the critical sliding surfaces of the upstream and downstream side slopes of the barrier and their sliding ranges were shown in Figs. 3.25 and 3.26.

(3) Stability after an earthquake

It was assumed that during the earthquake, the side slopes of both the upstream and downstream of the barrier dam were destabilized and slide with subsequent new barrier dam section. Based on the new cross section of barrier dam, the barrier stability analysis was carried out under the assumption that the upstream water level reached 750 m with little water in the downstream and saturated barrier dam.

The analysis showed that after the formation of a new cross section by VII-intensity earthquake, the minimum stability safety factor was 1.200 in the upstream of the barrier dam, which would not be destabilized again. And the minimum stability safety factor was 0.779 in the downstream, which would be destabilized. However, even if the barrier dam was destabilized along its critical destabilization sliding surface ($K = 0.980$–1.0), the barrier dam could still form a water retaining barrier section. From the perspective of stability analysis, it would not cause the barrier dam collapse. The stability analysis of the upstream and downstream side slopes after an VII-intensity earthquake and landslide was shown in Fig. 3.27.

After a new section formed by an VIII-intensity earthquake, the minimum stability factor was 1.217 for the upper reach of the barrier and 1.102 for the lower reach. The stability analysis of the upstream and downstream side slopes after an VIII-intensity earthquake and landslide was shown in Fig. 3.28.

(2) Barrier dam permeability stability analysis

In the plane seepage analysis, the calculation was mainly based on the parameters determined by engineering analogy, which were shown in Table 3.11. The program used for the calculation was the 2D seepage calculation program for earth and rock dams named UNSST2 developed by Nanjing Hydraulic Research Institute. UNSST2 (Unsteady Seepage and Stability, 2D) was a program for saturated porous media seepage, which was suitable for anisotropic seepage fields with irregular boundaries. It could be used to study various types of earth and rock dams, lock and dam foundations, channel leakage, as well as mine drainage or seepage issues of ash dams and tailing dams.

The calculation still assumed that there was no discharge channel arranged and the most unfavorable conditions were the upstream water level and downstream water level to be 750.00 m and 669.5 m, respectively. The geological longitudinal section in the trench core of the barrier dam and the section after the VIII-intensity earthquake and landslide were used to calculate the stable seepage flow, the infiltration slope drop and the single-width seepage of each layer of the barrier dam.

(1) Calculation results of geological longitudinal section of the trench core.

Table 3.12 showed the infiltration slope drop of the soil material at each layer of the barrier during the stable seepage period at the normal water storage level in the

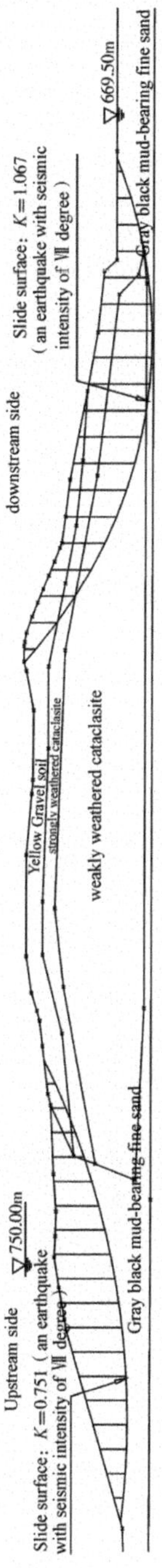

Fig. 3.25 Citical instability analysis diagram of upstream and downstream slope in the case of a VII-intensity earthquake

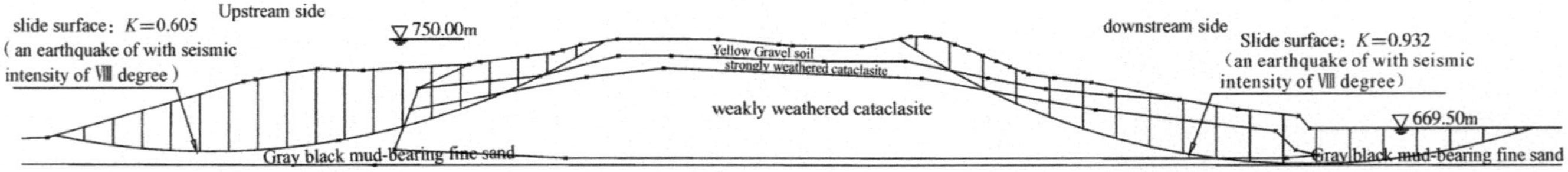

Fig. 3.26 Citical instability analysis diagram of upstream and downstream slope in the case of a VIII-intensity earthquake

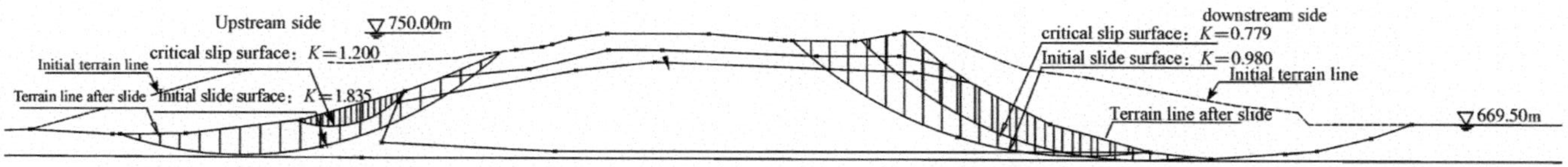

Fig. 3.27 Instability analysis diagram of upstream and downstream slope in the case of a VII-intensity earthquake

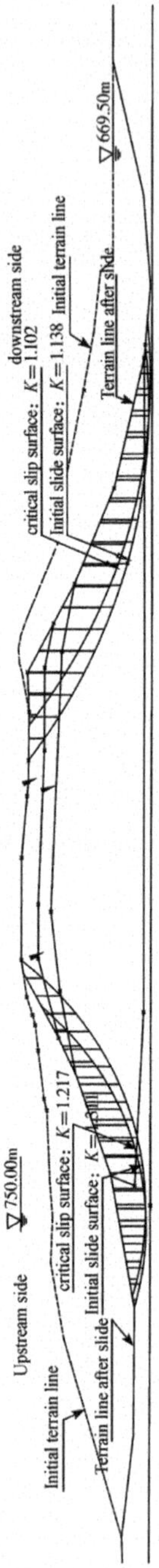

Fig. 3.28 Instability analysis diagram of upstream and downstream slope in the case of a VIII-intensity earthquake

Table 3.11 Stability analysis results of POWERCHINA Chengdu Engineering Corporation

Part of side slope	Upstream water level at 750 m elevation, little water in the downstream	Simultaneous superimposed 7° earthquake	Simultaneous superimposed 8° earthquake	Slope stability after superimposed 7° earthquake and landslide	Slope stability after superimposed 8° earthquake and landslide
Upstream side slope	1.009	0.751	0.605	1.200	1.217
Downstream side slope	1.345	1.067	0.932	0.779	1.102

Table 3.12 Soil infiltration slope drop of barrier at each layer with geological longitudinal section of trench core at normal water storage level during stable seepage period

Layer	Infiltration slope drop	Allowable slope drop *J*
① Grayish black mud-bearing fine sand	0.19	0.05
② Cataclasite	0.22	0.25
③ Crushed stone	0.22	0.10
④ Gravelly soil	1.15	0.20

geological longitudinal section of the trench core, and the single-width seepage rate was $2.490 \times 10^{-4} m^3/(s\ m)$. The distribution of seepage equipotential line was shown in Fig. 3.29.

From the calculation results, it could be seen that the seepage volume of the barrier was small during the stable seepage period in the geological longitudinal section of the trench core. Except for the permeability slope of fractured rock, which was less than the allowable slope drop, the permeability slope drop of soil material in all layers was greater than the allowable slope drop.

(2) Calculated results of the profile after a VII-intensity earthquake and landslide

Table 3.13 showed the infiltration slope drop of soil material at each level of the barrier during the stable seepage period at normal water storage level after VII-intensity earthquake and landslide, and the single-width seepage volume was $2.821 \times 10^{-4} m^3/(s\ m)$. The distribution of seepage isopotential lines was shown in Fig. 3.30.

From the calculation results, after earthquake and landslide, for the relative impermeable layer ④, the volume decreases, the seepage volume at smaller quantity level increases. For the soil layer ①, ② and ④, the infiltration slope drop increases. For the soil layer ③, the infiltration slope drop decreases. The infiltration slope drop of each soil layer is greater than the allowable slope drop.

(3) Evaluation of barrier dam stability analysis

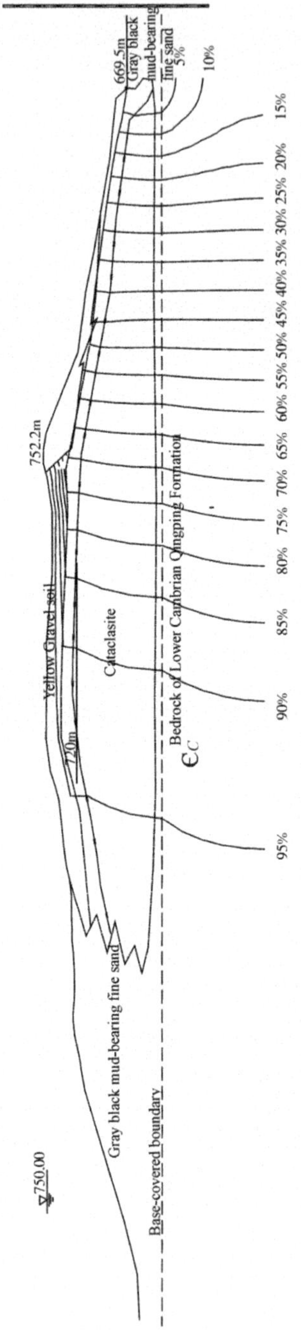

Fig. 3.29 Distribution of seepage equipotential line of geological longitudinal section of trench core

Table 3.13 Infiltration slope drop of soil material at each level of the barrier during the stable seepage period at normal water storage level after VII-intensity earthquake and landslide

Layer	Infiltration slope drop	Allowable slope drop *J*
① Grayish black mud-bearing fine sand	0.54	0.05
② Cataclasite	0.63	0.25
③ Crushed stone	0.12	0.10
④ Gravelly soil	1.28	0.20

Through the stability analysis of the two-dimensional longitudinal section of the upstream and downstream dam slopes, it could be concluded that the calculation was basically consistent with the macro geological judgment. That is to say, the overall barrier dam was relatively stable and the barrier dam was basically stable when the water level reached elevation of 750 m. In the view of stability analysis, the barrier dam might be partially destabilized and suffer landslide if aftershocks occurred. However, the barrier dam could still form a water retaining barrier section after the landslide but would not cause the collapse.

Through the analysis of the geological longitudinal section of the barrier dam and the plane stability of the section after the VIII-intensity earthquake and landslide, the seepage volume of the barrier was smaller, but the infiltration slope drop for most soil layers was greater than the allowable slope drop, which did not meet the requirements of permeability stability. The reason for the calculated larger infiltration slope drop lied in great the water level difference (nearly 100 m) between the upstream and the downstream of the barrier lake in the calculated working condition. And there was no impermeable body in the geological profile inferred from the analysis, which was insufficient to prevent seepage and extend the seepage path. The proposed boundary line of each layer was different from the actual boundary line, and the proposed permissible infiltration slope drop of each layer was conservative. The actual situation was that the highest water level at the upstream reservoir reached 743.10 mas the overtopping flow reached the discharge channel. And before that, in addition to find several spot outflows at the foot of the downstream dam slope at an elevation of around 669 m, with the rising of water level in front of the barrier, another outflow with a stable discharge of 1–2m^3/s was found at the elevation of 700 m on the dam slope. The outflow water was clear without turbidity and proved it presented no indication of pipe collapse in the barrier dam.

3.4.4.3 Comprehensive Comment

(1) The secondary landslides appeared at the right bank of upstream and downstream of the barrier dam, but did not reach the river valley. Under the strong rainfall or other induced factors, the sliding might happen again.

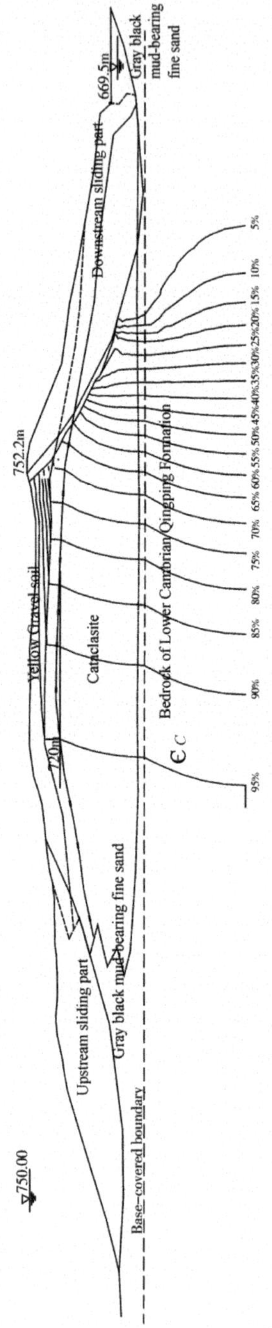

Fig. 3.30 Distribution of seepage equipotential line of geological longitudinal section after VII-intensity earthquake and landslide

(2) Due to the huge impulse of the landslide, the mud-bearing fine sand deposited in the reservoir area of Kuzhu Dam was uplifted and partially accumulated at the upstream of dam slope, which played a certain role in preventing infiltration. The possibility was small for seepage collapse in the whole barrier dam, but it could not be ruled out the existence of partial overhanging of the cataclasite rock. When the water level rose to a certain elevation, there was a possibility of concentrated leakage and might cause local collapse of the downstream dam slope.
(3) As the barrier was mainly composed of cataclasite rock, the overall stability was well. However, there was a possibility of local dam collapse when the water level rose to a certain elevation.
(4) In the absence of drilling and testing, it was effective for emergency rescue to select experts with rich practical experience, strong capabilities of observation and analysis to observe the field, deduce the composition of the barrier material and carry out geological zoning.

3.5 Evaluation and Analysis of Discharge Channel Stability

3.5.1 Evaluation of Discharge Channel Stability

There were three design schemes from the high, the medium and to the low standard for the discharge channel. In the high standard scheme, the crest elevation was 742 m, the bottom width was 13 m, and the total length was 695 m. The longitudinal slope of the upstream gentle section was 0.6%, while the longitudinal slope of the downstream steep section was designed in two sections with slope drops of 24% and 16%, respectively.

With the significant improvement of the actual construction conditions compared to the original plan, the actual construction formed a flat section inlet at an elevation of 740 m, the outlet at an elevation of 739 m, and the width of the channel bottom was 7–10 m.

According to the geological construction work of excavating the discharge channel, the bottom and slope of the channel in the section of pile number 0 + 000–0 + 270 were composed of gravel soil, which was consistent with the previous judgment. Moreover, in the section of pile number0 + 270–0 + 330, the bottom and slope of the channel were composed of undisintegrated hard fractured rock, and the upper part of the slope was gravel soil with a thickness of about 3.0 m (Fig. 3.31).

Geophysical test results showed that the top elevations of the cataclasite in the section of pile number 0 + 000–0 + 100 and 0 + 100–0 + 180 were, respectively, 725–735 m and 710 m. The gravel soil in the section of pile number 0 + 180–0 + 290 was deep. The bottom elevation of the cataclasite in the section of pile number 0 + 290–0 + 330 was 720–730 m, under which was the gravel soil. The gravelly rock was thicker in the outlet section of the discharge channel (Fig. 3.32).

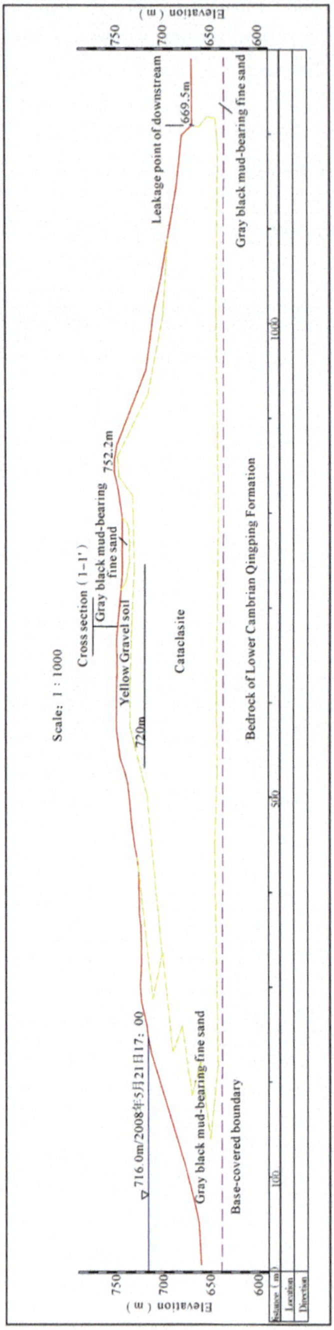

Fig. 3.31 Engineering geological longitudinal section of right channel of Tangjiashan barrier dam

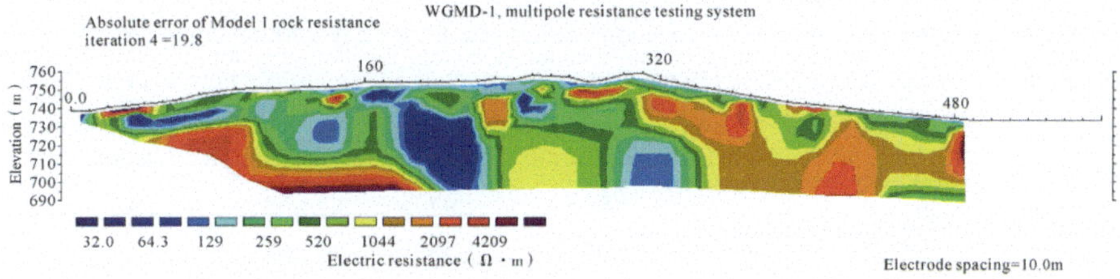

Fig. 3.32 Geophysical detecting profile of constructing excavation discharge channel (part of warm color is with high resistance like cataclasite, and part of cold color is with low resistance like gravel soil)

Because of the weak scouring resistance of the gravel soil, the cataclasite in the outlet section of the drainage channel would help slow down the down-cutting speed of the water flow and control the elevation of the entire channel bottom, which would reduce the impact of the lake water discharge on the downstream.

The discharge channel was excavated according to the terrain and presented a bow-shape bending to the right in plain. The overtopping water mainly scoured upon the right bank. And the right bank of the discharge channel side slope was mainly composed of gravel soil, which had weak scouring resistance and may be scoured to the landslide bed. The slope on the left side of the discharge channel was weakly scoured by the overtopping flow, and the slope was generally stable except for the gravel soil distributed on the slope which would be eroded away by the overtopping flow. The preliminary estimation was that the opening width of the discharge channel might reach 340 m after being scoured.

After the flow passing the discharge channel, the gravel soil at the bottom channel and the cataclasite formed by the strongly weathered rock would be washed away by the water very quickly. After reaching the top of the cataclasite formed by the weakly weathered rocks, the down-cutting speed of the discharge channel would be greatly slowed down. After the flow was stabilized, the bottom of the diversion channel might be lowered to an elevation of about 720 m.

3.5.2 *Stability Analysis of the Side Slope of the Discharge Channel*

In the design of the open channel drainage project, stability analysis was carried out for the slopes of all three programs.

3.5.2.1 Computational Analysis Results

The following were the results of the side slope stability analysis of the actual construction of the discharge channel section. The bottom of the channel was 10 m

Table 3.14 Safety factor of channel slope after construction for each working condition

Location	Water flow level of 739 m		Water flow level of 743 m		Water flow level of 752 m	
	Normal operation	7° seismic action	Normal operation	7° seismic action	Normal operation	7° seismic action
Left bank	2.06	1.71	1.80	1.47	1.52	1.16
Right bank	2.34	1.89	1.94	1.55	1.66	1.24

wide, and the water level of the lake was 752 m at the possible highest level, 743 m at the planned evacuation level and 739 m at the channel bottom, which corresponded to the least favorable conditions during the construction period of the slope. The stability under seismic action was calculated by the proposed static method, and the seismic action was mainly considered as the horizontal action, and the peak acceleration velocity of 0.1 g at 7° is considered.

The results of the safety factor of the channel slope after construction under each working condition were shown in Table 3.14 and Figs. 3.33, 3.34, 3.35, 3.36, 3.37 and 3.38.

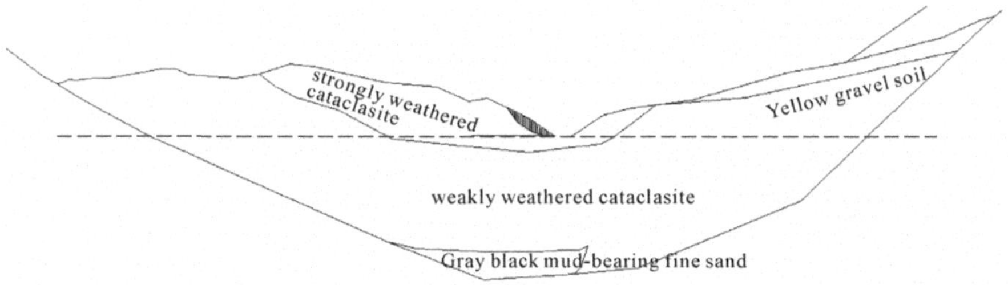

Fig. 3.33 Minimum safety factor $F_s = 2.06$ for the left bank circular arc sliding search method after excavation of the channel (water level elevation of 739 m)

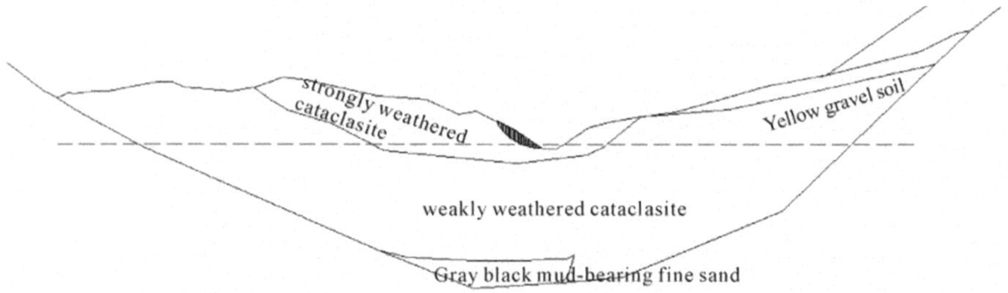

Fig. 3.34 Minimum safety factor $F_s = 1.8$ for the left bank circular arc sliding search method after channel excavation (water level elevation of 743 m)

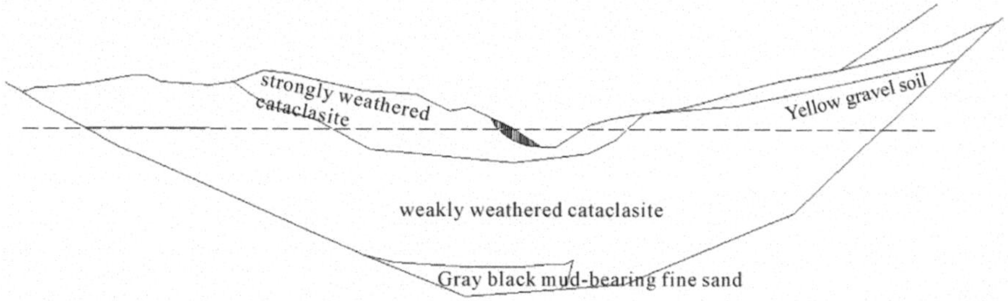

Fig. 3.35 Minimum safety factor $F_s = 1.52$ for the left bank circular arc sliding search method after excavation of the channel (water level elevation of 752 m)

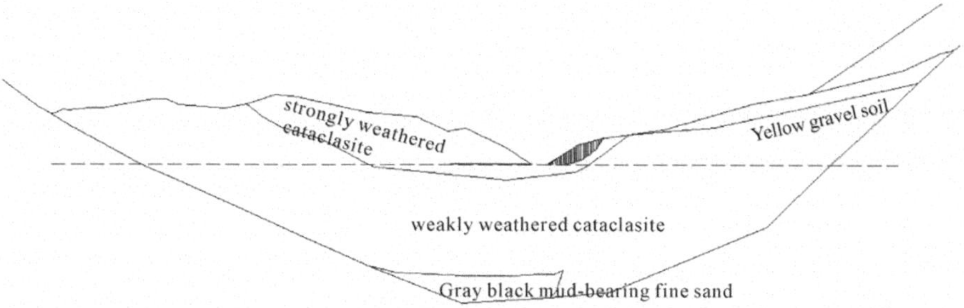

Fig. 3.36 Minimum safety factor $F_s = 2.34$ for the right bank circular arc sliding search method after excavation of the channel (water level elevation of 739 m)

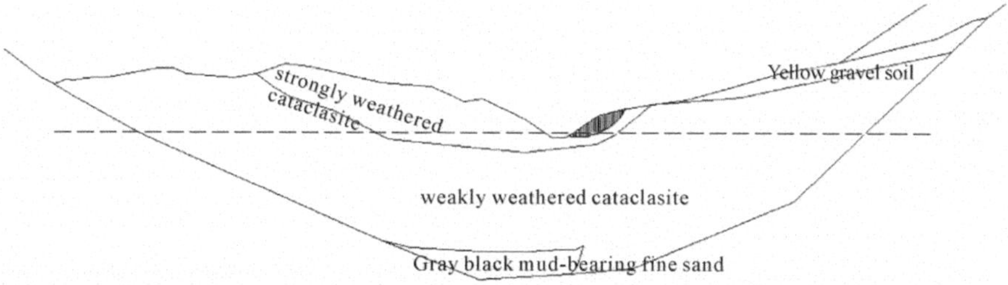

Fig. 3.37 Minimum safety factor $F_s = 1.94$ for the right bank circular arc sliding search method after excavation of the channel (water level elevation of 743 m)

3.5.2.2 Comprehensive Comments

(1) The right side of the barrier dam was a low-lying construction, and the upper part of the barrier dam was composed of the residual gravel soil on the top of the original slope with a thickness of 15 m. Therefore, the excavation from the right side required disposal of little excavated volume and thus the construction was more convenient.

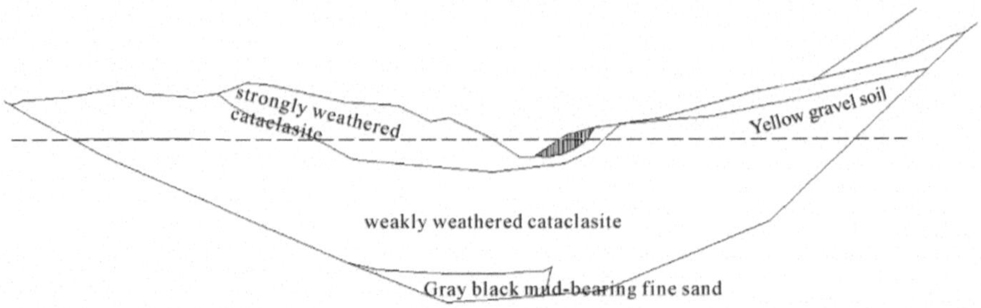

Fig. 3.38 Minimum safety factor $F_s = 1.66$ for the right bank circular arc sliding search method after excavation of channel (water level elevation of 752 m)

(2) The calculation results showed that the channel slope was in a stable state after the excavation with high standard plan, and it was great important for effective water discharge in time and water level reduction in front of the barrier dam.
(3) The cataclasite in the outlet section of the discharge channel would help slow down the down-cutting speed of the water flow and control the elevation of the entire bottom base of the discharge channel.
(4) The discharging water would mainly scour the right bank, and the right bank slope of the discharge channel was mainly composed of gravel soil, which has a weak scouring resistance, and may be washed away to the landslide bed. The left bank slope of the discharge channel was less scoured by the water flow, and the side slope was generally stable.

Chapter 4
Analysis on Flood Evolution of Tangjiashan Barrier Lake Dam-Break

Given the severity of consequences of the Tangjiashan barrier lake dam-break and the uncertainty of rainfall during the emergency, as well as the potential impact of secondary disasters induced by the upper Malingyan ancient landslide, it is of great necessity to conduct analysis on the flood evolution of the Tangjiashan barrier lake dam-break, so as to provide scientific basis for the barrier lake emergency handling engineering design scheme and the downstream-dam-personnel safe transfer and refuge program.

4.1 Analysis on Barrier Dam-Break Possibility and Scouring Form of Barrier Dam During the Emergency

In the process of emergency handling of the Tangjiashan barrier lake, according to the material composition and structural distribution characteristics of barrier dam, water resource experts analyzed and judged the barrier dam-break possibility and scouring form of the Tangjiashan barrier dam, which served as the basis for the formulation of the barrier dam-break calculation scheme.

4.1.1 Analysis on Scouring Resistance Capacity of Mixture Forming Barrier Dam

The empirical formulae for incipient velocity of general granular particles adopted at home and abroad were as follows:

Q. Yang and W. Yang, *Emergency Handling of the "5.12" Tangjiashan Barrier Lake*,
https://doi.org/10.1007/978-981-99-1706-8_4

(1) Sediment incipient velocity—Shamov Formula

$$U_c = 1.14\sqrt{\frac{\gamma_s - \gamma}{\gamma}\mathrm{gd}}\left(\frac{h}{d}\right)^{\frac{1}{6}}$$

(2) Sediment incipient velocity—Tang Cunben Formula

$$U_c = 1.53\sqrt{\frac{\gamma_s - \gamma}{\gamma}\mathrm{gd}}\left(\frac{h}{d}\right)^{\frac{1}{6}}$$

(3) Sediment incipient velocity—Zhang Ruijin Formula

$$U_c = 1.34\sqrt{\frac{\gamma_s - \gamma}{\gamma}\mathrm{gd}}\left(\frac{h}{d}\right)^{\frac{1}{7}}$$

(4) Scouring resistant velocity for rock foundation:

$$U_c = (5 \sim 7)\sqrt{d}$$

(5) Incipient velocity—Izbash Formula

$$U_c = K\sqrt{\frac{\gamma_s - \gamma}{\gamma}2\mathrm{gd}}$$

in which:

γ_s bulk density of the sediment particle (or rock block), N/m^3
γ bulk density of water, N/m^3
d grain size of the sediment particle (or dimension of the rock block), m
g acceleration of gravity, m/s^2
K comprehensive stability coefficient.
h water depth, m.

The results of calculation and analysis are shown in Tables 4.1 and 4.2. It could be seen that significant scouring would occur in the discharge channel for loose soil if the flow velocity exceeded 2 m/s, for cataclasite rock fragments if the flow velocity exceeded 5 m/s. According to preliminary judgment, the loose soil part of barrier dam was likely to have a barrier dam-break, and the cataclasite rock fragments would also be scoured in case of a fast flow velocity.

According to the gradation composition of the upper layer of barrier dam, in the process of water storage and overtopping, the barrier dam material infiltrated and inflated, its viscosity increases, in the meantime, its gradation was very uneven, so it was prone to form diluted debris flow. At this time, there would be three scouring forms: down-cutting scouring, making the river bed form deepen and widen; outer

Table 4.1 Analysis on scouring resistance capacity of mixture forming barrier dam (grain size: 20–200 mm)

Calculation formula		Grain size(mm)	Water depth(m)	Incipient velocity(m/s)
1	Shamov Formula	20	1	1.24
	Tang Cunben Formula	20	1	1.67
	Zhang Ruijin Formula	20	1	1.33
2	Shamov Formula	20	2	1.40
	Tang Cunben Formula	20	2	1.87
	Zhang Ruijin Formula	20	2	1.47
3	Shamov Formula	20	3	1.49
	Tang Cunben Formula	20	3	2.01
	Zhang Ruijin Formula	20	3	1.56
4	Shamov Formula	200	1	2.68
	Tang Cunben Formula	200	1	3.60
	Zhang Ruijin Formula	200	1	3.03
5	Shamov Formula	200	2	3.01
	Tang Cunben Formula	200	2	4.04
	Zhang Ruijin Formula	200	2	3.35
6	Shamov Formula	200	3	3.22
	Tang Cunben Formula	200	3	4.32
	Zhang Ruijin Formula	200	3	3.55

Table 4.2 Analysis on scouring resistance capacity of mixture forming barrier dam (grain size: 1.0–2.0 m)

Calculation formula	Grain size(m)	Scouring resistant (incipient) velocity(m/s)
Scouring resistant velocity for rock foundation	1.0	7.00
Incipient velocity—Izbash Formula	1.0	7.51
Scour resistant velocity for rock foundation	1.5	8.57
Incipient velocity—Izbash Formula	1.5	9.20
Scour resistant velocity for rock foundation	2.0	9.90
Incipient velocity—Izbash Formula	2.0	10.63

bank scouring in channel bends, making the water front expand; retrogressively scouring, making the temporary falling sill drawback. These three scouring forms coexisted, which would speed up the down-cutting scouring of the barrier dam.

4.1.2 *Analysis on Possibility of Abrupt Failure and Progressive Failure*

The Tangjiashan barrier dam had a total volume of 20.37 million m^3 and a surface area of about 307,200 m^2. Judging by a deluvial deposit cover (overburden layer) of about 15 m thick, it was estimated that the barrier dam had a deluvial deposit cover of 4.6 million m^3; taking 10 m-depth highly weathered material of 2.5–3 million m^3 into consideration, it aggregated to a total volume of 7–7.5 million m^3 and would be prone to scouring. The lower part of the barrier dam, with a total volume of 12 million m^3, consisted mainly of broken or cracked rocks. Rock accounted for approximately 65%, while soil (including riverbed overburden and intensely weathered zone) accounted for about 35%. In the meantime, the lower barrier dam was relatively dense, the lower part had stronger scouring resistant capacity than that of the upper part, while the left bank is stronger than that of the right bank.

Unlike those dams formed by soil landslides, the Tangjiashan barrier dam would not encounter soil flow failure. As the middle and lower part of its upstream slope was covered by the muddy silt clay deposited in the Kuzhu reservoir and had therefore relatively good anti-seepage property, plus the fact that the rapid landslide had strong compaction, the lower part of the barrier dam had a relatively dense structure, and piping failure would not easily occur. Only several spot outflows were observed at the toe of the downstream slope at an elevation of around 669 m. With the increase of the reservoir level, another outflow with a stable discharge of around 1–2 m^3/s was found at the elevation of 700 m on the downstream slope. The outflow water was clear without turbidity phenomenon and showed no indication of piping failure in the barrier dam.

The most probable failure scenario of the Tangjiashan barrier dam would be the scouring failure by flow of large discharge. From the analysis of the structure characteristics of the barrier dam, the scouring resistance capacity of its broken and cracked rock mass was relatively strong, while that of the rock fragment soil layer was relatively weak. The first to be scoured should be the broken rock layer on both sides of the artificial discharge channel within the low-lying area on the right side of the barrier dam. The lower boundary of the initial scouring failure was located in the contact zone between the upper broken rock soil layer and the lower cracked rock mass, at an elevation of 710–730 m, and decreased gradually from the left bank to the right bank.

The Tangjiashan barrier dam was huge in scale, forming by rapid bedrock landslide. It possessed dense lower part, there was no possibility of soil flow failure and little possibility of piping failure, the whole body was relatively stable, and thus an abrupt failure of the whole body could be basically ruled out.

Under the influence of flood scouring, the barrier dam could encounter retrogressively scouring. This process may be rapid and gradual, but abrupt failure would not occur. Before discharge of the Tangjiashan barrier dam, it came to the following conclusions on the barrier lake breach process after its water discharge.

(1) **The material composition of the barrier dam would be favorable for preventing it from a rapid failure**

The Tangjiashan barrier dam was formed by the rock landslide occurred along the bedding plane on the dip slope mountain body. The rock mass was composed of thin-to-medium thick layers of feldspar mica siltstone, siliceous rock, marl and mudstone in the Qingping Formation of Cambrian system, figured with soft–hard integrated rock formation. According to the field and aerial images, it was supposed that the landslide bedding slip along the bedding plane was just like cooking with spatula, which pushed the riverbed overburden and landslide leading edge mass to the left bank and upstream and downstream, and part of them was felled on the left surface of the landslide body.

As there were 12 million m^3 of broken and cracked rocks in the lower part of the barrier dam, the overall barrier dam was in a stable status, which was favorable for preventing it from a rapid failure. The surface of the barrier dam and the upper part of the right bank abundant with deluvial deposit cover, as well as the steep part of the slope, which possessed a poor scouring resistance capacity, were similar to the soil landslide and thus were likely to have rapid failure.

(2) **The broken and cracked rocks observed at two locations along the route of the discharge channel would be favorable for slowing down the breaching speed**

The barrier dam consisted mainly of broken and cracked rocks. According to the geological phenomenon revealed by excavation, there were broken and cracked rocks which could be observed or foreseen at two locations. When one located at the gentle outlet section and the other one was near the pile numbered 0 + 170 (hump), it would slow down the breaching speed. Unless it was over-topped by flows of high head and large discharge, an abrupt full breach or 50% breach would be very unlikely to occur.

(3) **The lower limit of the strongly weathered rock zone, i.e., 720 m, might be probably the lowest breach bottom elevation**

Once the loose part in the upper part of the barrier dam encountered flow of large discharge, there would be a rapid breach. When the flow velocity was higher, weathered and near-disintegrating broken and cracked rocks would also be scoured. Significant scouring would occur for loose medium if the flow velocity exceeded 2 m/s and for cataclasite rock fragments medium if the flow velocity exceeded 5 m/s. When the flow velocity was much greater than the scouring incipient velocity, rapid scouring would occur. The scenario of "scouring-stability-re-scouring-re-stability" is a basic law. As long as there were floods more serious than previous ones, the cycle of "scouring-stability-re-scouring-re-stability" would continue. The scouring would not stop until a stable slope ratio or relatively intact rock mass appeared.

According to the field investigation, the geological investigation data of surrounding areas and referred to the weathering status of similar rock strata, it was estimated that there might be a deluvial deposit cover (flood deposited overburden) of about 20 m thick on the right bank and a 10 m zone of strongly weathered rock below it. Considering the scouring resistance of the broken and cracked rocks which formed the lower part of the dam, the lower limit of the strongly weathered rock zone, i.e., 720 m, might be probably the lowest bottom elevation.

The following empirical formulae were adopted to calculate the potential breach width, respectively, developed by China Academy of Railway Sciences, Yellow River Conservancy Commission and Xie Renzhi.

Empirical formula developed by China Academy of Railway Sciences:

$$b_m = K\left(W^{1/4}B^{1/7}H_0^{1/2}\right)$$

Empirical formula developed by Yellow River Conservancy Commission:

$$b_m = K\left(W^{1/2}B^{1/2}H_0\right)^{1/2}$$

Empirical formula developed by Xie Renzhi:

$$b_m = KWH_0/(3E)$$

in which,

K coefficient related to the barrier dam material
W storage of the reservoir, m^3
B width of barrier dam, m
H_0 maximum water depth of the reservoir in the vicinity of the dam, m

E cross-sectional area of the valley (below the crest elevation) at the barrier dam site, m^2.

By using these three formulae, the breach depth of the Tangjiashan barrier dam was estimated at 200 m, 400 m and 120 m, respectively.

Based on the material composition of the barrier dam around the right-side channel, the breach might reach a lower limit of the strongly weathered cracked rock zone, i.e., 720 m, and the collapsed materials were rubble soil and strongly weathered cracked rock. Based on the material composition of the water flowing section, the breach width would be 340 m at the elevation of 752.5 m. It was predicted that the left side of the breach would be relatively steeply sloped with a probable angle of 35 ° and the right side gently sloped with a probable angle of 8 °, in a trapezoid shape. According to the material composition of the barrier dam, it was initially concluded that the breach width was 340 m, which was larger than the result calculated by the formulae developed by China Academy of Railway Sciences and by Xie Renzhi and slightly smaller than the result by formula developed by Yellow River Conservancy Commission.

4.2 Analysis on Discharge Capacity from Barrier Lake

When the water started to flow through the discharge channel, the retrogressively scouring would deepen and widen the water flow section, so as to improve the water flow capacity of the discharge channel. Since the water flow section in the process of water scouring could not be predicted in advance, based on the analysis of the influence of water souring on the discharge capacity and considering the available date of discharge section, as well as giving priority to safety, it was targeted at the determined discharge section for its discharge capacity calculation, which would provide relevant reference for the formulation of drainage design plan. In addition, the flood diversion calculation was carried out under the assumption that the discharge section did not change during the flood relief, in order to get the breaching flood levels under different design floods, which would provide relevant basis for the formulation of breaching flood calculation scheme.

4.2.1 Calculation Method

4.2.1.1 Method of Weir Flow Formula

The minimum flow section along the discharge channel was selected as generalization section of the trapezoidal barrier dam; in the meantime, the elevation of the barrier dam bottom was set as the elevation of the discharge channel bottom.

Adopting the weir flow formula with the channel end as water fall:

$$Q = \varepsilon_1 \mathrm{mn}\left(b' + 0.8H \cot\theta\right)2g^{1/2} H_0^{3/2}$$

in which,

H_0 full head corresponding to certain flow, m
ε_I coefficient of lateral contraction.
m coefficient of flow, in terms of safety, taking the smaller value 0.35 for $m\varepsilon_I$.

4.2.1.2 Method of Channel Water Surface Curve

From upstream to downstream, the discharge channel was divided into three sections based on the bottom slope ratio, i.e., $i = 0.006$, $i = 0.24$ and $i = 0.16$. The first section was gently sloped and served as control section for the discharge capacity of discharge channel, its channel flow pattern, which could be rapid flow or tranquil flow, depended on the flow volume. The following two sections were steeply sloped with rapid flow, which had no influence on the discharge capacity of discharge channel. Therefore, the discharge flow of the first section was calculated by taking the first end section as the downstream boundary section. As the following two sections carried rapid flow, the water depth at section 0 + 326 served as the critical water depth, which was used as the boundary condition to calculate the water surface curves under different flow volumes.

(1) Calculation of critical water depth

$$1 - \frac{\alpha Q^2 B}{g A^3} = 0$$

(2) Energy equation

$$z_1 + \frac{\alpha_1 v_1^2}{2g} = z_2 + \frac{\alpha_1 v_2^2}{2g} + h_{w1-2}$$

When the flow gradually increased, some calculated sections (such as 0 + 290, 0 + 100) will appear as rapid flows. At this time, the water depth of this section was taken as the boundary condition to continue calculate the water surface curves toward the upstream.

After calculating the water surface curve of the discharge channel, the water level in the reservoir and the incipient section of the discharge channel should meet with the energy equation. Moreover, assuming that the flow velocity in the reservoir was 0 and the local head loss coefficient between two sections was 0.25, thus the relationship between the water level in the reservoir and the flow volume in the discharge channel could be deduced.

4.2.1.3 Model of MIKE11

MIKE11 was used to establish the one-dimensional watercourse model. The first section was gently sloped and served as control section for the discharge capacity of discharge channel, and its channel flow pattern depended on the flow volume (rapid flow or tranquil flow). The following two sections were steeply sloped with rapid flow, which had no influence on the discharge capacity of discharge channel. However, as the flow pattern of the first section was uncertain, the water depth of section 0 + 326 was less than the critical water depth, and the specific value should be determined by calculation. The model covered a range from 0 + 000 to 0 + 426, a lower water level was fixed at the downstream section, the flow pattern was judged by the program, and then the water surface curve could be derived.

After calculating the water surface curve of the discharge channel, the water level in the reservoir and the incipient section of the discharge channel should meet with the energy equation. Moreover, assuming that the flow velocity in the reservoir was 0 and the local head loss coefficient between two sections was 0.25, thus the relationship between the water level in the reservoir and the flow volume in the discharge channel could be deduced.

4.2.2 Calculation of Discharge Capacity from Barrier Lake

4.2.2.1 Comparison Among Three Calculation Methods

The design section of the discharge channel was in trapezoidal shape, with bottom elevation of 742 m, bottom width of 13 m and side slope of 1:1.5. The discharge capacity was calculated, respectively, by the abovementioned three methods, and the results are shown in Table 4.3 and Fig. 4.1.

As shown in Table 4.3 and Fig. 4.1, in the case of flow volume below 1000 m^3/s, there was little difference among the flow results calculated by these three methods under the same water level condition. In the case of flow volume above 1000 m^3/s, the calculation results of weir flow formula method and channel water surface curve method were close, while that of MIKE11 model was relatively larger. In the case of flow volume below 2500 m^3/s, the discharge capacity calculated by weir flow formula method was larger than that calculated by channel water surface curve method. In the case of flow volume above 2500 m^3/s, the results calculated by the earlier-mentioned two methods were close.

It could be seen from the comparative analysis, in the case of small flow volume, these three methods would reach close results; in the case of large flow volume, the discharge capacity obtained by MIKE11 model was larger, while the result obtained by channel water surface curve method was slightly smaller than that obtained by weir flow formula method. Since the submerged weir flow formula method used one section for channel generalization, its accuracy was poor. In terms of actual situation

Table 4.3 Calculation results of discharge capacity from barrier lake corresponding to the scheme for 742 m bottom elevation

Method of weir flow formula		Method of channel water surface curve		Model of MIKE11	
$Q(m^3/s)$	H (m)	$Q(m^3/s)$	H(m)	$Q(m^3/s)$	H(m)
22	743	200	746.43	200	746.16
67	744	400	748.37	400	747.99
134	745	600	749.81	600	749.04
221	746	800	750.84	800	749.89
329	747	1000	751.75	1000	750.63
460	748	1200	752.54	1200	751.26
614	749	1400	753.30	1400	751.84
792	750	1600	753.98	1600	752.35
996	751	1800	754.63	1800	752.82
1220	752	2000	755.19	2000	753.26
1481	753	2200	755.72	2200	753.67
1765	754	2400	756.22	2400	754.05
2077	755	2600	756.69	2600	754.40
2419	756	2800	757.16	2800	754.72
2791	757	3000	757.63	3000	755.02
3193	758	3200	758.07	3200	755.30
3641	759	3400	758.50	3400	755.58
4173	760	3600	758.93	3600	755.85
4780	761	3800	759.35	3800	756.12
5508	762	4000	759.75	4000	756.37

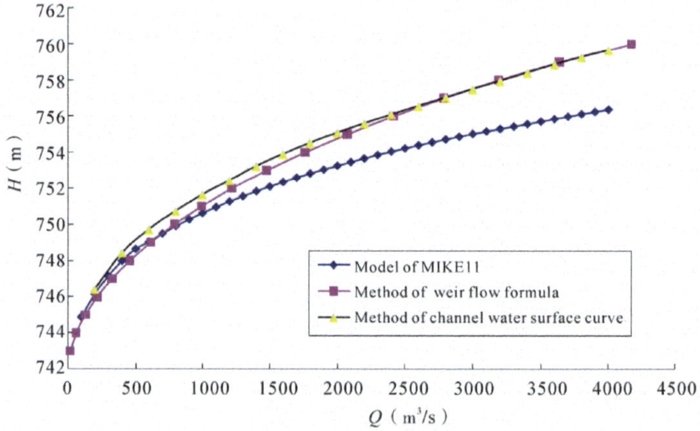

Fig. 4.1 Comparison diagram of discharge capacity calculated by three methods

of discharge channel and safety, the channel water surface curve method was adopted in flood regulation to calculate the discharge capacity.

4.2.2.2 Influence of Water Scouring on Discharge Capacity from Barrier Lake

As the scour effect of water flow on the discharge channel was actually a gradual process, the discharge channel gradually shortened and the bottom elevation gradually reduced in the scouring process. Meanwhile, due to the geological differences at different elevations, the bottom width of the discharge channel would also change. In order to reflect the influence of water scour on the discharge capacity of discharge channel, analyses were conducted under three scenarios of channel shortening, deepening and widening. The cross sections of the discharge channel were all generalized into trapezoidal sections.

(1) **Influence of discharge channel shortening on discharge capacity**

The scheme set the following values for the discharge channel: bottom elevation of 742 m, bottom width of 13 m, side slope ratio of 1:1.5 and the length of 326 m, 182 m, 100 m and 50 m, respectively. The discharge capacity of discharge channel for each scheme was calculated by channel water surface curve method, and the results are shown in Table 4.4 and Fig. 4.2.

As shown Table 4.4 and Fig. 4.2, when the length of discharge channel turned gradually from 326 to 100 m, its discharge capacity improved with slight increase. When the length of discharge channel turned gradually from 100 to 50 m, its discharge capacity improved with relatively great increase.

(2) **Influence of discharge channel bottom elevation reduction on discharge capacity**

The scheme set the following values for the discharge channel: bottom width of 13 m, side slope ratio of 1:1.5, the length of 326 m and the bottom elevation of 742 m, 738 m, 734 m and 730 m, respectively. The discharge capacity of discharge channel for each scheme was calculated by channel water surface curve method, and the results are shown in Table 4.5 and Fig. 4.3.

As shown in Table 4.5 and Fig. 4.3, when the bottom elevation of discharge channel kept reducing, its discharge capacity improved with relatively great increase.

(3) **Influence of discharge channel bottom width change on discharge capacity**

The scheme set the following values for the discharge channel: side slope ratio of 1:1.5, length of 326 m, bottom elevation of 738 m and 734 m, respectively, and bottom width of 13 m and 7 m, respectively. The discharge capacity of discharge channel for each scheme was calculated by channel water surface curve method, and the results are shown in Table 4.6 and Fig. 4.4.

As shown in Table 4.6 and Fig. 4.4, with the same water level, the larger of the bottom width, the larger of the discharge flow.

Table 4.4 Comparison table of discharge capacity in case of gradual length shortening

Flow volume(m^3/s)	Scheme on discharge channel length(m)			
	Water level			
	326 m	182 m	100 m	50 m
200	746.43	746.40	746.29	746.18
400	748.37	748.32	748.19	748.03
600	749.81	749.74	749.56	749.39
800	750.84	750.76	750.60	750.43
1000	751.75	751.63	751.45	751.23
1200	752.54	752.41	752.23	751.91
1400	753.30	753.12	752.97	752.50
1600	753.98	753.80	753.66	753.07
1800	754.63	754.43	754.31	753.57
2000	755.19	755.02	754.92	754.06
2200	755.72	755.58	755.50	754.51
2400	756.22	756.12	756.06	754.97
2600	756.69	756.64	756.59	755.27
2800	757.16	757.13	757.10	755.60
3000	757.63	757.60	757.59	755.92
3200	758.07	758.06	758.06	756.22
3400	758.50	758.50	758.51	756.52
3600	758.93	758.93	758.95	756.79
3800	759.35	759.35	759.36	757.07
4000	759.75	759.75	759.76	757.33

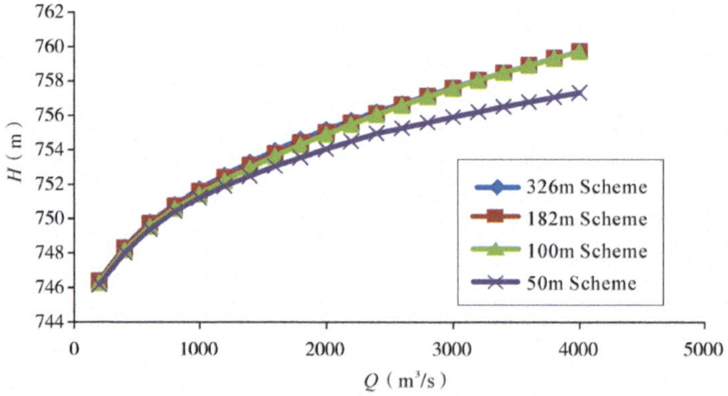

Fig. 4.2 Comparison diagram of discharge capacity in case of gradual length shortening

Table 4.5 Comparison table of discharge capacity in case of gradual bottom elevation reduction

Flow volume(m^3/s)	Scheme on discharge channel bottom elevation(m)			
	Water level			
	742 m	738 m	734 m	730 m
200	746.43	742.43	738.42	734.42
400	748.37	744.37	740.36	736.36
600	749.81	745.82	741.79	737.81
800	750.84	747.01	743.00	739.01
1000	751.75	748.06	744.03	740.05
1200	752.54	748.99	744.95	740.97
1400	753.30	749.84	745.78	741.82
1600	753.98	750.54	746.55	742.59
1800	754.63	751.21	747.26	743.31
2000	755.19	751.82	747.93	743.98
2200	755.72	752.41	748.56	744.62
2400	756.22	752.98	749.17	745.23
2600	756.69	753.53	749.73	745.80
2800	757.16	753.95	750.25	746.35
3000	757.63	754.43	750.78	746.87
3200	758.07	754.88	751.24	747.38
3400	758.50	755.32	751.72	747.87
3600	758.93	755.75	752.16	748.33
3800	759.35	756.16	752.60	748.79
4000	759.75	756.52	753.01	749.23

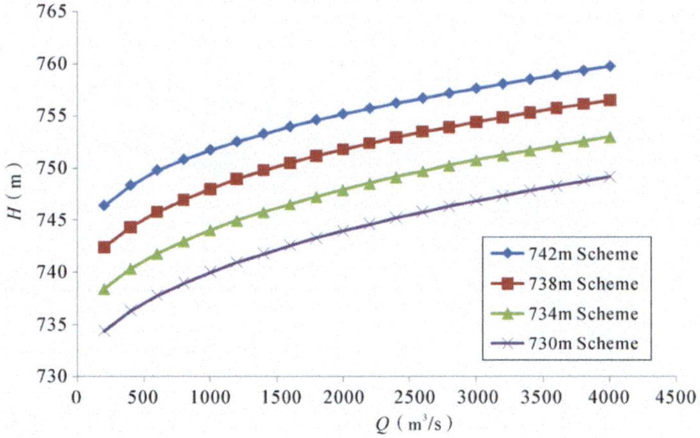

Fig. 4.3 Comparison diagram of discharge capacity in case of gradual bottom elevation reduction

Table 4.6 Comparison table of discharge capacity with different bottom width schemes

Flow volume (m^3/s)	Bottom elevation of discharge channel(m)			
	Water level of 738 m		Water level of 734 m	
	Bottom width of 7 m	Bottom width of 13 m	Bottom width of 7 m	Bottom width of 13 m
200	743.47	742.43	739.47	738.42
400	745.58	744.37	741.58	740.36
600	747.13	745.82	743.13	741.79
800	748.41	747.01	744.40	743.00
1000	749.50	748.06	745.47	744.03
1200	750.41	748.99	746.43	744.95
1400	751.16	749.84	747.31	745.78
1600	751.87	750.54	748.11	746.55
1800	752.53	751.21	748.85	747.26
2000	753.12	751.82	749.52	747.93
2200	753.71	752.41	750.18	748.56
2400	754.26	752.98	750.77	749.17
2600	754.79	753.53	751.32	749.73
2800	755.30	753.95	751.84	750.25
3000	755.78	754.43	752.34	750.78
3200	756.19	754.88	752.83	751.24
3400	756.60	755.32	753.28	751.72
3600	757.00	755.75	753.76	752.16
3800	757.39	756.16	754.23	752.60
4000	757.77	756.52	754.55	753.01

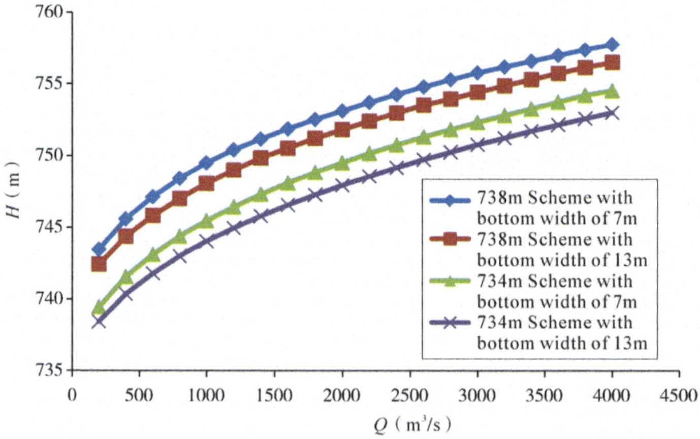

Fig. 4.4 Comparison diagram of discharge capacity with different bottom width schemes

As for the discharge capacity of the discharge channel, from the safety aspect, considering the material composition and geological condition, and without regard to scouring of water flow, the research results of the discharge capacity obtained by the above three schemes were smaller. On this basis, the flood regulation was carried out, and the results were relatively safe.

4.2.2.3 Discharge Capacity of Natural, Designed and Constructed Sections

(1) **Discharge capacity without discharge channel**

Without discharge channel, the reservoir stated overflowing at an elevation of 752.2 m (i.e., lowest elevation of the barrier dam crest), and the water flow section was a natural canyon section. Its discharge capacity was calculated by wide-top weir formula, and the results are shown in Table 4.7.

(2) **Discharge capacity of discharge channel designed section**

The designed section of the discharge channel was shaped in trapezoid, with bottom elevation of 742 m, bottom width of 13 m, side slope of 1:1.5 and length of 326 m. The discharge capacity of the discharge channel designed section is obtained by channel water surface curve method, and the calculation results are shown in Table 4.8. It can be seen that when the water level was 752 m, the discharge capacity of the discharge channel designed section was about 1063 m^3/s.

Table 4.7 Calculation results of discharge capacity without discharge channel

Water level (m)	Flow volume (m^3/s)	Water level (m)	Flow volume (m^3/s)
752.2	0.0	759.5	1579.0
752.5	4.5	760.0	1836.0
753.0	12.0	760.5	2145.0
753.5	37.5	761.0	2454.0
754.0	63.0	761.5	2818.5
754.5	115.0	762.0	3183.0
755.0	167.0	762.5	3605.0
755.5	251.5	763.0	4027.0
756.0	336.0	763.5	4510.5
756.5	457.5	764.0	4994.0
757.0	579.0	764.5	5541.0
757.5	742.0	765.0	6088.0
758.0	905.0	765.5	6701.0
758.5	1113.5	766.0	7314.0
759.0	1322.0	766.5	7995.5

Table 4.8 Calculation results of discharge capacity of discharge channel designed section

Water level (m)	Flow volume (m^3/s)	Water level (m)	Flow volume ($3^3/s$)
746.5	207.2	753.5	1458.8
747.0	258.8	754.0	1606.2
747.5	310.3	754.5	1760.0
748.0	361.9	755.0	1932.1
748.5	418.1	755.5	2117.0
749.0	487.5	756.0	2312.0
749.5	556.9	756.5	2519.1
750.0	636.9	757.0	2731.9
750.5	734.0	757.5	2944.7
751.0	835.2	758.0	3168.2
751.5	945.1	758.5	3400.0
752.0	1063.3	759.0	3633.3
752.5	1189.9	759.5	3875.0
753.0	1321.1		

(3) **Discharge capacity of discharge channel constructed section**

The constructed section of the discharge channel was shaped in trapezoid, with bottom elevation of 740 m, bottom width of 7 m, side slope of 1:1.5 and length of 326 m. The discharge capacity of the discharge channel constructed section was obtained by channel water surface curve method, and the calculation results are shown in Table 4.9. It can be seen that when the water level was 752 m, the discharge capacity of the discharge channel constructed section was about 1176 m^3/s.

4.2.3 *Calculation of Discharge Channel Flood Regulation*

4.2.3.1 Basic Data

(1) **Storage capacity curve**

The result of storage capacity curve developed by the POWERCHINA Chengdu was adopted.

(2) **Design flood data**

The design flood hydrograph of the dam site adopted the result developed by the Hydrology Bureau of Changjiang Water Resources Commission, including flood hydrograph with $P = 2\%$, $P = 5\%$, $P = 10\%$ and $P = 20\%$, respectively.

Table 4.9 Calculation results of discharge capacity of discharge channel constructed section

Water level (m)	Flow volume (m^3/s)	Water level (m)	Flow volume (m^3/s)
745.5	202.8	752.5	1309.6
746.0	250.2	753.0	1447.9
746.5	297.6	753.5	1588.7
747.0	345.0	754.0	1737.3
747.5	392.4	754.5	1893.5
748.0	453.2	755.0	2063.0
748.5	516.5	755.5	2251.0
749.0	579.7	756.0	2450.0
749.5	657.6	756.5	2659.6
750.0	742.4	757.0	2875.6
750.5	834.8	757.5	3100.0
751.0	943.5	758.0	3327.3
751.5	1057.1	758.5	3561.9
752.0	1176.2	759.0	3800.0

4.2.3.2 Flood Regulation Calculation of Static Reservoir Capacity

Assuming that the reservoir storage and water level changed in a straight line during the period of Δt, the water balance function was present as following:

$$\frac{I_b + I_e}{2} - \frac{Q_b + Q_e}{2} = \frac{V_e}{\Delta t} - \frac{V_b}{\Delta t}$$

in which:

I_b, I_e reservoir inflow at the beginning and the end during Δt, m^3/s
Q_b, Q_e discharge flow at the beginning and the end during Δt, m^3/s.
V_b, V_e storage volume at the beginning and the end during Δt, m^3.

Relationship between reservoir discharge and water level in front of the dam: $q = f(Z_{\text{dam}})$.

By combining the above two equations with trial algorithm, parameters such as water level in front of the dam and the discharge amount in the whole flood regulation process could be obtained.

4.2.3.3 Calculation Scheme and Flood Regulation Results

(1) **Calculation schemes**

It was supposed that the section form of the discharge channel remained unchanged during flood discharge, and the flood regulation calculations were carried out under three water flow section scenarios: without discharge channel, designed section for discharge channel, constructed section for discharge channel, encountering $P = 5\%$, $P = 10\%$ and $P = 20\%$, respectively.

The specific calculation schemes were as following:

Scheme 1: without discharge channel, encountering floods with $P = 5\%$, $P = 10\%$ and $P = 20\%$, respectively;
Scheme 2: discharge channel with designed section (high standard scheme), encountering floods with $P = 5\%$, $P = 10\%$ and $P = 20\%$, respectively;
Scheme 3: discharge channel with constructed section, encountering floods with $P = 5\%$, $P = 10\%$ and $P = 20\%$, respectively

(2) **Flood regulation results**

Based on the basic data and flood regulation principle, the results are present in Table 4.10 under these three schemes. The calculation results showed that the highest water lever ranges for them in the cases of $P = 5\%$, $P = 10\%$ and $P = 20\%$ were 759.09–761.26 m, 751.72–755.42 m and 750.85–754.75 m, respectively.

4.3 Calculation Analysis on Barrier Dam-Break Flood

The calculation results of barrier dam-break flood were directly related to the well-being of the downstream people and the population of the emergency evacuation. The initial water level of the barrier dam-break, the development process of the breach and the duration of the breach formation played an important role on the barrier dam-break flood; however, currently, the analysis of barrier dam-break floods was still in the exploration stage. Therefore, many institutions adopted several methods to conduct calculation and comparison analysis in the event of barrier dam-break flood, in order to maximize the reliability of the results.

4.3.1 Calculation Model for Barrier Dam-Break Flood

CISPDR, POWERCHINA Chengdu, CRSRI and IWHR adopted multiple methods to carry out parallel calculation and comparison analysis, including empirical formula method, one-dimensional barrier dam-break and flood propagation mathematical model, two-dimensional barrier dam-break and one-dimensional flood evolution mathematical model, Denmark MIKE11, USA DAMBRK barrier dam-break flood evolution calculation model and so on. Table 4.11 shows the five representative models adopted by CISPDR, River Research Department and Hydraulics Department of CRSRI, POWERCHINA Chengdu and IWHR.

The flood evolution model covered the watercourses from Tangjiashan to Tongkou on Tongkou River and Jiangyou to Fujiangqiao in Mianyang on Fujiang River. In

Table 4.10 Flood regulation results for three schemes

Calculation scheme	Scheme 1			Scheme 2			Scheme 3		
Initial water level (m)	752.2			742			740		
Length of discharge channel(m)	326			326			326		
Bottom width (m)				13			7		
Flood frequency (%)	5	10	20	5	10	20	5	10	20
Peak discharge (m^3/s)	3920	3920	2190	3920	3040	2190	3920	3040	2190
Maximum flood level (m)	761.26	760.23	759.09	755.42	753.69	751.72	754.75	752.88	750.85
Maximum discharge capacity (m^3/s)	2650	1980	1370	2090	1510	990	1970	1410	910

Table 4.11 Calculation models for barrier dam-break flood

No.	Institution	Modeling approach
Model 1	CISPDR	MIKE11 barrier dam-break model
Model 2	River Research Department of CRSRI	The self-developed BRESZHU model was used to calculate the breach flow process, and the MIKE11 HD module was used to calculate the flood evolution of the downstream dam
Model 3	Hydraulics Department of CRSRI	The weir flow empirical formula was used to calculate breach flow based on the reservoir water level and the shape of the crest section, a one-dimensional unsteady flow based on the Preissmann four-point implicit scheme was used to conduct flood evolution downstream of the dam, and a planar two-dimensional model was used for some schemes
Model 4	POWERCHINA Chengdu	The transplanted and improved DAMBRK barrier dam-break model was used, which was developed originally by NWS
Model 5	IWHR	Several empirical formulae were used to calculate the breach flow and the downstream flood propagation process of the barrier dam-break dam flood

the upper reaches, the barrier dam-break flood and upstream inflow of Fujiang River were taken as the boundary conditions. In the lower reaches, the water level-discharge relation at Fujiangqiao gauging station was taken as the boundary condition. The calculation range of Fujiang River stretched a 31 km river section from Jiangyou to Fujiangqiao in Mianyang. In the calculation of impact on Chongqing City, the lower boundary was extended to the small river dam 392 km downstream from the barrier lake. The calculation range of Tongkou River stretched a 59 km river section from Kaiping Town to Huikou. The river section for Tangjiashan to Fujiangqiao adopted the 21 large cross-section topography provided by POWERCHINA Chengdu, and the rest was obtained according to the above-water DEM aerial survey in 1988. The calculation ranges of Models 1 to 4 were basically consistent with the accepted topographic data. The value of roughness coefficient along each section for Model 1 was 0.04, and the values of roughness coefficient for Models 2 and 3 ranged from 0.035 to 0.05.

4.3.1.1 Model 1 (CISPDR)

The MIKE11 hydraulic calculation model was a one-dimensional river network hydrodynamic calculation model developed by Denmark Hydraulics Institute (DHI). The barrier dam-break model for MIKE11 took "Barrier Dam-break Structure" module to simulate the breaching process of dam body.

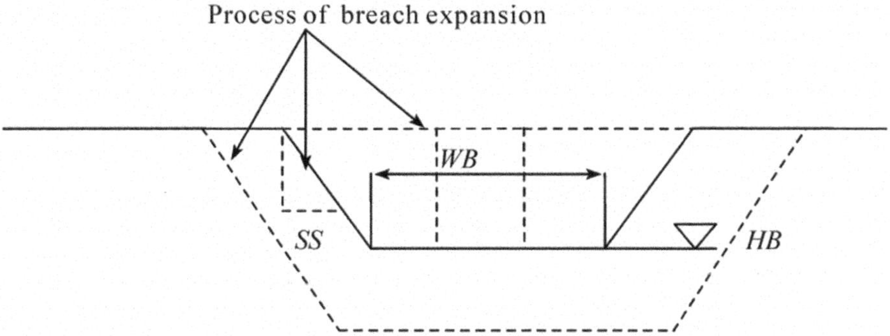

Fig. 4.5 Schematic diagram process of breach expansion and control parameters

A dam-break structure was similar to a broad-crested weir, but differed in two aspects:

(1) The shape of the dam would change with time, the breach became larger, the dam crest became smaller, and thus the relationship between water level and discharge of critical flow at dam crest and breach could not be calculated in advance.
(2) Since the relationship between water level and discharge of critical flow at dam crest and breach differed from each other, the crest overflow and the breach flow should be calculated separately.

It was assumed that the breach was an isosceles trapezoid for Model 1. The shape of the breach was mainly described by breach bottom elevation (HB), bottom width (WB) and side slope (SS, horizontal/vertical, and the left slope angle was equal to the right one). The extension of the breach was described by time series files. HB, WB and SS were determined by linear interpolation among the given time points within the time series. In this study, the elevation and the width of the breach bottom both changed linearly with time, while the breach side slope kept unchanged, so as to simulate the gradual breaching process of the barrier dam. The process of breach expansion and its control parameters are shown in Fig. 4.5.

The reservoir in the upstream of the barrier lake was simulated by adding additional volume to the upstream section by water level area curve in the mathematical model. The storage capacity curve was calculated by using the 1:2000 topographic map measured before the earthquake.

4.3.1.2 Model 2 (River Research Department of CRSRI)

The BRESZGU model independently developed by CRSRI was adopted for the calculation. This model was a mathematical model of earth-rock dam failure based on physical mechanism, which was coupled with reservoir water balance-breaching flow change-breach pattern development-dam body scouring process. It was permissible

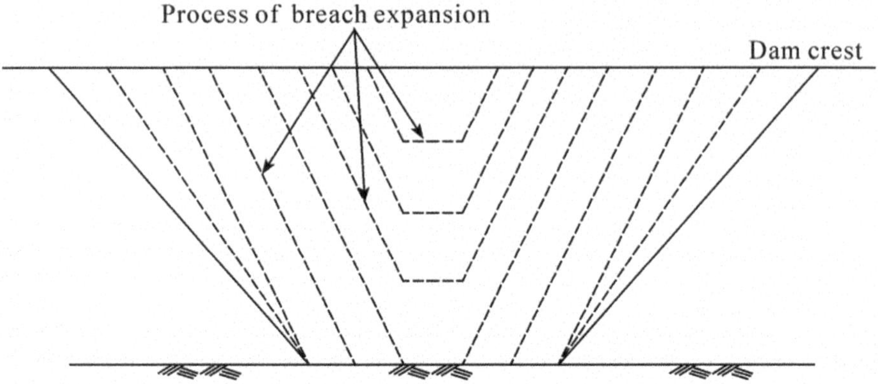

Fig. 4.6 Generalized map of breach expansion process of Tangjiashan barrier dam

to simulate directly the breach development process and flow process based on the upstream inflow, the storage capacity curve and the set breaching conditions, which are shown in Fig. 4.6.

For the evolution process of barrier dam-break flood toward downstream, a one-dimensional model was used to solve the Saint–Venant equations, for which the one-dimensional model adopted the MIKE11 module, and its equation was discretized in six-point center implicit format, the algorithm was Abbot-Ionescu method (tri-diagonal matrix method), and the solution was direct.

4.3.1.3 Model 3 (Hydraulics Department of CRSRI)

The empirical formula of weir flow and dynamic balance equation of reservoir water volume were used for flood regulation. Under the condition that the barrier lake did not encounter breach, based on the combination of the upstream inflow and the outflow of the barrier lake, it was capable to solve storage change within the reservoir, and to figure out the reservoir water level by storage capacity-water level relation. The breach flow was obtained by the empirical formula of weir flow according to the reservoir water level and crest section shape. Different empirical formulae were used for different barrier dam-break modes, such as instantly full-barrier dam-break, instantly partial barrier dam-break, subperiods instantly partial barrier dam-break and gradual barrier dam-break. The change of reservoir storage capacity was derived from the dynamic balance equation of reservoir volume.

When calculating the flow process at the barrier lake breach, different calculation methods were used for different barrier dam-break modes.

(1) **Mode of instantly full-barrier dam-break**

The generalized typical flow process curve method was adopted to calculate the barrier dam site flow process, on the basis of instantly full-barrier dam-break flow Q_m, discharge flow before barrier dam-breakQ_0, available discharge flow before barrier dam-break W and assumed emptying time T. The maximum flow empirical formula in the case of instantly full-barrier dam-break was presented as: $Q_m = \lambda B(g)^{0.5} H_0^{1.5}$.

(2) **Mode of instantly partial barrier dam-break**

The generalized typical flow process curve method was adopted to calculate the barrier dam site flow process, on the basis of instantly partial barrier dam-break flow Q_m, discharge flow before barrier dam-breakQ_0, available discharge flow above residual dam elevation before barrier dam-break W and assumed emptying time T.

The maximum flow of the instantly partial barrier dam-break was calculated by the empirical equation(shown as below), which was developed by the Water-way Experiment Station, U.S. Army Corps of Engineering.

$$Q_m = 0.296b(g)^{0.5}(BH_0/bh)^{0.28}h^{1.5}$$

in which,

b	breach width, m
B	dam length, m
h	water depth at the breach top, m
H_0	water head before the dam including marching water flow head, m
G	gravity acceleration, m/s^2.

(3) **Mode of subperiods instantly partial barrier dam-break**

The discharge process of each period was calculated by divided period according to the residual storage capacity and breach shape at the end of the previous period. The calculation method was the same as that of the instantly partial barrier dam-break mode.

(4) **Mode of gradual barrier dam-break**

In line with the water level change of the barrier lake and the development process of the breach scour (breach shape, barrier dam-break duration), the weir flow empirical formula was used to figure out the gradual barrier dam-break process at barrier dam site, for which the method for calculating water level was the same as that for flood regulation.

In view of the evolution process of the barrier dam-break flood toward downstream, one and two-dimensional unsteady flow models were adopted to solve the Saint–Venant equations, and the model solution was four-point eccentric implicit difference method.

The control equation of the planar two-dimensional mathematical model was as follows:

(1) Continuity equation

$$\frac{\partial h}{\partial t} + \nabla \cdot \left(\overset{r}{u}\, h\right) = 0$$

(2) Momentum equation

$$\frac{\partial \vec{u}}{\partial t} + \vec{u} \cdot \nabla \vec{u} + g\left(\nabla \varsigma + \vec{S}_f\right) - \nabla\left(\nu \nabla \vec{u}\right) = 0$$

in which,

$\vec{u}$ velocity vector, $\vec{u} = (u, v)$
$\vec{S}_f$ riverbed friction slope vector, $\vec{S}_f = \left(S_{fx}, S_{fy}\right)$, and $\vec{S}_f = \frac{1}{\rho g h} \vec{\tau}_{zb}$.

For the control equation, its spatial discretization adopted the control volume method, temporal discretization adopted the prediction-explicit calibration scheme, and irregular quadrilateral grid discretized the computation area. The model fully considered the influence of channel planform change on the flood evolution and the role played by floodplain and channel storage. The local velocity distribution under the dam derived from the model could provide a basis for the stability analysis of the barrier dam.

4.3.1.4 Model 4 (POWERCHINA Chengdu)

The calculation method adopted the transplanted and improved DAMBRK barrier dam-break flood mathematical model originally developed by USA National Weather Service. Before the introduction of the earlier-mentioned model, its developers tested it by using the investigation data of the 5 barrier dam-break flood events occurred in the USA. After the introduction of the model, Tianjin Water Resources and Hydropower Survey, Design and Research Institute of the Ministry of Water Resources, tested it by using the investigation data of the barrier dam-break events in the Banqiao Reservoir and Shimantan Reservoir.

4.3.1.5 Model 5 (IWHR)

The maximum breach flow was calculated, respectively, by broad-crested weir flow calculation formula developed by Tsinghua University, Shockledge partly gradual barrier dam-break maximum breach flow formula.

(1) **Broad-crested weir flow calculation formula developed by Tsinghua University**

$$Q = \delta \varepsilon m b \sqrt{2g} H_0^{\frac{3}{2}}$$

in which,

b breach width, m
H_0 effective water level at which the maximum breach flow occurs, m.

(2) **Empirical formula**

According to barrier dam-break cases at home and abroad, many researchers set up some empirical formulae, most of which established relations based on the reservoir storage, water depth and maximum barrier dam-break flow of the dam-broken reservoir. However, most of the accepted barrier dam-break cases occurred in small-sized reservoirs with storage capacity of several million cubic meters. According to verification and analysis, these formulae did not fit for the Tangjiashan barrier lake case.

Based on the theoretical weir flow formulae, the empirical formulae could be obtained by empirical coefficients, among which the typical Shockledge partly gradual barrier dam-break maximum breach flow formula could be shown as:

$$Q_{\max} = 0.9 \times \left(\frac{H - h}{H - 0.827}\right) B(H - h)\sqrt{H}$$

in which,

H total water depth, m
h residual dam height, m.

(3) **Water balance function**

Regardless of the inflow to the barrier lake and according to the triangular discharge process, the water balance relationship was shown as:

$$Q_{\max} = 2W/\Delta t$$

in which,

W total reservoir storage capacity, m^3/s
Δt emptying time, relating to reservoir storage capacity.

(4) **Calculation model of breach flow unsteady process**

According to the reservoir dynamic water balance equation, the assumed rate of breach scouring and expansion dynamic development and the calculation formula of breach dynamic outflow, the calculation model of breach flow unsteady process was jointly developed to analyze the breach flow process.

Reservoir dynamic water balance equation:

$$\Delta W = (Q_{\text{out}} - Q_{\text{in}})\Delta t$$

Breach dynamic flow calculation:

$$Q_b = c_v k_s C_d \sqrt{2g}\left[\frac{2}{3} b_s (h_i - h_b)^{1.5} + \frac{8}{15} z (h_i - h_b)^{2.5}\right]$$

The breach bed scouring rate depended on the breach flow intensity and the souring resistance property of the barrier dam material composition, and the accepted empirical formula was as follows:

$$S_r = 0.218\frac{\gamma}{\gamma_s - \gamma}\frac{nU}{h^{0.67}d^{0.25}}\left[\frac{\gamma n^2 U^3}{h^{0.333}} - 0.1\frac{\gamma}{g}\left(\frac{\gamma}{\gamma_s - \gamma}gd\right)^{1.5}\right]$$

in which,

S_r scouring rate per unit area
γ_s, γ bulk density of soil and water, N/m^3
n roughness coefficient.
U flow velocity, m/s.
H water depth, m
d scouring resistance representative grain size of scouring material, m
g gravity acceleration, m/s^2.

(5) **Calculation of downstream propagation process of barrier dam-break flood**

The formula of barrier dam-break flood maximum flow at the downstream $L_{(m)}$ away from the barrier dam site was:

$$Q_L = \frac{W}{\frac{W}{Q_{\max}} + \frac{L}{V_{\max}K}}$$

in which,

Q_L barrier dam-break flood maximum flow at the downstream control section $L_{(m)}$ away from the barrier dam site, m^3/s
W total reservoir storage capacity, m^3
$Q_{\max}$ maximum flow at barrier dam site, m^3/s
L distance between control section and reservoir barrier dam site, m
$V_{\max}$ maximum flow velocity of extraordinary flood (without data, 3.0–5.0 m/s for mountainous area, 2.0–3.0 m/s for hilly area, 1.0–2.0 m/s for plain area)
K empirical coefficient (1.1–1.5 for mountainous area, 1.0 for hilly area and 0.8–0.9 for plain area).

4.3.2 Calculation Analysis of First-Time Barrier Dam-Break Flood

4.3.2.1 Calculation Scheme of Barrier Dam-Break Flood

When formulating the emergency evacuation plan, in order to make it foolproof, the water conservancy group of the Earthquake Relief Headquarters of the State Council determined three typical breach bottom elevations, i.e., 720 m, 695 m and 663 m. For better understanding, they were called 1/3-barrier dam-break, 1/2-barrier dam-break and full-barrier dam-break. The breaching duration was set as 1, 2, 3, 4, 5 and 6 h, and the initial water level was set to 752 m. The breach shape and breaching duration of these three typical breaching modes were as follows:

(1) **1/3-barrier dam-break**

The breach was shaped in trapezoid, with a breach depth of 32 m, bottom elevation of 720 m, crest width of 340 m and bottom width of 35 m. The gradual barrier dam-break duration, respectively, carried 1, 2 and 3 h. The 1 h scheme was recommended.

(2) **1/2-barrier dam-break**

The breach was shaped in trapezoid, with a breach depth of 57 m, bottom elevation of 695 m, crest width of 340 m and bottom width of 35 m. The gradual barrier dam-break duration, respectively, carried 1, 2 and 3 h. The 2 h scheme was recommended.

(3) **Full-barrier dam-break**

The breach was shaped in trapezoid, with a breach depth of 89 m, bottom elevation of 663 m, crest width of 340 m and bottom width of 100 m. The gradual barrier dam-break duration, respectively, carried 1, 2, 3, 4 5, and 6 h. The 2 h scheme was recommended.

(4) **In addition to the above three typical barrier dam-break modes, the following computing scheme was conducted**:

(1) **Combined scheme of different breach shape, breaching duration, initial water level and other factors**

Models 1, 2, 3 and 4 conducted in parallel the barrier dam-break flood algorithm by the combined scheme of different breach shape, breaching duration, initial water level and other factors.

(2) **Joint algorithm of flood regulation and barrier dam-break flood**

In order to reflect the barrier dam-break situation in the cases of upstream reservoir inflow floods with different frequencies, Models 1, 2, 3 and 4 conducted joint algorithm of flood regulation and barrier dam-break flood.

(3) **Barrier dam-break flood calculation in the case of encountering Fujiang frequency flood**

In order to reflect the influence of the barrier dam-break induced by Fujiang floods with different frequencies on the Mianyang City and the cities and towns in the downstream of Mianyang, Models 1, 2, 3 and 4 conducted the barrier dam-break simulation in the cases of different reservoir inflow floods encountered by the Tangjiashan barrier lake and analyzed the impact of the serious floods with the same frequency as that on Fujiang River on the Mianyang City and the cities and towns in the downstream of Mianyang.

(4) **Calculation for influence of breaching duration extension on barrier dam-break flood evolution**

Given that the barrier dam was composed of the landslide of the whole mountain, the natural flood propagation time was 4–5 h, the discharge channel was protected artificially, and thus the actual breaching duration might be prolonged. In order to study the influence of breaching duration extension on the breaching flow volume and its flood propagation, Model 1 conducted the barrier dam-break flood algorithm for the typical schemes featuring breaching duration of 9 h, 12 h, 18 h and 24 h, respectively.

(5) **Comparison calculation of influence on barrier dam-break flood with or without discharge channel under the same condition**

In order to study the role played by the discharge channel excavation on the barrier dam-break flood relief, Model 1, Model 2 and Model 3 were used to conduct the comparison calculation and analysis of the influence on barrier dam-break flood with or without discharge channel under the same condition.

(6) **Calculation analysis of flood retention by Tongkou HPP**

The Tongkou HPP was located about 24 km downstream from the Tangjiashan barrier lake, in order to study its flood retention action on the barrier dam-break flood, Model 1 conducted calculation analysis of flood retention by Tongkou HPP.

(7) **Sensitivity analysis of calculation conditions such as breach change process and watercourse roughness**

① Sensitivity analysis of section development process: Model 1 barrier dam-break calculation conducted analysis on breach development process and compared the barrier dam-break flood calculation results between the two scenarios where section width and breach bottom elevation changed linearly and in line with the Weber Function.
② Sensitivity analysis of watercourse roughness: the watercourse roughness was an important parameter for the calculation analysis of barrier dam-break flood evolution. The greater the watercourse roughness, the greater the resistance of the watercourse, then the obvious the flood retention effect, and the greater the attenuation effect of the flood peak. Therefore, Model 1 simulated the Fujiangqiao flood peak discharge and its appearing time in different roughness values of the barrier lake downstream watercourse.

(8) **Simulated calculation of three schemes by planar two-dimensional mathematical model**

These three schemes all considered the initial barrier dam-break occurred in the case of the pre-excavated bottom elevation of 740 m, the upstream flood frequency of 5% and the reservoir water level of 752 m. Three schemes of simulated calculation were carried out in the scenarios of breaching to 720 m elevation within 1 h, to 695 m elevation within 2 h and to 663 m elevation with 2 h. In the preliminary stage, it

Table 4.12 Calculation schemes for three typical barrier dam-break patterns

No.	Breach bottom elevation (m)	Upstream inflow (m^3/s)	Fujiang inflow (m^3/s)	Breaching duration (h)	Initial water level (m)
1–1	720 (1/3-barrier dam-break)	160	1400	1	752
1–2		160	1400	2	752
1–3		160	1400	3	752
1–4	695 (1/2-barrier dam-break)	160	1400	1	752
1–5		160	1400	2	752
1–6		160	1400	3	752
1–7	663	160	1400	1	752
1–8	663 (full-barrier dam-break)	160	1400	2	752
1–9		160	1400	3	752
1–10		160	1400	4	752
1–11		160	1400	5	752
1–12		160	1400	6	752

Notes in the cases of breach bottom elevation at 720 and 695 m, Model 4 also calculated the working conditions with breaching duration of 4 h, 5 h and 6 h, respectively

was assumed that the section possessed dry riverbed from downstream dam to the Fujiangqiao and in regardless of the storage action of Tongkou HPP.

The working conditions obtained by Models 1–4 are shown in Tables 4.12, 4.13, 4.14, 4.15, 4.16, 4.17, 4.18, 4.19, 4.20 and 4.21. The following focused on the analysis of the calculation results of three typical barrier dam-break patterns with 752 m initial water level, as well as some atypical barrier dam-break patterns. Meanwhile, the results of various calculation conditions were systematically analyzed, so as to summarize the rules and draw the main understandings and conclusions.

4.3.2.2 Analysis on Calculation Results for Barrier Dam-Break Floods

The three typical barrier dam-break patterns embodied, respectively, as: 1/3-barrier dam-break, gradual breaching duration of 1 h; 1/2-barrier dam-break, gradual breaching duration of 2 h; and full-barrier dam-break, gradual breaching duration of 2 h.

Tables 4.22, 4.23, 4.24, 4.25, 4.26 and 4.27 show, respectively, the results of flood peak discharge, propagation time and water level at the barrier dam site and along the section from downstream to Mianyang City, which were obtained by the different barrier dam-break models in the cases of three typical breach patterns with a 752 m initial water level. Figures 4.7, 4.8, 4.9, 4.10, 4.11 and 4.12 present the comparison figures of flood peak discharge and propagation time of each model.

It can be seen from the analysis results that, in the calculation condition of 1/3-barrier dam-break recommended scheme, the flood peak discharge values at

Table 4.13 Calculation schemes for barrier dam-break floods at different initial water levels

No.	Breach bottom elevation (m)	Upstream inflow (m^3/s)	Fujiang inflow (m^3/s)	Breaching duration (h)	Initial water level (m)
2–1	720	160	1400	1–3	760
2–2	695	160	1400	1–3	760
2–3	663	160	1400	1–6	760
2–4	720	160	1400	1–3	742
2–5	695	160	1400	1–3	742
2–6	663	160	1400	1–6	742
2–7	720	160	1400	1–3	730
2–8	695	160	1400	1–3	730
2–9	663	160	1400	1–3	730
2–10	663	160	1400	1–3	724
2–11	663	160	6000	1–3	724
2–12	690	20%	1400	1–3	720
2–13	663	20%	1400	1–3	720

Table 4.14 Calculation schemes for barrier dam-break floods at different initial water levels (Seven working conditions by Models 2 and 3, sensitivity analysis calculation schemes)

No.	Breach bottom elevation (m)	Discharge channel bottom elevation (m)	Frequency of upstream dam flood (%)	Breaching duration (h)	Initial water level (m)
2–14	720	740	5	1	742
2–15			5	Decided by calculation	740.2
2–16			5	Decided by calculation	752
2–17		740	5	2	742
2–18			5	Decided by calculation	740.2
2–19		740	5	3	742
2–20			5	Decided by calculation	740.2

dam site obtained by each model were 26,440 m^3/s, 43,200 m^3/s, 33,771 m^3/s and 28,503 m^3/s, respectively, among which the maximum value was about 63.4% larger than the minimum value. It showed that there were greater differences in the results of flood peak discharge at the dam site derived from each model. When the barrier dam-break flood traveled to the Fujiangqiao located about 68.9 km downstream of the dam, the flood peak discharge values at dam site obtained by each model were 10,861 m^3/s, 12,800 m^3/s, 12,107 m^3/s and 12,860 m^3/s, respectively, slightly smaller

Table 4.15 Calculation schemes for barrier dam-break floods with different Fujiang inflow (Nine working conditions by Models 1, 2 and 3, sensitivity analysis calculation schemes)

No.	Breach bottom elevation (m)	Upstream inflow (m^3/s)	Fujiang inflow (m^3/s)	Breaching duration (h)	Initial water level (m)
3–1	720	160	1400	1	752
3–2	695	160	1400	2	752
3–3	663	160	1400	2	752
3–4	720	160	1400	1	752
3–5	695	160	1400	2	752
3–6	663	160	1400	2	752
3–7	720	160	1400	1	752
3–8	695	160	1400	2	752
3–9	663	160	1400	2	752

Table 4.16 Calculation schemes for barrier dam-break floods with breaching duration extension (13 working conditions by Model 1, sensitivity analysis calculation schemes)

No.	Breach bottom elevation (m)	Upstream inflow (m^3/s)	Fujiang inflow (m^3/s)	Breaching duration (h)	Initial water level (m)
4–1	663	160	1400	9	760
4–2	663	160	1400	12	760
4–3	663	160	1400	18	760
4–4	663	160	1400	24	760
4–5	663	160	1400	9	752
4–6	663	160	1400	12	752
4–7	663	160	1400	18	752
4–8	663	160	1400	24	752
4–9	720	160	1400	6	752
4–10	720	160	1400	9	752
4–11	720	160	1400	12	752
4–12	720	160	1400	18	752
4–13	720	160	1400	24	752

than the 13,500 m^3/s flood control design discharge of Mianyang urban river channel, among which the maximum value was about 18.4% larger than the minimum value. It showed that with barrier dam-break flood evolution toward downstream, the differences tended to decrease along the route among the flood peak discharge values by different models.

In the calculation condition of 1/2-barrier dam-break recommended scheme, the flood peak discharge values at dam site obtained by each model were 40,470 m^3/s, 47,200 m^3/s, 47,252 m^3/s and 53,051 m^3/s, respectively, among which the maximum

Table 4.17 Calculation schemes for barrier dam-break floods with discharge channel (31 working conditions by Models 1, 2 and 3, it was supposed that scouring did not occur along discharge channel with water flow, when the more than 10-year return period flood occurred upstream, the highest water level before the dam would exceed 752 m based on the flood regulation results)

No.	Breach bottom elevation (m)	Discharge channel bottom elevation (m)	Frequency of upstream dam flood or precipitation (%, mm)	Breaching duration (h)	Initial water level (m)
5–1	720	747	20	1	752
5–2			10	1	752
5–3			5	1	752
5–4		745	20	1	752
5–5			10	1	752
5–6			5	1	752
5–7		742	20	1	752
5–8			10	1	752
5–9	720	742	5	1	752
5–10	695	747	20	2	752
5–11			10	2	752
5–12			5	2	752
5–13		745	20	2	752
5–14			10	2	752
5–15			5	2	752
5–16		742	20	2	752
5–17			10	2	752
5–18			5	2	752
5–19	663	747	20	2	752
5–20			10	2	752
5–21			5	2	752
5–22		745	20	2	752
5–23			10	2	752
5–24			5	2	752
5–25		742	20	2	752
5–26			10	2	752
5–27			5	2	752
5–28	695	742	5	2	754
5–29	695	742	Daily precipitation10	2	752
5–30	695	742	Daily precipitation12	2	752

(continued)

Table 4.17 (continued)

No.	Breach bottom elevation (m)	Discharge channel bottom elevation (m)	Frequency of upstream dam flood or precipitation (%, mm)	Breaching duration (h)	Initial water level (m)
5–31	695	742	Daily precipitation15	2	752

Table 4.18 Calculation schemes for flood retention at Tongkou HPP (12 working conditions by Model 1, sensitivity analysis calculation schemes)

No.	Breach bottom elevation (m)	Upstream inflow (m^3/s)	Fujiang inflow (m^3/s)	Breaching duration (h)	Initial water level (m)
6–1	720	160	1400	1–3	752
6–2	695	160	1400	1–3	752
6–3	663	160	1400	1–6	752

Table 4.19 Calculation schemes for barrier dam-break flood in the cases of different breach development processes (Model 1, breach development changed by Weber Function, 12 working conditions)

No.	Breach bottom elevation (m)	Upstream inflow (m^3/s)	Fujiang inflow (m^3/s)	Breaching duration (h)	Initial water level (m)
7–1	720	160	1400	1–3	752
7–2	695	160	1400	1–3	752
7–3	663	160	1400	1–6	752

Table 4.20 Calculation schemes for barrier dam-break flood with different roughness values (Model 1, 0.04, 0.05 as watercourse roughness in the downstream dam, 12 working conditions, sensitivity analysis calculation schemes)

No.	Breach bottom elevation (m)	Upstream inflow (m^3/s)	Fujiang inflow (m^3/s)	Breaching duration (h)	Initial water level (m)
8–1	720	160	1400	1–3	752
8–2	695	160	1400	1–3	752
8–3	663	160	1400	1–6	752

value was about 31.1% larger than the minimum value. The flood peak discharge values at Fujiangqiao obtained by each model were 18,708 m^3/s, 19,400 m^3/s, 21,169 m^3/s and 22,031 m^3/s, respectively, higher than the 5000–8000 m^3/s flood control design discharge of Mianyang urban river channel, and the maximum value was about 17.8% larger than the minimum value.

Table 4.21 Other calculation working conditions (Model 4, 17 working conditions, 720 m and 730 m schemes as secondary barrier dam-break flood calculation schemes, 752 m scheme as breach expansion scheme)

Initial water level(m)	Breach bottom elevation)	Breach crest width(m)	Breach bottom width(m)	Breaching duration(h)	Reservoir inflow discharge(m^3/s)
720	663	200	0	2	100
720	663	200	0	2	2100
720	663	450	100	2	100
720	663	450	100	2	2100
720	690	220	40	1	100
720	690	220	40	2	100
720	690	220	40	1	2100
720	690	220	40	2	2100
742	732	260	130	1	100
742	720	260	35	1	100
742	695	260	35	2	100
742	663	260	100	2	100
730	690	220	40	1	100
730	690	220	40	2	100
730	663	450	100	2	100
752	663	610	100	1	100
752	663	610	100	2	100

In the calculation condition of full-barrier dam-break recommended scheme, the flood peak discharge values at dam site obtained by each model were 65,686 m^3/s, 63,100 m^3/s, 69,577 m^3/s and 76,699 m^3/s, respectively, among which the maximum value was about 21.6% larger than the minimum value, and the calculation results of each model differed little from each other. The flood peak discharge values at Fujiangqiao obtained by each model were 27,610 m^3/s, 22,800 m^3/s, 30,461 m^3/s and 34,330 m^3/s, respectively, all of which were much higher than the flood control standard of Mianyang urban river channel.

For the 1/3-barrier dam-break recommended scheme, the peak discharge occurrence time at the dam site was consistent among all models with the same value of 1 h. At Tongkou HPP, the peak time calculated by each model was 1.8–2.1 h. At Fujiangqiao, the peak time calculated by each model was 5.3–6.0 h.

For the 1/2-barrier dam-break recommended scheme, the peak time at the dam site, Tongkou River and Fujiangqiao ranged successively 1.9–2.0 h, 2.4–2.5 h and 5.2–5.8 h.

For the full-barrier dam-break recommended scheme, the peak time at the dam site, Tongkou River and Fujiangqiao ranged successively 1.5–1.8 h, 2.2–2.3 h and 4.2–5.2 h.

Table 4.22 Flood peak discharge results calculated by 1/3-barrier dam-break scheme (initial water level of 752 m, breaching duration of 1 h)

Section location	Distance away from barrier dam site (km)	Model 1		Model 2		Model 3		Model 4	
		Flood peak discharge (m^3/s)	Flood peak propagation time(hour: minute)	Flood peak discharge (m^3/s)	Flood peak propagation time(hour: minute)	Flood peak discharge (m^3/s)	Flood peak propagation time(hour: minute)	Flood peak discharge (m^3/s)	Flood peak propagation time(hour: minute)
Barrier dam site	0	26,440	01:00	43,200	01:00	33,771	01:00	28,503	01:00
Beichuan	4.64	19,040	01:22	38,700	01:05	27,522	01:09	26,438	01:04
Tongkou HPP	24.40	15,550	02:06	28,400	01:47	21,727	01:59	21,144	01:54
Jiangjunshi Hydrology Station	36.10	15,189	02:26	26,700	02:00	19,726	02:19	19,270	02:22
Yongxingchang	40.30	14,720	02:36	25,500	02:15	19,017	02:39	18,461	02:39
Qinglian Town	46.51	13,805	03:02	24,000	02:40	17,446	03:09	16,984	03:11
Tongkou estuary	50.20	13,130	03:36	23,000	02:55	15,230	03:29	15,489	03:32
Fujiangqiao	68.88	10,861	05:20	12,800	05:53	12,107	05:59	12,860	05:59

Table 4.23 Flood peak discharge results calculated by 1/2-barrier dam-break scheme (initial water level of 752 m, breaching duration of 2 h)

Section location	Distance away from barrier dam site (km)	Model 1		Model 2		Model 3		Model 4	
		Flood peak discharge (m^3/s)	Flood peak propagation time(hour: minute)	Flood peak discharge (m^3/s)	Flood peak propagation time(hour: minute)	Flood peak discharge (m^3/s)	Flood peak propagation time(hour: minute)	Flood peak discharge (m^3/s)	Flood peak propagation time(hour: minute)
Barrier dam site	0	40,470	02:00	47,200	01:55	47,252	02:00	53,051	02:00
Beichuan	4.64	36,967	02:06	46,800	02:03	44,155	02:00	51,668	02:01
Tongkou HPP	24.40	31,307	02:32	42,400	02:26	39,381	02:29	45,013	02:26
Jiangjunshi Hydrology Station	36.10	31,080	02:48	41,000	02:40	37,461	02:49	41,925	02:40
Yongxingchang	40.30	29,920	02:54	40,200	02:48	36,516	02:59	40,871	02:48
Qinglian Town	46.51	27,709	03:14	38,500	03:05	34,176	03:19	36,851	03:10
Tongkou estuary	50.20	24,882	03:38	37,600	03:20	30,528	03:29	32,859	03:26
Fujiangqiao	68.88	18,708	05:12	19,400	05:45	21,169	05:29	22,031	05:23

Table 4.24 Flood peak discharge results calculated by full-barrier dam-break scheme (initial water level of 752 m, breaching duration of 2 h)

Section location	Distance away from barrier dam site (km)	Model 1		Model 2		Model 3		Model 4	
		Flood peak discharge (m^3/s)	Flood peak propagation time(hour: minute)	Flood peak discharge (m^3/s)	Flood peak propagation time(hour: minute)	Flood peak discharge (m^3/s)	Flood peak propagation time(hour: minute)	Flood peak discharge (m^3/s)	Flood peak propagation time(hour: minute)
Barrier dam site	0	65,686	01:44	63,100	01:44	69,577	01:29	76,699	01:31
Beichuan	4.64	64,129	01:48	62,700	01:50	66,785	01:49	75,285	01:51
Tongkou HPP	24.40	57,630	02:14	58,900	02:13	62,511	02:09	70,776	02:09
Jiangjunshi Hydrology Station	36.10	57,683	02:16	56,500	02:27	60,263	02:29	68,115	02:20
Yongxingchang	40.30	57,900	02:22	55,200	02:34	59,082	02:29	67,234	02:26
Qinglian Town	46.51	53,673	02:40	53,200	02:49	55,652	02:49	62,825	02:42
Tongkou estuary	50.20	44,470	03:02	51,600	03:00	49,179	02:59	57,525	02:54
Fujiangqiao	68.88	27,610	04:12	22,800	05:14	30,461	04:39	34,330	04:26

Table 4.25 Flood peak water level results calculated by 1/3-barrier dam-break scheme (initial water level of 752 m, breaching duration of 1 h)

Section location	Distance away from barrier dam site (km)	Model 1	Model 2	Model 3	Model 4	Maximum difference value of model water level(m)
		Water level (m)	Water level (m)	Water level (m)	Water level (m)	
Barruer dam site	0			665.57	676.69	
Beichuan	4.64	639.4	638.2	642.56	640.95	4.36
Tongkou HPP	24.40	561.8	558.3	567.14	567.6	9.30
Jiangjunshi Hydrology Station	36.10	538.4	535.1	537.84	538.37	3.30
Yongxingchang	40.30	525.2	524.7	526.30	526.66	1.96
Qinglian Town	46.51	506.2	502.9	507.85	507.95	5.05
Tongkou estuary	50.20	495.9	494.9	496.72	497.09	2.19
Fujiangqiao	68.88	466.8	464.9	467.27	467.5	2.60

Table 4.26 Flood peak water level results calculated by 1/2-barrier dam-break scheme (initial water level of 752 m, breaching duration of 2 h)

Section location	Distance away from barrier dam site (km)	Model 1	Model 2	Model 3	Model 4	Maximum difference value of model water level(m)
		Water level (m)	Water level (m)	Water level (m)	Water level (m)	
Barrier dam site	0			670.05	683.46	
Beichuan	4.64	644.4	645.3	647.43	647.18	3.03
Tongkou HPP	24.40	568.0	570.3	573.86	575.96	7.96
Jiangjunshi Hydrology Station	36.10	544.5	545.4	541.88	543.79	3.52
Yongxingchang	40.30	527.3	527.0	528.73	529.84	2.84
Qinglian Town	46.51	508.6	508.4	510.53	511.23	2.83
Tongkou estuary	50.20	498.2	497.2	499.21	500.14	2.94
Fujiangqiao	68.88	469.8	468.6	469.14	469.99	1.39

In conclusion, there existed little difference of peak time at the dam site among all models. The peak time at the dam site was basically the same as the breaching duration in the patterns of 1/3-barrier dam-break and 1/2-barrier dam-break, while the former value was slightly smaller than the latter value for the full-barrier dam-break pattern. The peak time difference among the main control stations downstream of the dam tended to increase along the route, and the peak time difference at the Fujiangqiao ranged from 0.6 h to 1.0 h. The reason might be: the peak time at the

Table 4.27 Flood peak water level results calculated by full-barrier dam-break scheme (initial water level of 752 m, breaching duration of 2 h)

Section location	Distance away from dam site (km)	Model 1	Model 2	Model 3	Model 4	Maximum difference value of model water level(m)
		Water level (m)	Water level (m)	Water level (m)	Water level (m)	
Dam site	0			675.53	688.91	
Beichuan	4.64	649.3	651.3	652.93	651.85	3.63
Tongkou HPP	24.40	575.8	578.9	580.34	582.94	7.14
Jiangjunshi Hydrology Station	36.10	553.0	549.4	545.66	548.32	7.34
Yongxingchang	40.30	529.9	532.8	531.02	532.49	2.90
Qinglian Town	46.51	511.7	509.6	512.93	514.30	4.70
Tongkou estuary	50.20	501.1	501.5	501.56	503.10	2.00
Fujiangqiao	68.88	473.0	470.9	470.71	472.44	2.29

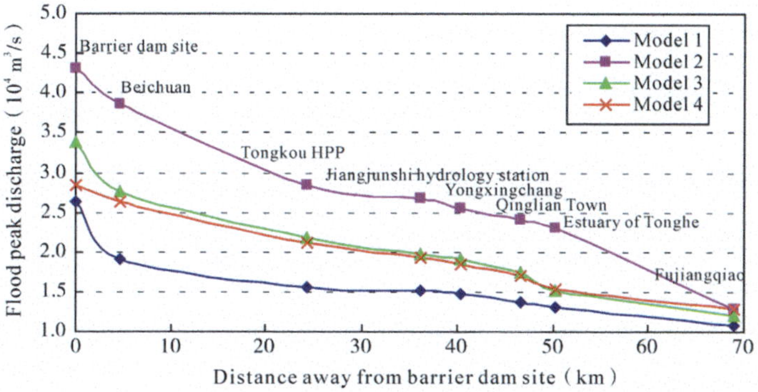

Fig. 4.7 Flood peak discharge change along the way by 1/3-barrier dam-break scheme of each model

dam site depended mainly on the breaching duration, when the barrier dam-break flood evolved toward the downstream, barrier dam-break flow process, flood peak discharge, roughness value differed among the various model calculations, which resulted in the difference of peak time.

Tables 4.25, 4.26 and 4.27 show the peak water level results obtained by each model. It could be seen that there were some differences among the values of peak water level in each model. As calculated by each model, the values of water level difference in the recommended schemes of 1/3-barrier dam-break, 1/2-barrier dam-break and full-barrier dam-break were 2.60 m, 1.39 m and 2.29 m, respectively. The reason might be: (1) each model used different numerical formats, which made the numerical discretization error of the model different; (2) the model calculation results

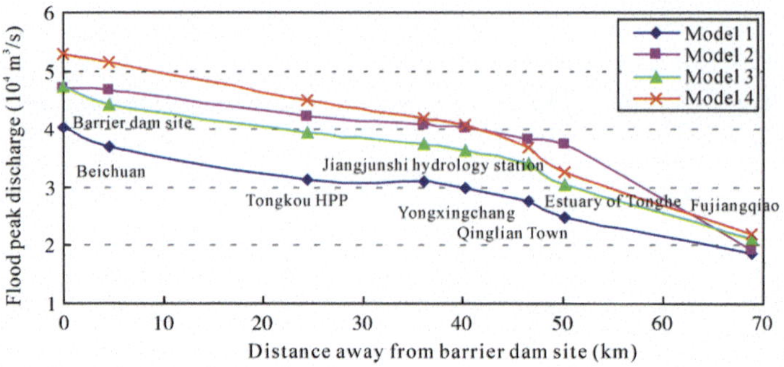

Fig. 4.8 Flood peak discharge change along the way by 1/2-barrier dam-break scheme of each model

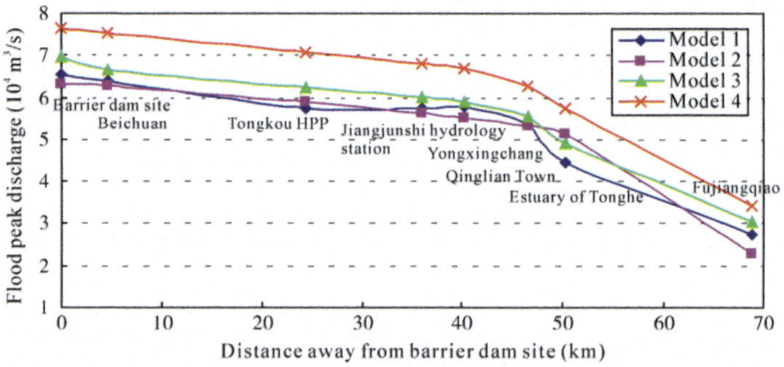

Fig. 4.9 Flood peak discharge change along the way by full-barrier dam-break scheme of each model

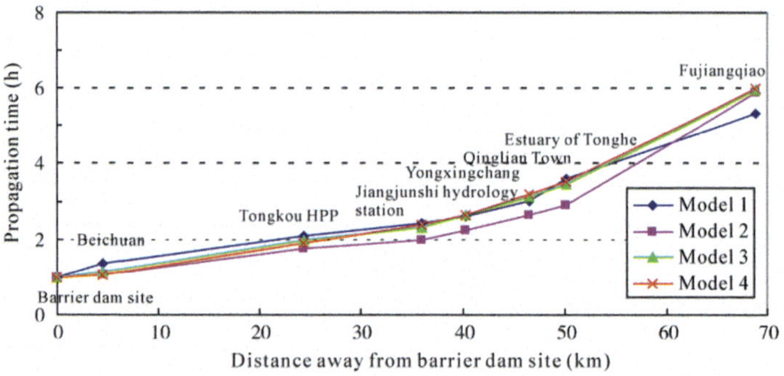

Fig. 4.10 Flood peak propagation time by 1/3-barrier dam-break scheme of each model

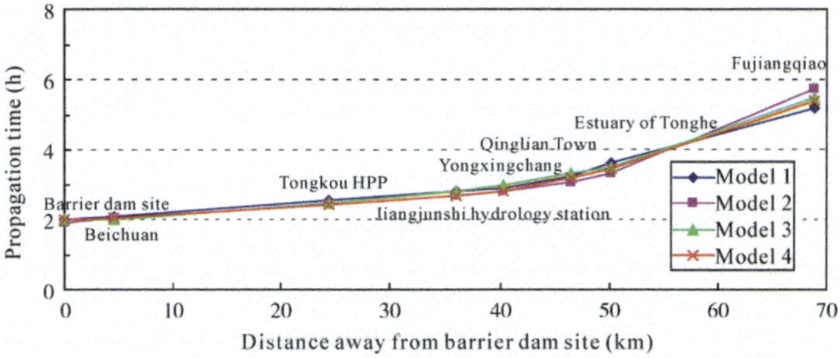

Fig. 4.11 Flood peak propagation time by 1/2-barrier dam-break scheme of each model

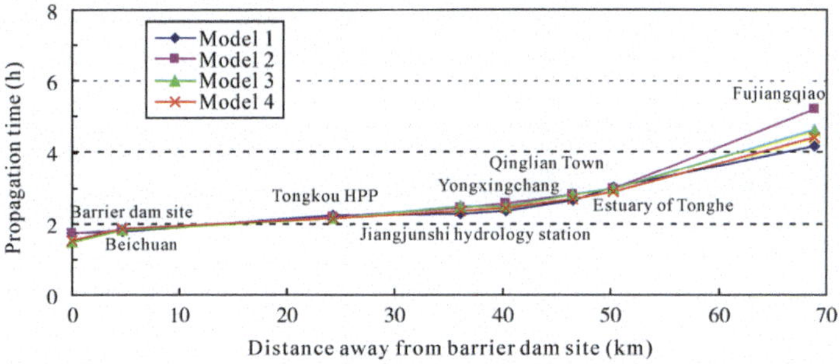

Fig. 4.12 Flood peak propagation time by full-barrier dam-break scheme of each model

of water level were sensitive to the roughness values, and the values of roughness for each model were not completely consistent; (3) there were also differences among the barrier dam-break peak discharge and flow process obtained by each model.

In general, for the recommended schemes of 1/3-barrier dam-break, 1/2-barrier dam-break and full-barrier dam-break, the peak discharges all exceeded the river safety discharge limit in the upstream reach of the Tongkou River and also approached or exceeded the river safety discharge limit in the downstream reach. Therefore, while taking engineering measures to eliminate risks, it was necessary to make evacuation plan in the downstream area to avoid potential risk of barrier dam-break. Meanwhile, when formulating the evacuation plan, it was preferable to select the maximum values of each model for the peak flow and water level, so as to ensure the evacuation plan foolproof.

4.3.2.3 Calculation Results by Planar Two-Dimensional Model

Given the fact that the one-dimensional model had the characteristics of high computational efficiency and relatively few data requirements, the barrier dam-break flood mainly adopted one-dimensional model. Considering the curved river channel and complex plane morphology of the lower reaches of the barrier lake, the planar two-dimensional model was used by Model 3 to calculate the barrier dam-break flood in some working conditions. Some calculation results are shown in Figs. 4.13, 4.14, 4.15 and 4.16.

The analysis of calculation results by Model 3 showed that, by adopting one-dimensional and planar two-dimensional unsteady flow to conduct the calculation for barrier dam-break flood evolution, the maximum discharge obtained by one-dimensional and planar two-dimensional models was close at the sections of dam site and Beichuan. At the section from Tongkou HPP to Fujiangqiao, the width of water flow section changed sharply along the river channel, the channel widened in the river reaches of Beichuan, Xiaohekou, Shimashan and confluence with obvious storage effect, and the maximum discharge calculated by planar two-dimensional model was a little bit smaller than that calculated by one-dimensional model. It showed that the evolution of barrier dam-break flood toward downstream was impacted by the

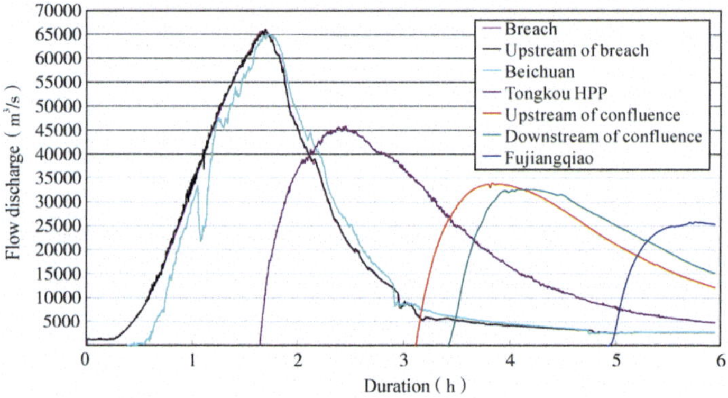

Fig. 4.13 Discharge hydrograph of each section for 663–740-20 scheme

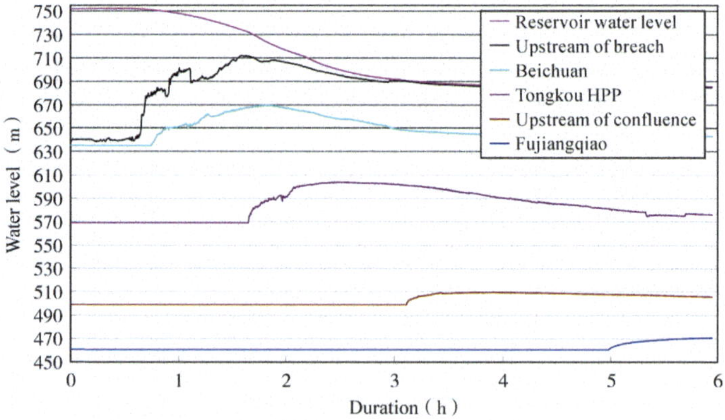

Fig. 4.14 Water level hydrograph of each station for 663–740-20 scheme

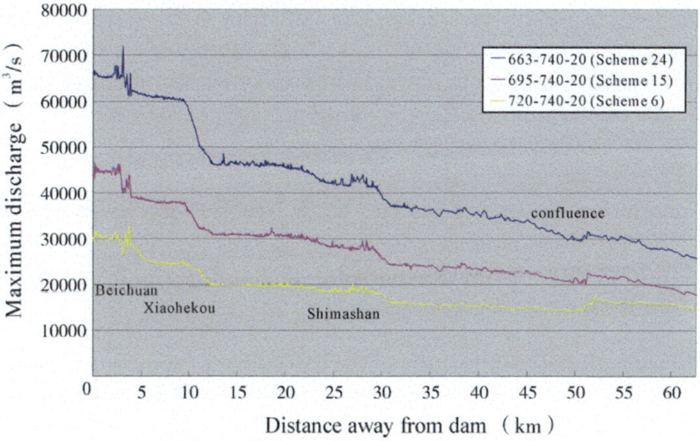

Fig. 4.15 Maximum discharge along the route for each scheme

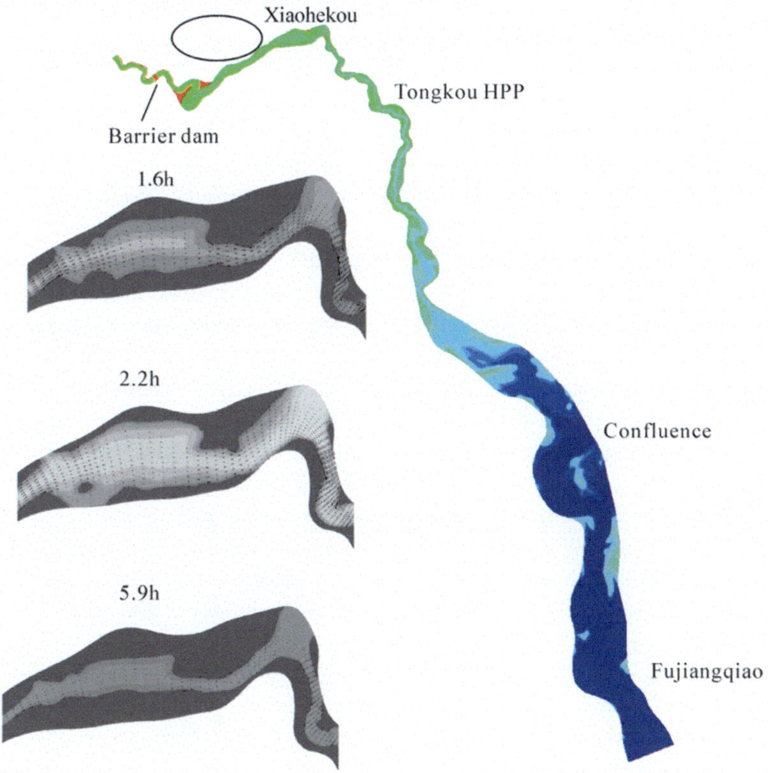

Fig. 4.16 Water depth and flow field distribution at Xiaohekou section

plane morphology of the lower reaches and boundary conditions, so it was necessary to appropriately make the calculation sections more intense in the case of one-dimensional unsteady flow model simulation, in order to reflect the influence of channel boundary conditions on flood evolution as much as possible.

4.3.2.4 Calculation Results of Barrier Dam-Break Flood by Model 5

(1) **Calculation results of barrier dam-break flood in the first stage of emergency handling**

According to the material composition of the barrier dam, based on the general law of sediment scouring and the calculation results derived from the barrier dam material composition and its responding incipient velocity, it was estimated that about 1/3 of the barrier dam materials could remain stable during the barrier dam-break flood scouring. At the same time, given the influences of the downstream Kuzhu Dam, it was supposed that a residual dam of 27 m could be maintained. In line with an overflow crest elevation of 750 m, the bottom elevation of local river was 667 m, and thus the barrier dam-break effective water depth reached 56 m. In addition, the average breach width was about 220 m with the maximum discharge based on the original topography of the local river.

Based on the above parameters, the values of breach maximum discharge calculated respectively by Tsinghua broad-crested weir formula and by Shockledge Formula were 32,000 m^3/s and 30,900 m^3/s of Model 5. Combined with analysis results of domestic and foreign cases, referring to the characteristics of the barrier dam material composition and the other available data, it was deemed that the reasonable emptying time was 3-5 h. By taking the emptying time as 4 h after the formation of breach maximum discharge and using the water balance equation of storage capacity and discharge flow process, the breach maximum discharge was about 35,000 m^3/s.

The most critical question for calculating the barrier dam-break flood process based on the unsteady barrier dam-break flood calculation model was to determine the breach forming rate. According to the emptying time—breach development time relation and the barrier dam material composition, it was initially supposed that the final breach was formed within 4 h of scouring, and the scouring rate varied according to the isosceles triangle scheme. The barrier dam-break flood flow process at the dam site by unsteady barrier dam-break flood calculation model is shown in Fig. 4.17, with maximum discharge of about 36,000 m^3/s.

In general, according to the secure outsourcing principle, breach maximum discharge of 36,000 m^3/s was to be recommended. The propagation of maximum discharge all along the downstream was calculated by the former formula, and its results could be seen as that of the barrier dam crest at elevation of 750 m marked in the Fig. 4.18, and the maximum discharge was 31,900 m^3/s when evolving to Beichuan County, 18,000 m^3/s to Qinglian Town, 16,500 m^3/s to the downstream Fujiang River and 12,900 m^3/s to Mianyang City. For the calculation, based on the assuming conditions of overtopping static storage, the calculation excluded the impact of the barrier

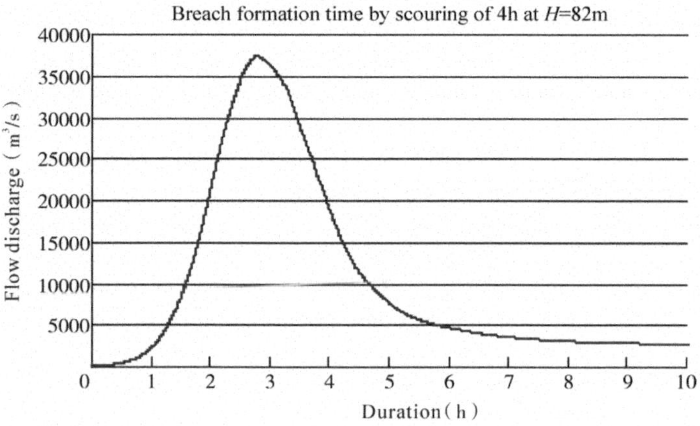

Fig. 4.17 Barrier dam-break flood flow process at the dam site by unsteady barrier dam-break flood calculation model

lake in the downstream river channel, facilities of water conservancy and change of the channel morphology.

(2) **Calculation results of barrier dam-break flood in the second stage of emergency handling**

Based on the geological data observed in the field, it was preliminarily judged that the landslide could be divided into four layers. The barrier dam surface of the first

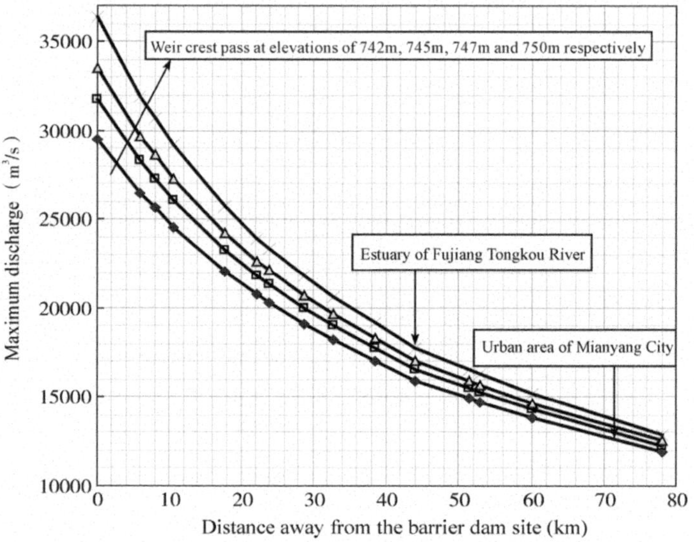

Fig. 4.18 Calculation results of barrier dam-break flood discharge all along the route

layer went to an elevation of 735 m, composed of silty soil with loose particles and poor scouring resistance capacity. The second layer was strong-weathered, judged as cracked stone particles with scouring resistance capacity to some extent. The third layer was weak-weathered with relatively strong in a general way. By analyzing, the pattern of breach scouring formation and development embodied mainly as surface gradual scouring and retrogressively scouring on the downstream surface of the barrier dam. The scouring from water surface to a 735 m elevation might occur in a fast speed, while relatively gentle from 720 to 735 m and hard to scour from 670 to 720 m. It could be inferred that, in terms of the whole barrier dam, it could not break within short time, in addition of the existed Kuzhu Dam, thus it was favorable to the stability of the Tangjiashan barrier dam, it was reliable for the barrier dam to maintain a residual barrier dam at 720 m elevation, and it should be safe in the case of stable breach forming in 3 h scouring.

Figures 4.19, 4.20 and 4.21 reflected successively the elevations of different residual barrier bodies and the maximum breach discharges with different forming times at different initial water levels of 747, 745 and 742 m. It was showed in Fig. 4.20 of the breach maximum discharge calculated by model of 720 m-residual dam elevation and 745 m-reservoir initial water level, in which the maximum discharge was 17,600 m^3/s with a stable breach scouring time of 1 h, 14,000 m^3/s with a stable breach scouring time of 3 h, 12,200 m^3/s with a stable breach scouring time of 5 h and 11,000 m^3/s with a stable breach scouring time of 7 h. By comparing the occurrence time of breach maximum discharge—the breach forming time, and verifying by water balance equation, it was supposed that a breach forming time of 3 h was relatively reasonable. Meanwhile, further calculations were carried out for the maximum discharges of corresponding characteristic-dimensional breach by adopting the Tsinghua broad-crested weir formula and the Shockledge Formula, proving the reliability of the calculation results.

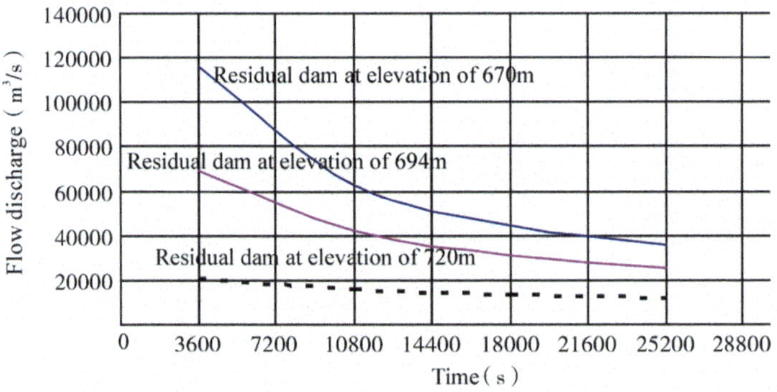

Fig. 4.19 Breach maximum discharges with different breach forming times at different residual weir elevations (water level of 747 m)

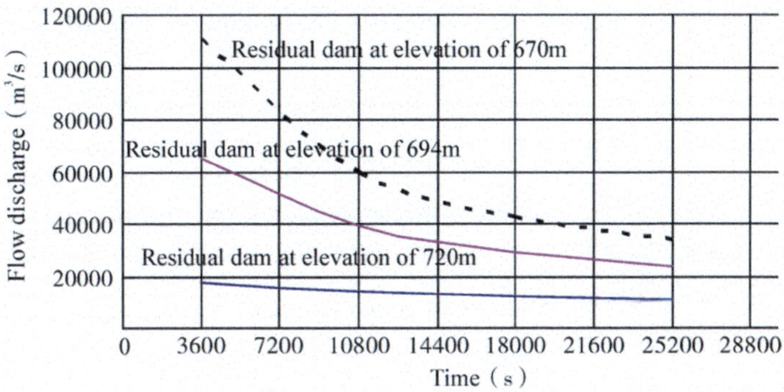

Fig. 4.20 Breach maximum discharges with different breach forming times at different residual weir elevations (water level of 745 m)

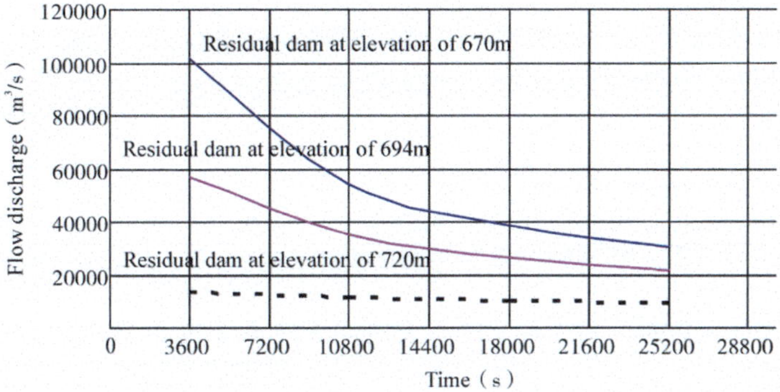

Fig. 4.21 Breach maximum discharges with different breach forming times at different residual weir elevations (water level of 742 m)

(3) Preliminary analysis of the role played by engineering measures of emergency handling

From the comparison of barrier dam-break maximum discharges with different water flow elevations, it could be seen that reducing the water flow elevation was an important way to reduce downstream safe risk. When the reservoir water level reduced from 752 to 747 m, the breach maximum discharge reduced by 29.31%. When the reservoir water level reduced from 752 to 745 m, the breach maximum discharge reduced by 38.79%. When the reservoir water level reduced from 752 to 742 m, the breach maximum discharge reduced by 51.77%.

4.3.3 Analysis of Secondary Barrier Dam-Break Flood

At the construction and completion of the discharge channel, it was deemed that the bottom elevation might decrease to an elevation of 720 m after the water flow scouring of the discharge channel. If there was a large inflow process in the upper reaches, secondary barrier dam-break flood would occur. In order to analyze the influence of barrier dam-break after water flow, the calculation was conducted for secondary barrier dam-break flood.

4.3.3.1 Secondary Barrier Dam-Break at Elevation of 720 m

When formulating the emergency contingency plan, it was preliminarily judged that the possible breach bottom elevation was around 720 m. Another barrier dam-break flood occurred after the formation of the first breach bottom at an elevation of 720 m, the impact on the upper reaches was taken into consideration in the cases of the following three scenarios, and the calculations were conducted for the secondary barrier dam-break flood.

(1) **Calculation working conditions**

Working condition 1: the secondary breach was shaped in isosceles triangle, with crest elevation of 720 m, width of 200 m, bottom elevation of 663 m and breaching duration of 2 h.

Working condition 2: the secondary breach was shaped in isosceles trapezoid, with crest elevation of 720 m and width of 220 m, bottom elevation of 690 m and width of 40 m and breaching duration of 1 h.

Working condition 3: the secondary breach was shaped in isosceles trapezoid, with crest elevation of 720 m and width of 220 m, bottom elevation of 690 m and width of 40 m and breaching duration of 2 h.

Working condition 4: the secondary breach was shaped in isosceles trapezoid, with crest elevation of 720 m and width of 450 m, bottom elevation of 663 m and width of 100 m and breaching duration of 2 h.

For each working condition, the inflow into Tangjiashan barrier lake was design flood with $P = 20\%$, approaching flow of Fujiang River was 1400 m^3/s.

(2) **Calculation results**

Table 4.28, 4.29, 4.30 and 4.31 show Model 1 calculation results of the four working conditions at an initial water level of 720 m, and the corresponding peak discharges at barrier dam site were 16,784 m^3/s, 13,085 m^3/s, 9649 m^3/s and 30,673 m^3/s, respectively. As it evolved to Fujiangqiao, the discharges corresponding to each working condition were 8737 m^3/s, 6297 m^3/s, 6441 m^3/s and 8662 m^3/s, all smaller than 13,500 m^3/s, which was the design flood standard of the urban area of Mianyang City. This indicated that the secondary barrier dam-break flood at an elevation of 720 m posed no threat to Mianyang.

Table 4.28 Flood elements of downstream main sections for working condition 1 of secondary barrier dam-break

Section location	Distance away from barrier dam site(m)	Maximum discharge(m^3/s)	Occurrence time(hour: minute)	Highest water level(m)
Barrier dam site	0	16,874	02:00	
Beichuan	4.64	15,541	02:08	637.7
Tongkou HPP	24.40	12,807	02:42	560.3
Jiangjunshi Hydrology Station	36.10	12,063	03:00	537.0
Yongxingchang	40.30	11,698	03:12	524.6
Qinglian Town	46.51	10,994	03:36	505.5
Estuary of Tonghe	50.20	10,632	04:06	495.2
Fujiangqiao	68.88	8737	05:54	465.9

Table 4.29 Flood elements of downstream main sections for working condition 2 of secondary barrier dam-break

Section location	Distance away from barrier dam site(m)	Maximum discharge(m^3/s)	Occurrence time(hour: minute)	Highest water level(m)
Barrier dam site	0	13,085	01:00	
Beichuan	4.64	9965	01:20	634.4
Tongkou HPP	24.40	7798	02:18	557.1
Jiangjunshi Hydrology Station	36.10	7404	02:42	534.1
Yongxingchang	40.30	7176	02:54	523.6
Qinglian Town	46.51	6738	03:26	504.2
Estuary of Tonghe	50.20	7222	04:00	494.0
Fujiangqiao	68.88	6297	05:54	464.8

4.3.3.2 Full-Barrier Dam-Break at Initial Water Level of 724 m

(3) Calculation working conditions

Working condition 1: full-barrier dam-break at initial water level of 724 m, breaching duration of 1 h, flow discharge of 1500m^3/s in Fujiang River;

Working condition 2: full-barrier dam-break at initial water level of 724 m, breaching duration of 1 h, flow discharge of 6000 m^3/s in Fujiang River;

Working condition 3: full-barrier dam-break at initial water level of 724 m, breaching duration of 2 h, flow discharge of 1500 m^3/s in Fujiang River;

Working condition 4: full-barrier dam-break at initial water level of 724 m, breaching duration of 2 h, flow discharge of 6000 m^3/s in Fujiang River.

Table 4.30 Flood elements of downstream main sections for working condition 3 of secondary barrier dam-break

Section location	Distance away from barrier dam site(m)	Maximum discharge(m^3/s)	Occurrence time(hour: minute)	Highest water level(m)
Barrier dam site	0	9649	02:00	
Beichuan	4.64	8908	02:10	633.7
Tongkou HPP	24.40	7473	03:04	556.9
Jiangjunshi Hydrology Station	36.10	7206	03:26	534.0
Yongxingchang	40.30	7040	03:38	523.6
Qinglian Town	46.51	6709	04:08	504.2
Estuary of Tonghe	50.20	7297	04:42	494.1
Fujiangqiao	68.88	6441	06:34	464.9

Table 4.31 Flood elements of downstream main sections for working condition 4 of secondary barrier dam-break

Section location	Distance away from barrier dam site(m)	Maximum discharge(m^3/s)	Occurrence time(hour: minute)	Highest water level(m)
Barrier dam site	0	30,673	01:52	
Beichuan	4.64	26,103	02:02	641.6
Tongkou HPP	24.40	18,269	02:30	562.8
Jiangjunshi Hydrology Station	36.10	16,274	02:50	539.2
Yongxingchang	40.30	15,278	02:58	525.2
Qinglian Town	46.51	13,348	03:22	506.0
Estuary of Tonghe	50.20	11,720	03:48	495.4
Fujiangqiao	68.88	8662	05:34	465.9

(4) **Calculation results**

Table 4.32 shows the calculation results of peak discharge under the four working conditions by Model 1. It could be seen from Table 4.32 that the peak discharges at Fujiangqiao corresponding to the four working conditions were 12,218 m^3/s, 16,431 m^3/s, 12,167 m^3/s and 16,595 m^3/s, respectively, all of which exceeded or were close to the design flood control standard of the urban areas of Mianyang. This indicated that in the case of encountering flood on the mainstream Fujiang, there would be an obvious increase of peak discharge at Fujiangqiao section, bringing threat to the urban areas of Mianyang.

Table 4.32 Peak discharge of downstream main sections for full-barrier dam-break scheme at initial water level of 724 m

Section	Working condition 1		Working condition 2		Working condition 3		Working condition 4	
	Peak discharge (m^3/s)	Time (hour: minute)	Peak discharge (m^3/s)	Time (hour: minute)	Peak discharge(m^3/s)	Time (hour: minute)	Peak discharge(m^3/s)	Time (hour: minute)
Barrier dam site	50,163	00:56	50,163	00:56	24,933	01:34	24,933	01:34
Beichuan	45,163	01:02	45,163	01:02	24,369	01:42	24,369	01:42
Tongkou HPP	40,243	01:28	40,243	01:28	23,443	02:10	23,443	02:10
Jiangjunshi Hydrology Station	38,516	01:36	38,516	01:36	22,836	02:22	22,836	02:22
Yongxingchang	29,596	01:52	29,596	01:52	22,118	02:36	22,118	02:36
Qinglian Town	24,583	02:16	24,405	02:14	20,129	02:58	19,998	02:58
Estuary of Tonghe	23,123	02:18	22,818	02:20	19,159	03:02	19,046	03:08
Fujiangqiao	12,218	04:22	16,431	04:14	12,167	05:04	16,595	04:56

4.3.4 Related Understanding of Calculation Analysis on Barrier Dam-Break Flood

By combining three typical barrier dam-break patterns in the different condition values of breaching duration, initial water level and Fujiang inflow for each model, as well as calculation analysis results of atypical barrier dam-break pattern, the laws and understandings obtained by each model were almost the same, which were listed as following.

4.3.4.1 Influences of Initial Water Level, Breach Shape and Its Development Process on Barrier Dam-Break Flood

The barrier dam-break flood peak discharge at barrier dam site depended on the initial water level (which decided the available discharge water volume), breaching duration, breach shape and its development process and other factors. The higher the initial water level, the shorter the breaching duration, the faster the development of breach width and depth, and the larger the peak discharge at the barrier dam site. Therefore, by implementing measures as excavating river channels could effectively reduce initial water level and available discharge water volume, obviously reduce the peak discharge at the barrier dam site and relieve the threat of barrier dam-break flood to the downstream flood control.

By taking the calculation results of Model 1 as an example, an analysis was carried out for the dam site peak discharge of three typical barrier dam-break schemes under different initial water levels and breaching duration values. It could be seen from Tables 4.33, 4.34 and 4.35 that: (1) in the condition of the same breaching duration, the higher the initial water level, the larger the available discharge flow, and the larger the peak discharge at the barrier dam site. (2) In the condition of the same initial water level, the shorter the breaching duration and the larger the peak discharge at the barrier dam site.

Table 4.33 Peak discharge at barrier dam site for 1/3-barrier dam-break scheme (unit: m^3/s)

Initial water level (m)	Breaching duration		
	1 h	2 h	3 h
760	35,940	24,508	18,361
752	26,440	20,289	16,320
742	15,544	11,955	9453

Table 4.34 Peak discharge at barrier dam site for 1/2-barrier dam-break scheme (unit: m^3/s)

Initial water level (m)	Breaching duration		
	1 h	2 h	3 h
760	68,373	44,222	30,721
752	58,935	40,470	28,520
742	40,417	26,624	19,447
730	25,601	17,196	12,152
720	14,776	9048	6277

Table 4.35 Peak discharge at barrier dam site for full-barrier dam-break scheme (unit: m^3/s)

Initial water level (m)	Breaching duration					
	1 h	2 h	3 h	4 h	5 h	6 h
760	136,234	69,692	48,452	37,475	30,710	26,090
752	132,537	65,686	44,497	34,085	28,011	23,859
742	95,030	48,730	33,517	25,749	20,981	17,737
730	62,408	31,580	22,055	17,102	14,036	11,941
720	42,058	21,554	14,697	11,196	9057	7608

4.3.4.2 Influences of Encountering Fujiang Frequency Flood on Barrier Dam-Break Flood

The barrier dam-break was likely to be accompanied by rainfall flood; therefore, in the calculation schemes of Model 1, Model 2 and Model 3, the cases were considered for encountering 5-year return period flood, 10-year return period flood and 20-year return period flood on the mainstream Fujiang River by the barrier dam-break floods, and analyzes were conducted for the increase and arriving time of peak discharge at each station. The calculation results of Model 1 are shown in Tables 4.36, 4.37 and 4.38.

The calculation results obtained by each model were basically consistent: (1) the larger the peak discharge on the mainstream Fujiang, the faster the discharge propagation of barrier dam-break flood, and the greater the increase of peak discharge to some extent. (2) In the case of flood on the mainstream Fujiang and 1/3-barrier dam-break, 1/2-barrier dam-break and full-barrier dam-break occurred simultaneously with the Tangjiashan barrier lake breach, it would aggravate the downstream flood control pressure to some extent; for example, the increase values of peak discharge at Tongnan obtained by Model 1 were, respectively, 3000–4000 m^3/s, 5000–6000 m^3/s and 6000–7000 m^3/s.

Table 4.36 Situation of encountering 20-year return flood on mainstream Fujiang by barrier dam-break flood

Location	Distance away from barrier dam site(km)	1/3-barrier dam-break1h		1/2-barrier dam-break2h		Full-barrier dam-break2h	
		Peak discharge increase(m^3/s)	Occurrence time (h)	Peak discharge increase(m^3/s)	Occurrence time (h)	Peak discharge increase(m^3/s)	Occurrence time (h)
Fujiangqiao	68.88	9725	4.20	18,343	3.07	29,402	2.40
Santai	160.88	5859	11.47	9671	9.87	12,692	8.63
Shehong	206.88	5223	14.03	8567	12.33	10,963	11.07
Suining	216.66	5142	14.80	8448	13.03	10,759	11.77
Tongnan	363.1	3395	28.07	5484	25.83	6689	24.13
Xiaoheba	391.88	3389	28.70	5418	26.43	6571	24.73

Table 4.37 Situation of encountering 10-year return flood on mainstream Fujiang by barrier dam-break flood

Location	Distance away from barrier dam site(km)	1/3-barrier dam-break1h		1/2-barrier dam-break2h		Full-barrier dam-break2h	
		Peak discharge increase(m^3/s)	Occurrence time (h)	Peak discharge increase(m^3/s)	Occurrence time (h)	Peak discharge increase(m^3/s)	Occurrence time (h)
Fujiangqiao	68.88	9634	4.23	18,246	3.07	29,205	2.40
Santai	160.88	5820	11.70	9599	10.03	12,446	8.93
Shehong	206.88	5214	14.33	8506	12.57	10,786	11.47
Suining	216.66	5136	15.07	8333	13.33	10,589	12.20
Tongnan	363.1	3320	29.23	5137	26.93	6144	25.43
Xiaoheba	391.88	3326	29.83	5112	27.53	6093	26.03

Table 4.38 Situation of encountering 5-year return flood on mainstream Fujiang by barrier dam-break flood

Location	Distance away from barrier dam site(km)	1/3-barrier dam-break1h		1/2-barrier dam-break2h		Full-barrier dam-break2h	
		Peak discharge increase(m^3/s)	Occurrence time (h)	Peak discharge increase(m^3/s)	Occurrence time (h)	Peak discharge increase(m^3/s)	Occurrence time (h)
Fujiangqiao	68.88	9513	4.27	18,105	3.10	28,858	2.43
Santai	160.88	5750	12.00	9506	10.23	12,255	9.17
Shehong	206.88	5145	14.77	8408	12.90	10,619	11.77
Suining	216.66	5075	15.50	8250	13.67	10,367	12.53
Tongnan	363.1	3194	30.07	4982	28.00	5899	26.63
Xiaoheba	391.88	3204	30.63	4972	28.57	5875	27.23

4.3.4.3 Sensitivity Analysis on Breach Change Process

Model 1 conducted the sensitivity analysis on breach change process. Figure 4.22 described the change of the breach width and elevation in accordance with the linear and Weber Functions at initial water level of 752 m and in the case of 1/3-barrier dam-break in 1 h; Fig. 4.23 shows the change process of breach area in the case of 1/3-barrier dam-break. As seen from Figs. 4.22 and 4.23, the expansion rates of breach width and elevation in the Weber Function were faster than that in the linear change during the initial period, while slower than that during the later period. In terms of area change, the area expansion by Weber Function distribution first gradually increases, then gradually slows down, while by linear change, the area expansion tended to gradually increase.

The section elements were distributed by linear change and Weber Function, simulation was carried out for breach section development process, the peak occurrence of sections distributed by Weber Function appeared earlier than that by linear change, and the values of peak discharge were larger than that by linear change (Table 4.39).

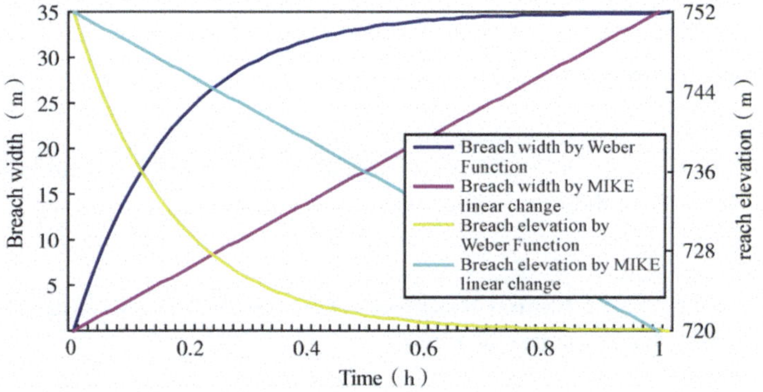

Fig. 4.22 Development process of breach section elements for 1/3-barrier dam-break

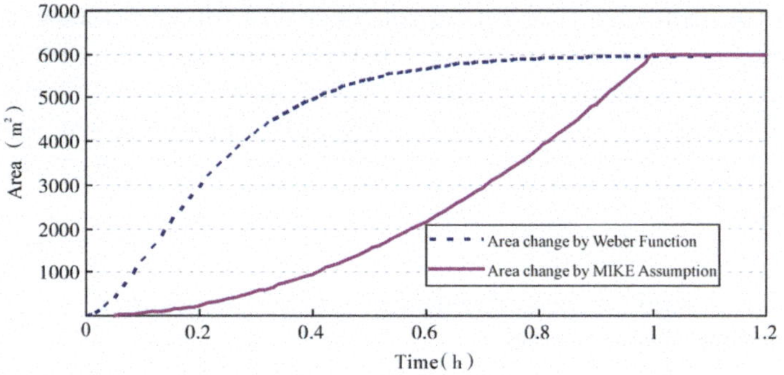

Fig. 4.23 Change process of breach area for 1/3-barrier dam-break

Table 4.39 Influences of section development process of 1/3-barrier dam-break scheme on the barrier dam-break flood

Section elements	Breaching duration		
	1 h	2 h	3 h
Peak discharge of section by linear change (m^3/s)	26,440	20,289	16,320
Peak discharge of sction by Weber Function change (m^3/s)	23,906	20,552	15,510
Peak occurrence time of section by linear change(hour: minute)	01:00	02:00	03:00
Peak occurrence time of section by Weber Function change(hour: minute)	00:28	00:36	00:58

Table 4.40 Peak discharge at Fujiangqiao section under different roughness values for 1/3-barrier dam-break scheme

Initial water level of 752 m, 1/3-barrier dam-break	Breaching duration		
	1 h	2 h	3 h
Peak discharge with $n = 0.04(m^3/s)$	10,861	10,733	10,477
Peak discharge with $n = 0.05(m^3/s)$	9595	9524	9376
Peak occurrence time with $n = 0.04$ (hour: minute)	05:20	06:06	06:52
Peak occurrence time with $n = 0.05$ (hour: minute)	06:16	07:00	07:46

The reason was that the section elements varied by Weber Function, and the section area expansion rate in the initial stage was faster than that by linear change.

4.3.4.4 Influences of Channel Roughness Change in Downstream Dam on Barrier Dam-Break Flood

In the calculation analysis on the barrier dam-break flood evolution, river channel roughness was a vital parameter. The greater the river channel roughness, the greater the river resistance, thus the greater the flood retention effect of the river channel, and the greater the attenuation effect on the peak flood flow. Model 1 compared the peak discharge and peak occurrence time at Fujiangqiao section under different values of roughness for the downstream channel of the barrier lake. The river roughness increased from 0.04 to 0.05, the peak discharge at Fujiangqiao section decreased by 8.5%-17% under different working conditions, and the peak occurrence time lagged 40–60 min (Tables 4.40, 4.41 and 4.42).

4.3.4.5 Effect Analysis on Reducing Barrier Dam-Break Risk of Discharge Channel

After the implementation of the discharge channel of Tangjiashan emergency dredging project, it was capable to reduce the initial water level from 752 m before the

Table 4.41 Peak discharge at Fujiangqiao section under different roughness values for 1/2-barrier dam-break scheme

Initial water level of 752 m, 1/2-barrier dam-break	Breaching duration		
	1 h	2 h	3 h
Peak discharge with $n = 0.04(m^3/s)$	18,968	18,708	18,089
Peak discharge with $n = 0.05(m^3/s)$	16,433	16,262	15,828
Peak occurrence time with $n = 0.04$ (hour: minute)	04:28	05:12	06:00
Peak occurrence time with $n = 0.05$ (hour: minute)	05:16	06:02	06:46

Table 4.42 Peak discharge at Fujiangqiao section under different roughness values for full-break scheme

Initial water level of 752 m, full-barrier dam-break	Breaching duration					
	1 h	2 h	3 h	4 h	5 h	6 h
Peak discharge with $n = 0.04(m^3/s)$	28,247	27,610	25,958	23,842	21,785	19,943
Peak discharge with $n = 0.05(m^3/s)$	23,440	23,072	22,255	20,941	19,597	18,247
Peak occurrence time with $n = 0.04$ (hour: minute)	03:26	04:12	04:58	05:40	06:20	06:56
Peak occurrence time with $n = 0.05$ (hour: minute)	04:14	05:00	05:44	06:24	07:01	07:36

project implementation to 742 m, decrease the barrier dam-break discharge volume of 78 million m^3 and greatly alleviating the threat of barrier dam-break flood. The calculation results of each model showed that by excavating the discharge channel, the barrier dam-break water head and available discharge volume could be reduced, so as to effectively reduce the water level and peak discharge.

By taking the calculation results of Model 1 as an example, analysis was conducted for the effect of reducing barrier dam-break flood by excavating discharge channel (Tables 4.43 and 4.44). After excavating the discharge channel, the initial water level reduced from 752 to 742 m, and for the corresponding three calculation schemes in the cases of 1/3-barrier dam-break, 1/2-barrier dam-break and full-barrier dam-break, the peak discharge at the barrier dam site decreased averagely by 41.5, 32.5 and 25.7%, and those at the Fujiangqiao section decreased by 36.7, 30.4 and 26.6%.

4.3.4.6 Flood Retention Effect of Tongkou HPP

Tongkou HPP is located 24.40 km downstream of the Tangjiashan barrier lake, with storage capacity of 36million m^3. The calculation of Model 1 showed that in the case of emptying the whole reservoir and opening all discharge orifices, if the Tangjiashan barrier lake suffered breach by the patterns of 1/3-barrier dam-break, 1/2 and full-barrier dam-break at an initial water level of 752 m, thanks to the reservoir flood retention effect, it was capable to reduce the peak discharge by 3000 m^3/s (Table

Table 4.43 Peak discharge change at barrier dam site with initial water level reduced by discharge channel

Scheme	1/3-barrier dam-break			1/2-barrier dam-break		
	Duration of1h	Duration of2h	Duration of3h	Duration of1h	Duration of2h	Duration of3h
Discharge at initial water level of 752 m(m^3/s)	26,440	20,289	16,320	58,935	40,470	28,520
Discharge at initial water level of742m(m^3/s)	15,544	11,955	9453	40,417	26,624	19,447
Percentage of discharge decreased (%)	41.2	41.1	42.1	31.4	34.2	31.8
Scheme	Full-barrier dam-break					
	Duration of1h	Duration of2h	Duration of3h	Duration of4h	Duration of5h	Duration of6h
Discharge at initial water level of 752 m(m^3/s)	132,537	65,686	44,497	34,085	28,011	23,859
Discharge at initial water level of742m(m^3/s)	95,030	48,730	33,517	25,749	20,981	17,737
Percentage of discharge decreased (%)	28.3	25.8	24.7	24.5	25.1	25.7

4.45) at Tongkou HPP and by 800 m^3/s at the downstream Fujiangqiao section, as well as postpone the peak arriving time. Therefore, Tongkou HPP played a role in peak cutting and hysteresis on the Fujiangqiao section.

4.3.4.7 Influences of Prolonged Breaching Duration on Barrier Dam-Break Flood

When formulating emergency contingency plan, in order to study on the influences of breaching duration on the breach discharge and its evolution, Model 1 conducted calculation analyzes on barrier dam-break floods for full-break-dam scheme with duration of 9, 12, 18 and 24 h at initial water level of 760 m and 752 m (Tables 4.46, 4.47 and 4.48).

It is showed by the calculation that: (1) breaching duration played a relatively important role on the peak discharge at barrier dam site section, but as the barrier

Table 4.44 Peak discharge change at Fujiangqiao section with initial water level reduced by discharge channel

Scheme	1/3-barrier dam-break			1/2-barrier dam-break		
	Duration of1h	Duration of2h	Duration of3h	Duration of1h	Duration of2h	Duration of3h
Discharge at initial water level of 752 m(m^3/s)	10,861	10,733	10,477	18,968	18,708	18,089
Discharge at initial water level of742m(m^3/s)	6880	6795	6638	13,244	13,014	12,563
Percentage of discharge decreased (%)	36.7	36.7	36.6	30.2	30.4	30.5
Scheme	Full-barrier dam-break					
	Duration of1h	Duration of2h	Duration of3h	Duration of4h	Duration of5h	Duration of6h
Discharge at initial water level of 752 m(m^3/s)	28,247	27,610	25,958	23,842	21,785	19,943
Discharge at initial water level of742m(m^3/s)	20,611	20,116	19,031	17,588	16,095	14,620
Percentage of discharge decreased (%)	27.0	27.1	26.7	26.2	26.1	26.7

Table 4.45 Maximum peak discharge before and after flood regulation at Tongkou HPP

Scheme	Calculation working conditions		
	1/3-barrier dam-break scheme	1/2-barrier dam-break scheme	Full-barrier dam-break scheme
Peak discharge before flood regulation (m^3/s)	15,500	31,300	57,630
Peak discharge after flood regulation(m^3/s)	12,600	28,000	54,800
Reduced peak discharge (m^3/s)	2900	3300	2830

Table 4.46 Peak discharge for full-barrier dam-break scheme at initial water level of 760 m (unit: m^3/s)

Section location	Breaching duration									
	1 h	2 h	3 h	4 h	5 h	6 h	9 h	12 h	18 h	24 h
Barrier dam site	136,234	69,692	48,452	37,475	30,710	26,090	18,119	13,960	9633	7380
Fujiangqiao	34,301	32,979	30,293	27,467	24,810	22,472	17,332	13,961	10,461	8484

Table 4.47 Peak discharge for full-barrier dam-break scheme at initial water level of 752 m (unit: m^3/s)

Section location	Breaching duration									
	1 h	2 h	3 h	4 h	5 h	6 h	9 h	12 h	18 h	24 h
Barrier dam site	132,537	65,686	44,497	34,085	28,011	23,859	16,679	12,918	8990.5	6938
Fujiangqiao	28,247	27,610	25,958	23,842	21,785	19,943	15,602	12,828	9663.1	7959

Table 4.48 Peak discharge for 1/3-barrier dam-break scheme at initial water level of 752 m (unit: m^3/s)

Section location	Breaching duration							
	1 h	2 h	3 h	6 h	9 h	12 h	18 h	24 h
Barrier dam site	26,440	20,289	16,320	9689	6673	5110	3498	2667
Fujiangqiao	10,861	10,733	10,477	9054	7567	6313	4832	4032

dam-break flood evolved toward downstream, the effect diminished rapidly, while breaching duration had little impact on the maximum discharge and highest water level of several important cities and towns in the downstream dam. However, in the case of short breaching duration, the highest water level occurrence time at each section was advanced obviously. (2) In the full-barrier dam-break scheme at initial water level of 760 m, if the breaching duration was prolonged to 12 h with Fujiang inflow of 1400 m^3/s, then the peak discharge at the barrier dam site was almost of the same order as that at the section of Fujiangqiao. (3) In the full-barrier dam-break scheme at initial water level of 752 m, if the breaching duration was prolonged to 18 h with Fujiang inflow of 1400 m^3/s, then the peak discharge would reduce to about 9660 m^3/s at the section of Fujiangqiao, guaranteeing basically the flood control safety of Mianyang City. (4) In the 1/3-barrier dam-break scheme at initial water level of 752 m, if the breaching duration was prolonged to 6 h with Fujiang inflow of 1400 m^3/s, then the peak discharge would be 9054 m^3/s at the section of Fujiangqiao, which was below the channel flood control standard of the urban areas of Mianyang.

The above results fully indicated that the prolongation of breaching duration would significantly reduce the flood peak discharge at the barrier dam site and downstream, so as to effectively reduce the threat of barrier dam-break flood to the downstream.

4.3.4.8 Influences of Different Discharge Channel Sections on Barrier Dam-Break Flood

The calculation of Models 1, 2 and 3 showed that, in the case of the same inflow flood and barrier dam-break process, the bottom elevation of different discharge channels had little influence on the flood peak discharge downstream of the dam, and the influence degree was generally less than 3%. Under the condition of the same discharge channel bottom elevation and barrier dam-break process, the floods of different return periods in the upper reaches of Tongkou River had little influence on the flood peak discharge in the lower reaches of the dam, and the influence degree was generally less than 4%.

4.4 Calculation Analysis Pattern of Barrier Dam-Break Flood

Based on the knowledge and experience obtained in the calculation of the barrier dam-break flood of Tangjiashan barrier lake, it listed the patterns for the calculation analysis of the barrier dam-break flood of Tangjiashan barrier lake, including what basic data should be collected, how to formulate the calculation conditions and the analysis of the calculation results, so as to provide reference for the calculation of barrier dam-break flood of the barrier lake in the future.

4.4.1 Basic Data to Be Collected

The basic data to be collected for the calculation of barrier dam-break flood of the barrier lake mainly included:

(1) **Topographic data**

Such as the topographic map of the reservoir and the corresponding water level storage capacity curve of the barrier dam site, the barrier dam site and its downstream river cross section and other data. In terms of the characteristics of barrier lake emergency handling, remote sensing image data and digital processing should be fully used if possible to obtain the data quickly.

(2) **Hydrologic data**

The hydrologic data included historical statistics on the basin area of the barrier dam site, the level of the control hydrology station, the discharge, the rainstorm and the hydrograph of flood of different return periods. If there was afflux of large tributary in the lower reaches, it was necessary to collect the flood process of the downstream tributaries and calculate the water level-flow relationship curve at the outlet boundary section. In the absence of such data, then the assumption could be carried out by the uniform flow formula of open channel combined with hydraulic elements of cross section.

(3) **Roughness coefficient of river channel of the downstream barrier dam**

Generally speaking, it was difficult to collect the measured hydrological data of rivers in the area where the barrier lake is located to verify the roughness coefficient of the mathematical model. The more feasible method was to refer to the values of roughness of other similar rivers in the same basin. At the same time, in order to ensure the reliability of the calculation results, the roughness sensitivity should be calculated and analyzed as much as possible.

(4) **Flood control capacity of river channel of the downstream barrier dam**

The flood control capacity of the river downstream of the dam included the flood control standard of the downstream towns, the corresponding discharge and flood level, the grade and elevation of the river embankments on both sides, the protection area and the population involved and other data. This was one of the basic data for barrier dam-break flood risk and disaster assessment.

(5) **Basic information of important facilities of the downstream barrier dam**

If there were other cascaded reservoirs under construction or already built downstream of the dam, then it was necessary to collect the dam body structure form, elevation-storage curve, discharge capacity and operation mode of these reservoirs. If there were important transportation facilities such as railways and bridges downstream of the dam, then it was necessary to collect the distances between these facilities and the barrier lake, as well as the structure and elevation of the buildings.

4.4.2 Definite Solution Condition and Boundary Treatment

Generally, the boundary conditions of one-dimensional mathematical models should be represented by water level process line, flow process line and water level-flow relation curve. For tranquil flow, flow process line should be taken as the boundary condition of inflow section, and water level process line or water level-flow relation curve should be taken as the boundary condition of outflow section. Two of these conditions should be given at the inflow section of the rapid flow.

For the calculation of barrier dam-break flood, it was difficult to give the boundary conditions according to the above requirements because the flow pattern of the rapid and tranquil flow in each section of the reach (including the outlet section) was variable. It was feasible to take the flow process line as the boundary of inflow section and the water level-flow relationship curve as the boundary condition of outflow section. When the mathematical model of barrier dam-break flood could simulate the performance of the rapids and slow flow interphase in the reach, reasonable calculation results could be obtained by using such boundary conditions.

Since the barrier dam-break water level and discharge were larger than the measured data range, the measured water discharge curve should be extended. The steps of the extension method were as follows: (1) according to the trend of the existing water level-flow relationship curve, the approximate range of the extension curve was given. (2) The water level and flow relationship curve with higher water level under the same discharge was used to calculate flood evolution. The calculated results showed that the water level of the outlet section was higher than that of several sections upstream. (3) By gradually reducing the water level under the same flow, when the water level of the outlet section was no longer higher than the water level of the upstream section under the calculated flow volumes, then it was the upper limit of the water–flow curve of the outlet section used for extension calculation. This upper limit was taken for safety reasons.

4.4.3 Preparation of Calculation Working Conditions and Results Analysis

4.4.3.1 Preparation of Calculation Working Conditions

According to the available research results, the harm of barrier dam-break flood depended on the barrier dam-break pattern, the initial water level (which determined the collapsible storage capacity and the size of the upstream and downstream water head), the breach size and the breaching duration.

(1) **Determination of barrier dam-break pattern**

Dam-break patterns could be divided into abrupt full failure and gradual failure. Gradual failure could be divided into gradual full failure and gradual partial failure, and partial failure could be divided into vertical partial failure and lateral partial failure. According to the statistical data of the barrier dam-break of earth-rock dam or barrier lake, the barrier dam-break patterns of barrier lake were generally gradual failure.

(2) **Determination of initial water level**

It was assumed that the barrier dam-break occurred in non-flood season and the incoming flow was relatively small. The initial water level could be assumed to be the highest water level of the barrier lake. Assuming that the barrier dam-break occurred in the flood season, it was necessary to determine the flood return period that might be encountered, and the initial water level needed to be determined after the flood regulation for the incoming flood.

(3) **Determination of breach size**

The breach size mainly referred to the possible breach shape, the breach depth and the breach width. The shape of the breach could be approximately rectangular or ladder. The final size of the entrance should be based on the material composition, structure and geological conditions of the barrier dam. The process from the initial entrance shape to the final entrance shape could be approximately processed in linear.

According to the *DAMBRK Model Manual* compiled by the U.S. Weather Service Bureau, the average breach width of an earth dam after its failure ranged from dam height to three times dam height. The formation of the breach was a gradual process, and the failure time of the dam body could vary from several minutes to several hours, depending on the height of the dam, the material of the dam, the density of the material and the degree of overtopping. As the damage began to form at a point below the crest, the flow of water scoured the channel inside the dam, creating larger and larger openings and finally accelerating the failure due to the crest failure.

Due to the uncertainty of the actual wreck mechanism and the difficulty in determining the calculation parameters, the accurate model which simulated the scour, failure and expansion process could not meet the practical requirements. At present, simplified and widely applicable methods were used to figure out the duration from the beginning to the end of the failure and the final size and shape of the breach, based on experience and data and other wrecked dams. (1) Slope failure was determined by slope coefficient m, whose value range was $0 \leq m \leq 2$. (2) The final shape of the breach was determined by the slope coefficient m and the breach bottom width b. For example, m = 0,b > 0 was a rectangle; m > 0,b = 0 was a triangle; and m > 0,b > 0 was a trapezoid. In calculation, the 1 h initial breach bottom width was given, during the development process of the breach, that is, within the breaching duration (t), the bottom width of the breach expanded at a linear speed until the final bottom width (b) was reached and the elevation of the breach is scoured to the bottom of the reservoir or the bottom elevation of the discharge channel (Fig. 4.24).

In the simulation of dam-break, it was assumed that the breach calculation started when the reservoir water level exceeded the specified value h_c. If h_c was higher than the elevation of dam crest, it was the case of simulated overtopping failure. If h_c was lower than the elevation of dam crest, it was the case to simulate the situation of the failure caused by leakage or piping, and it was called non-overtopping failure.

Empirical formulae for estimating the breach width have also been put forward by relevant domestic institutions and scholars, such as the Scientific Research Institute

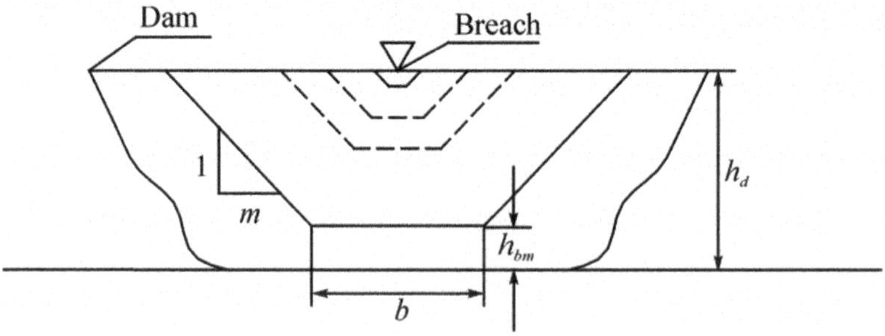

Fig. 4.24 Schematic diagram of breach expansion process

of the Ministry of Railways, the Yellow River Conservancy Commission and Xie Renzhi.

(4) **Determination of breaching duration**

The breaching duration was affected by many factors, such as the material composition of the barrier dam, the upstream inflow and the water level in front of the dam, so it was difficult to estimate in advance. According to the statistics of earth-rock dam failure, most of the failure duration was 0–3 h. In practical calculation, different breaching duration could be assumed to analyze the harm of barrier dam-break flood in various possible situations.

4.4.3.2 Analysis on Calculation Results

The analysis on the calculation results of barrier lake dam-break flood mainly included:

(1) Reliability, accuracy and rationality analysis on the calculated data.
(2) The changing process curves of water level, flow and velocity with time at the barrier dam site section and each characteristic section of the upstream and downstream river channel.
(3) The relationship of barrier dam-break flood submergence area with time; the time and advancing speed of barrier dam-break wave at each feature section, the maximum values of water level, flow rate and velocity at each feature section and their occurrence time and the changing process of hydraulic elements at each feature point when downstream towns were submerged.
(4) The law of barrier dam-break flood changing with breach entrance shape and breaching duration, the law of the barrier dam-break flood changing with the incoming flood of different return periods and the law of barrier dam-break flood changing at different water levels in front of different dam.

According to the population, social and economic development of the important cities and towns at downstream, as well as the existing standards of protective facilities, and according to the calculation results of the barrier dam-break flood, analysis was conducted for the impact degree on downstream residential areas and various building facilities and inundation scope, so as to estimate the possible losses caused by the barrier dam-break flood. Therefore, it provided the relevant basis for the formulation of the contingency plan of the barrier lake emergency handling and the evacuating plan for the downstream personnel.

4.4.4 Working Pattern for Calculating Barrier Dam-Break Flood of Tangjiashan Barrier Lake

4.4.4.1 Strengthening Hydrological Analysis by Establishing Water and Rain Forecasting System

In order to meet the needs of analysis and calculation and frontline rescue command, the Changjiang Water Resources Commission (CWRC) attaches great importance to the basic work of hydrological survey, report, analysis and calculation. On the basis of the extensive collection of relevant basic data, the Hydrology Bureau of CWRC has carried out a large number of hydrological measurements and reports and hydrological analysis and calculation by various technical means.

The monitoring of hydrologic elements along the Tangjiashan barrier lake should be strengthened. On the basis of making full use of the existing stations, more observation stations should be added. There are now Baishi and Piankou rain gauge stations in the basin upstream of the barrier lake, and it is supposed to add Tangjiashan, Fushun, Zhicheng and Youfang rain gauge stations, then a total of six rain gauge stations are stand by to monitor the rainfall regime in the basin. Beichuan hydrology station has automatic measuring and reporting equipment, which can normally monitor rainfall and water level. In order to ensure the reliability of upstream water level observation, Tongkou temporary water level automatic observation station is set up about 1.5 km downstream of Tongkou HPP, and Xiangshui temporary observation station is set up downstream of Tongkou HPP.

The communication scheme is formulated according to the analysis of the communication conditions of each monitoring station. ① The information of hydrological monitoring stations in the upper reaches of Tangjiashan is automatically collected and transmitted to the Mianyang branch center by Beidou satellite. ② The information of Beichuan hydrology station is automatically transmitted by Beidou satellite to the Mianyang branch center. ③ Walkie-talkie communication is adopted in the Tongkou and Xiangshui observation stations, and the observation information is transmitted to the Mianyang branch center by satellite telephone and voice. The abovementioned information is sent to the Hydrology Information Center of the Ministry of Water Resources and the Hydrology Bureau of CWRC through the current water situation network.

From May 20 to June 9, the Hydrology Bureau of CWRC overcame the difficulties of having no forecast plan and lack of basic hydrological data, gave full play to the subjective initiative of forecasters and kept improving the forecast. The daily forecast and analysis results were provided to the hydrologist of the headquarters, the hydrology group and the Hydrology Bureau of the Ministry of Water Resources (HBMWR) through SMS and email, which laid strong technical support for the front decision-making.

4.4.4.2 Organizing Several Institutions to Conduct Calculation and Analysis on Barrier Dam-Break Evolution in Parallel

Due to the complexity of breach mechanism and the uncertainty of influencing factors, the numerical simulation technology of barrier dam-break flood is not very mature at present. In the process of emergency handling of barrier lake, the calculation of barrier dam-break flood should be carried out in parallel by organizing several institutions and adopting several sets of mathematical models, so that the comprehensive results can provide reference for decision-making.

In view of the fact that the accuracy of the dam-beak flood calculation results is directly related to the safety of the lives and property of the downstream people and the number of the emergency evacuation population, the CWRC organized the technical backbones of the CISPDR, CRSRI, Hydrology Bureau of CWRC and other institutions to cooperate closely in the front and rear and adopted a variety of methods for calculation and comparative analysis, in order to maximize the reliability of the results. In this calculation, different empirical formula methods, self-developed one-dimensional barrier dam-break and flood evolution mathematical model, two-dimensional barrier dam-break and one-dimensional flood evolution mathematical model, MIKE11 barrier dam-break flood evolution calculation model and other methods were adopted successively to carry out parallel calculations and comparative analyses. In the calculation process, all institutions adopted uniform topographic data, lake capacity curve, calculation conditions and related parameters, so as to analyze and compare the results.

(1) **Breach calculation of various combination schemes with different beach elevation, shape, breaching duration, initial water level and other factors**

In order to make the emergency evacuation plan foolproof, the front headquarters hold conferences of consultation to determine three possible breach elevations and corresponding breach shapes in the cases of 1/3-barrier dam-break, 1/2-barrier dam-break and full-barrier dam-break according to the geological characteristics of the barrier dam, worked out the breaching duration of 1, 2, 3, 4, 5 and 6 h and set out corresponding initial water level with reference to the flood regulation results provided previously. CISPDR and CRSRI carried out the barrier dam-break flood calculation of each combination scheme in parallel and made a comparative analysis. On this basis, in accordance with the principle of safety priority, it was supposed to submit the calculation results of barrier dam-break flood at each section to the front headquarters.

(2) **Joint calculation of flood regulation and barrier dam-break flood**

Since the barrier dam shaped long along flow direction, the total length of the discharge channel was about 680 m, and it is a variable slope channel, how to accurately determine the discharge capacity of the drainage groove was the key to the calculation of flood regulation. To this end, it was supposed to use weir flow formula method with the channel tail water fall, channel water surface line method and MIKE11 model for comparative analysis, so as to determine the discharge curves of the three schemes.

In order to study the impact of upstream floods of different return periods on barrier lake dam-break flood and initial water level, the CWRC organized the CISPDR, Hydrology Bureau of CWRC and CRSRI to carry out barrier lake flood regulation by using a variety of methods. Such flood regulation studies were conducted for incoming flood processes of return periods of 20%, 10% and 5%, and the combined calculations of 28 schemes with different barrier dam-break schemes were carried out.

(3) **Barrier dam-break flood calculation for encountering Fujiang frequency flood**

In order to reflect the impact of barrier dam-break resulting from Fujiang River floods with different frequencies on Mianyang and the towns below Mianyang, CWRC organized CISPDR, CRSRI to analyze the impact on Mianyang and the towns below Mianyang due to the serious situation where Tangjiashan suffered barrier dam-break resulted from meeting with 10% incoming flood and the floods with the same frequency as Fujiang River floods through different mathematical models. Relevant calculation results was submitted to Chongqing Flood Defense Office.

(4) **Calculation of the influence of further extension of beaching time on barrier dam-break flood evolution**

Considering the geological composition of the barrier dam was the whole mountain landslide, the propagation time of natural flood was 4–5 h, and there were two exposed places of the broken and cracked rocks along the discharge channel, the actual breaching time might be prolonged. In order to study the influence of prolonged breaching time on barrier dam-break discharge and flood propagation, CWRC organized CISPDR and CRSRI to conduct the barrier dam-break flood calculation of typical schemes with breaching duration of 9, 12, 18 and 24 h. According to the calculation of 1/3 dam-break scheme, if the breaching duration extended to 9 h and the discharge at Fujiang River was still 1500 m^3/s, the peak flood discharge at Fujiangqiao in Mianyang would decrease to about 9000 m^3/s. This fully demonstrated that extending the breaching time would greatly alleviate the downstream threat.

(5) **Comparative calculation of the influence on barrier dam-break flood with or without discharge channel under the same condition**

In order to study the effect of discharge channel excavation on reducing barrier dam-break flood, CWRC organized CISPDR and CRSRI to make a comparative calculation for the influence on barrier dam-break flood with or without drainage flood. The influences of three discharge channel schemes were compared and calculated under the same conditions. Through comparative analysis, it came to the conclusion that the barrier dam-break flow could be reduced significantly after the discharge channel excavation.

(6) **Calculation analysis of flood retention in downstream reservoir**

Tongkou HPP is located in the lower reaches of Tangjiashan barrier lake. In order to study its flood retention effect on barrier dam-break flood, CWRC organized CISPDR and CRSRI to calculate and analyze for the flood retention effect in Tongkou HPP. Through calculation, Tongkou HPP had obvious flood retention effect on the downstream. But if Tongkou HPP suffered barrier dam-break after its flood retention, it would increase the impact on Mianyang. Therefore, the safety of Tongkou HPP should be given priority.

(7) **Calculation of the impact of barrier dam-break of small barrier lakes in the lower reaches**

According to the field data, there are small storage barrier lakes as Kuzhu dam and Yanyangtan in the downstream of Tangjiashan barrier lake. If the two barrier lakes suffered continuous failures when the flood peak of Tangjiashan barrier lake arrived, it would affect the Mianyang flood peak flow. The results showed that they had little influence on the peak discharge of barrier dam-break flood in Mianyang because of their small water storage capacity.

4.4.5 Conclusion

(1) According to the field investigation, the nearby regional geological survey data and the reference to the weathering condition of similar strata, the water resource experts concluded that the Tangjiashan barrier lake was of a large scale, the upstream dam slope was relatively gentle, the substructure was dense, the overall stability was good, there was no risk of overall failure, the barrier dam-break process was gradual change, and there would be no instant barrier dam-break. It was deemed that the elevation of 720 m was the most likely breach bottom elevation. The actual discharge process of Tangjiashan barrier lake was in the form of gradual barrier dam-break, and the inlet elevation of the channel bottom was about 714 m after discharge. The results of expert field research on the possibility of barrier dam-break and scour form were basically consistent with the actual situation.

(2) Under the condition of different discharge channel length, channel bottom elevation and channel bottom width, the sensitivity analysis of discharge channel discharge capacity change showed results as below. Under the condition of the same water level, the lower the discharge channel bottom elevation, the greater the discharge; that is, regardless of widening of channel scouring width, the effect of reducing the channel bottom elevation was more obvious than that of reducing the water level of the barrier lake. The peak discharge of barrier dam-break flood depended on the initial water level, the breaching duration, the breach shape and its expansion process. The higher the initial water level, the shorter the breaching duration, the faster the development of the width and depth

of the breach, and the greater the flood peak discharge at the barrier dam site. Therefore, by excavating discharge channels and other measures to effectively reduce the initial water level and the available discharge water, the peak flood discharge at the barrier dam site could be significantly reduced, and the threat of barrier dam-break flood to downstream flood control could be effectively mitigated.

(3) According to the calculation results of the barrier dam-break mathematical model, under the three typical barrier dam-break patterns of 1/3-barrier dam-break, 1/2-barrier dam-break and full-barrier dam-break, the peak discharges all exceed the river safety discharge limit in the upstream reach of the Tongkou River and also approach or exceed the river safety discharge limit in the downstream reach of the Tongkou River. Therefore, in addition to the engineering measures, we should also simultaneously make the evacuation plan in the downstream area to avoid the risk caused by barrier dam-break. Furthermore, people living in the upstream area of Tongkou River estuary should also be evacuated. The barrier dam-break flood would not pose a threat to Mianyang City if the secondary failure occurred after the first failure of Tangjiashan barrier dam reached 720 m.

(4) The water flow began to flow through discharge channel at 07:08 on June 7, the barrier dam reached the highest level of 743.23 m at 01:30 on June 10, and the maximum discharge of 6500 m^3/s appeared at 12:30 on June 10. The implementation of the emergency handling project reduced the barrier lake head by 9.0 m, reduced the water storage by 0.7 million m^3, reduced the peak flood discharge at the barrier dam site by about 3400 m^3/s and the peak flood discharge at the Fujiangqiao section by about 3000 m^3/s. Compared with the original 1/3-barrier dam-break scheme, the peak flood discharge at the main control sections downstream of the dam was reduced by 33.6–34.6%. Without the implementation of drainage project, i.e., if the barrier dam sufferer barrier dam-break when the lake level rose to 752 m, the flow at the section of Fujiangqiao would greatly exceed its allowable flow capacity, and the consequences were unimaginable. This showed that the implementation of the emergency handling project effectively reduced the threat of the barrier lake flood to the safety of downstream people's well-being and brought huge social and economic benefits.

(5) The new channel was formed after the completion of the barrier lake discharge. When encountering the 200-year return period flood, the highest lake level was lower than 725.44 m and the maximum discharge was greater than 5440 m^3/s. The new channel had the capacity to carry a 200-year return period flood, laying the foundation for the subsequent handling of the barrier lake.

Chapter 5
Study on the Anti-Scouring Stability of Discontinuous Wide Graded Mixture of Barrier Dam

The Tangjiashan barrier dam was formed in a very short time by earthquake-induced landslide. The material composition and scouring resistance ability can only be inferred by on-site simple surveying and mapping and the expertise in limited time. In order to review the on-site judgment during emergency handling and provide experience and theoretical basis for subsequent barrier lakes risk elimination, the "*Researches on Barrier Lake Emergency Handling Technology and Experiences*" program was arranged in the National Science and Technology Support Program during the 11th Five-Year Plan Period from 2009 to 2011. In this program, the scouring process and internal mechanism of barrier dam were studied through laboratory scouring experiments and dam-break tests to provide theoretical support for future disposal of similar disasters.

The geometric and physical characteristics of the material composition of Tangjiashan barrier dam were analyzed through on-site investigation and field sampling of Tangjiashan barrier dam residues. Then, the particle diameter range and bulk density of indoor experimental research were simulated and designed. The building alum stone and natural sand were selected to prepare different wide graded mixtures according to different proportion. And the flume tests on the incipient motion characteristics of incipient motion and scouring were carried out by those mixtures. The movement patterns of various mixtures under different flow rates, water depths and flow velocities were observed. The critical incipient motion flow rate was measured, and the relationship between the critical incipient motion velocity coefficient and particle diameter characteristics of those mixtures was analyzed. Combined with the material composition of Tangjiashan barrier dam, two kinds of wide graded mixtures were selected as material of laboratory tests. A newly built dam-break device and a newly developed breach morphology capture technology were used in the dam-break tests of wide graded barrier dam. The scouring resistance ability, the scouring characteristics and the dam-break mode of wide graded barrier dam mixtures were studied under the conditions of different barrier dam section shapes, different barrier dam materials, different inflow and discharge channel. The influence of barrier dam section shape on the barrier dam-break process and peak discharge was summarized.

Q. Yang and W. Yang, *Emergency Handling of the "5.12" Tangjiashan Barrier Lake*,
https://doi.org/10.1007/978-981-99-1706-8_5

Three typical modes and the scouring process mechanism of barrier dam-break were revealed.

The experiment increased the insight and knowledge of the incipient motion and scouring characteristics of wide graded mixture, accumulated the test data of the dam-break process of the wide graded barrier dam, studied the change process of the breach shape and hydraulic elements and analyzed the breach formation and development mechanism. These results provided scientific basis and technical support for the early warning and prediction of the failure of barrier dams and similar artificial dams.

5.1 Study on the Barrier Dam-Break

5.1.1 Research Status of Barrier Dam-Break

Generally, the research on barrier dam-break can be divided into two aspects: ① study on the evolution of dam-break flood and the propagation law of dam-break wave; ② study on the dam-break mechanism and breach development process. The research methods mainly include physical model test, numerical calculation and statistical analysis of historical data.

To the study on the evolution of dam-break flood and the propagation law of dam-break wave, which involves the transmission of hydraulic characteristics of long-distance and large-scale floods, mathematical models are mainly used for simulation, including one-dimensional models and two-dimensional models. In addition, physical models with a small scale of 1:100–1:500 are also used for experimental research. At present, there are many research results in this area.

The research on dam-breach mechanism and breach development process is complicated because it involves many factors such as hydraulics, soil mechanics and sediment movement mechanics. Breach development process is closely related to dam type, geometric profile, dam materials, gradation and compactness. And research in this area is still being explored. This research has recently become the focus of barrier dam-break. The breach development process of the dam-break is a process of interaction between water and soil or stone. The vertical development of the breach is dominated by water erosion and undercut, and horizontal expansion of the breach is dominated by gravity-induced collapse. The breach shape changes violently and develops nonlinearly, and it is highly three dimensional. At present, large-scale indoor physical model tests are mostly used. Under certain assumptions, a series of numerical calculation models have also been developed. In addition, relevant countries and institutions have also carried out field experiments, such as the European Union's IMPACT project, in which homogeneous earth dams and homogeneous rock-fill dams were applied and certain research results were achieved.

Scholars at home and abroad have made many beneficial explorations in dam-break research. At the end of 1950s, a series of flume tests of earth-rock dam-break were carried out in the USA, and large-scale model tests were also carried out on site.

In 1960s, Austria conducted a lot of laboratory tests on rock-fill dam-break. Since 1990s, European and American countries have gradually shifted the focus of dam-break research to dam-break mechanism, including the National Dam Safety Plan (NDSP) of the United States and the European Union's IMPACT project. Among them, five large-scale field tests (dam height 4–6 m) and 22 indoor small-scale tests were carried out in the European Union IMPACT project. In the test, the influencing factors such as different dam types, different dam geometries (different dam slopes, different dam crest widths), different dam-building materials (non-viscous materials, viscous materials), different material properties (particle size distribution, compaction degree, water content) and different dam-break forms (overtopping, piping) were studied in detail, and enormous results were obtained.

Since 1970s, some model tests of self-break dams have been carried out in China. More than 30 times of self-break tests have been carried out in Yahekou Reservoir. The formation time and undercut rate of the breach have been studied in detail, and the relationship between the breach formation time and undercut rate and the model scale has been summarized. In 1978, the field scouring experiment of Nanshan Reservoir was carried out. In this experiment, the water head and the dam-break duration and the scouring rate and the scouring process were measured. It was considered that the factors affecting the scouring rate of the dam mainly depended on the form and thickness of the cutoff wall besides the water head. Then, through a series of small-scale laboratory tests, the relationship between scouring rate and model scale was obtained. In 1993, the dam-break accident occurred in Qinghai Gouhou concrete face sand-gravel dam. In order to reveal the failure mechanism of this accident, a series of physical model tests were carried out by relevant departments.

The National Science and Technology Support Program during the 11th Five-Year Plan Period "*Research on Dam Break Test and Simulation Technology*" was conducted from three aspects: ① large-scale physical model test of dam-break; ② numerical simulation of dam-break process and ③ numerical simulation technology of dam-break flood and software development of dam-break risk map. In this project, the technical route of combining physical model with numerical calculation was adopted to study the dam-break mechanism of homogeneous cohesive soil dam. At the same time, the numerical calculation method was adopted to simulate the flood routing of dam-break. Through this project, rich achievements have been made, while improving the understanding level of dam-break research in China.

5.1.2 *Research Status of Incipient Motion and Scouring Characteristics of Wide Graded Mixture*

The scouring of the riverbed by natural rivers, the movement process of debris flow and the formation and dredging of barrier lakes all involve the movement of sediment. At present, some progress has been made in the research of homogeneous sediment movement in scientific research areas. However, due to the wide gradation of bed sand of mountainous rivers, complicated debris flow composition and commonly discontinuous particle size distribution, etc., the sediment movement involved in the scouring process of natural riverbed and the formation and dredging of barrier lake is quite different from that of homogeneous sediment. Therefore, in view of the particularity of the movement of continuous and discontinuous wide graded non-uniform sand, it is of great value to strengthen the study of its movement law and incipient motion velocity for the management of natural rivers and the prevention and treatment of flash flood disasters.

Research work on non-homogeneous sand mainly focused on its incipient motion velocity and incipient motion mechanism. Rongyu [1] considered that non-homogeneous sand was also subjected to an additional resistance force proportional to the average shear resistance of mixed sand besides its underwater self-weight of sediment, thus deriving the formula of incipient velocity of non-homogeneous sediment. Yuaner and Jianheng [2] clarified the complexity of the incipient motion mechanism of non-uniform sediment based on the near-bed flow structure and the experimental results of the flow around a sphere on the bed surface and established a simplified formula for calculating the incipient motion velocity that can reflect the exposure and concealment of the non-uniform sand bed. The coefficients of the formula were determined by experiments and field observations. Han and He [3] introduced the concept of comprehensive incipient motion velocity of non-homogeneous riverbed sand, with the law of low sediment transport rate of bed load, to calculate formula of comprehensive velocity. He et al. [4] adopted rolling incipient motion mode with additional mass force and relative exposure parameters to deduce the corresponding incipient motion formula. Tang and Fang [5] pointed out that the influence of the hiding-exposing action and impulse force should be considered at the same time in the incipient motion of non-homogeneous sand. Wang et al. [6] comprehensively considered the influence of water flow conditions, sediment particle characteristics of bed surface and sediment particle diameter composition in the incipient motion environment. The mathematical expressions of each influencing factors were given, and the formula of incipient motion velocity of non-homogeneous sediment was established. On the basis of theoretical analysis and previous experimental results, Mai and Wang [7] analyzed the motion characteristics of non-uniform sediment from the aspects of the starting condition and mechanism of non-uniform sediment, the calculation method of sediment carrying rate and the unbalanced sediment transport of suspended sediment. According to the formula of bed load transport rate of non-uniform sediment, the variation law of sediment transport rate and the influencing factors of sediment transport rate under the same incipient motion condition were analyzed. Haitaoxu et al. [8] established the formula of incipient motion velocity based on the drag force and thrust force of the non-uniform sand bed and concluded that compared with uniform sand of the same size, fine particles in non-uniform sand were more difficult to move, and coarse particles were easier to move.

Because of the randomness of non-uniform sediment movement, it was difficult to determine the criterion of sediment incipience, and the accurate measured data were not easy to obtain, so the research work has no satisfactory results. In addition, the problems such as riverbed scouring were not only related to sediment incipience, but also directly related to the movement of sediment in various degrees under different flows.

5.2 Material Composition Analysis of Barrier Dam

In order to know the material composition of Tangjiashan barrier dam, while collecting and analyzing data in the same time, technical personnel carried out site survey (Fig. 5.1) on April 2010, to fully investigate the scouring stability of bank slopes and the discharge channel of Tangjiashan barrier lake, the accumulation state of landslide mass and the particle size distribution of earth-rock materials and took back some soil and stone samples of the accumulation mass.

The bedrock on both banks of Tangjiashan barrier lake dam area was gray-black thin-to-medium-thick layered feldspar mica siltstone of Cambrian Qingping Formation, siliceous slate with thin-to-extremely thin marl and mudstone. The left bank was a slope in the counter direction, and the right bank was a dip slope. Bedrock fissures were relatively developed, and the rock mass was relatively fractured. The strongly weathered zone was 5–10 m thick. Underneath was weakly weathered zone. There were eluvium-deluvium gravel soil distributed on both banks, with a thickness of 5–30 m. The gravel soil was composed of silty loam, rock debris and cobblestones.

The barrier dam was composed of boulders, block gravel, eluvium-deluvium gravel soil and muddy silty fine sand deposit in Kuzhu reservoir area. The muddy silty fine sand of the original riverbed was scattered in the left area. See Table 5.1 for the stratification revealed by the surface survey and five boreholes. The material composition of the barrier dam had obvious stratification characteristics from top to bottom.

According to the material composition characteristics of the barrier dam, it could be divided into four layers from top to bottom: gravel soil layer, cobblestones layer, boulders layer (layer-like) and gray-black silty gravel layer.

(1) Layer ④: Gravel Soil Layer

It was distributed in the remaining part on the right side of the existing barrier dam, the left bank near the top of the discharge channel and the upstream slope of the barrier dam, which was the eluvium-deluvium of the original Tangjiashan surface. Gravel soil was consisted of silty loam, rock debris and cobblestone, of which silty

Fig. 5.1 Site survey on material composition of Tangjiashan barrier dam

Table 5.1 Stratification of riverbed

Location	Borehole number	Borehole elevation (m)	Borehole depth (m)	Barrier dam thickness (m)	Floor elevation-floor depth-layer thickness (m)			
					Layer ④	Layer ③	Layer ②	Layer ①
Discharge channel	ZK01	718.0	84.1	71.5	–	–	651.6–66.4–66.4	646.5–71.5–5.1
	ZK05	714.0	73.2	67.1	–	–	656.9–57.1–57.1	646.9–67.1–10
Left bank of barrier dam	ZK04	742.0	106.8	87.5	735.7–6.3–6.3	729.1–12.9–6.6	670.2–71.8–58.9	654.5–87.5–15.7
	ZK02	780.0	83.6	47.1	–	756.5–23.5–23.5	739.8–40.2–16.7	732.9–47.1–6.9
	ZK03	782.9	105.8	85.8	781.0–1.9–1.9	754.0–28.9–27.0	703.0–79.9–51.0	697.0–85.9–6.0

loam accounted for about 60%, rock debris accounted for 30–35% and cobblestone accounted for 5–10%. The particle diameter was less than 5 cm, and the cobblestone was highly weathered. The thickness of this layer was 5–15 m. The structure was loose, the scouring resistance capacity was poor, and many tensile cracks had been spotted in the range of 20–40 m near the discharge channel.

(2) Layer ③: Cobblestones Layer

Underneath layer ④, the cobblestones layer was distributed around the surface on the left side of the barrier dam. This layer was mainly composed of cobblestones, a small amount of boulders and a small amount of gravel soil. The cobblestones were mainly 6–40 cm in diameter, and the boulders were 1–2 m in diameter. Most of the cobblestones were strongly weathered, some of them were weakly weathered, and the strength of rocks was extremely high. There was overhead phenomenon on the surface of this layer. No drill was lost during drilling. Generally, the thickness was 10–30 m, the structure was loose, and it had certain scouring resistance capacity.

(3) Layer ②: Boulders Layer (Layer-Like)

Underneath layer ③, the boulders layer was continuously distributed from the inlet to the outlet near the foot of the left bank slope of the discharge channel. According to the surface survey, the layer on the surface generally appeared as a layered structure, with incomplete disintegration and obvious pressure cracks, mainly 0.2–1 cm. The thickness was mainly 5–15 cm. The boulders formed in the front section of the discharge channel had a maximum diameter of 8 m, and its lithology was mainly siltstone, which was weakly weathered fresh and had high strength. The layer-like surface occurrence was N30°–90° E/NW∠10°–45°, which showed that the dip angle gradually decreased from the right bank to the left bank and from the middle to the upstream. According to the geological mapping of the surface, the top elevation of this layer was low on the left side and high on the right side. The top elevation of the left side of discharge channel was 718–735 m, and the outcropping elevation of the middle section on the left side was 715 m.

It was revealed by drilling that the thickness of this layer was 50–67 m, with particle size of 2 cm mainly, and the proportion of particles with diameter greater than 6 cm accounted for 20–60% of the total thickness of this layer. This layer had dense structure, strong scouring resistance capacity and strong seepage resistance capacity. This was also the reason why the flood ended at this layer after discharge on June 11, 2008.

(4) Layer ①: Gray-black Silty Gravel Layer

This layer was the bottom material of the barrier dam, which was a sliding zone formed by the rapid decline of the barrier dam. The particles were fine, mixed with the silty sand of the original riverbed or the gravel soil of the bank slope. It was revealed by drilling that the thickness of this layer was 6–15.7 m, the gravel size was mainly less than 2 cm, accounting for about 60%, and the rest was sand and silt, with a small amount of medium-coarse sand. The bottom elevation of this layer was

646.5 m, and it was directly connected to the bedrock of the original river bed at the bottom of ZK01 and ZK05 boreholes.

The Tangjiashan barrier dam was composed of eluvium-deluvium gravel soil accumulated on the upper slope of the original hillside and cataclasite rocks formed by sliding, extruding and crushing of upper bedrock of the Cambrian Qingping Formation. According to the on-site geological survey of the barrier dam, the gravel soil accounted for about 14% and the cataclasite rocks accounted for about 86%.

Gravel soil: yellow consisted of silty loam and cobblestones, of which silty loam accounted for about 60%, gravel accounted for 30–35% and cobblestones (with particle diameter of 5–20 cm) accounted for 5–10%.

Cataclasite rocks: The composition was heterogeneous. The cataclasite rocks formed by soft rocks were mainly gravel, and the particle diameter was mostly less than 20 cm. The cataclasite rocks formed by hard rocks were mainly cobblestones, with the diameter of 1.0–3.0 m, and the diameter of boulders could reach several meters. The local rock mass was not disintegrated, and the layered structure of the original rock remained.

According to the on-site geological survey, the barrier dam was divided into three areas rely on different material composition.

Area I: Located on the right side of the barrier dam, it was lower in the middle and higher at two ends, with the lowest elevation of 747 m. The upper part of the barrier dam was gravel soil, about 15 m thick, thin on the left and thick on the right; the lower part was cataclasite rocks. The right groove was run through the upstream and downstream of the barrier dam in the middle. The original riverbed was located in this area, with an elevation of 664.7 m. Due to the impact of the sliding force of the landslide, most of the 20-m-thick silty fine sand deposited in the original Kuzhu reservoir was scoured away, and the thickness was reduced to about 8.0 m. The elevation of underlying bedrock surface was 735–738 m.

Area II: Located on the left side of the barrier dam, it was featured with high terrain, with the highest elevation of 793.9 m. It was composed of cataclasite rocks, 35–107 m thick and partially covered with gravel soil, 2–4 m thick. Under the barrier dam, the original left bank slope was eluvium-deluvium gravel soil and bedrock.

Area III: Located on the upstream of the barrier dam, it was composed of muddy silty fine sand deposited in Kuzhu reservoir, formed by the sliding and extruding of landslide and distributed on the upstream dam slope below the elevation of 730–740 m.

See Table 5.2 for the physical and mechanical parameters of various types of soil based on the material composition of the barrier dam and the experience of the site experts.

Combining the analysis from field survey and analysis from data, Tangjiashan barrier dam was formed by bank slope landslide, and its geological composition was mostly loose accumulation of soil and rocks, with strong material heterogeneity. The barrier dam materials were mostly silty loam, coarse particles, gravel, cobblestones and even boulders, with particle diameters ranging from less than 1 mm to several meters, which was wide particle size distribution. Compared with the man-made

Table 5.2 Physical and mechanical parameters of various types of soil in Tangjiashan barrier dam

Layer location	Density ρ (g/cm^3)		Internal friction angle φ ($''$)		Cohesion c (KPa)		Permeability coefficient K (cm/s)	Allowable gradient J
	Natural	Saturated water	Natural	Saturated water	Natural	Saturated water		
Gray-black muddy fine sand	1.5–1.7	1.6–1.8	17	12	10	5	1×10^{-4}	0.10
Gravel soil	1.7–1.8	1.8–1.9	25	20	30	10	1×10^{-4}	0.20
Strongly weathered cataclasite rocks	2.0	2.1	33	30	30	10	1×10^{-2}	0.60
Weakly weathered cataclasite rocks	2.1	2.2	37	34	60	30	1×10^{-3}	0.80

earth-rock dam, the barrier dam was irregular in shape, with uneven crest and overtopping flow often occurred in the weak area of the pass. In terms of the structure of barrier dam, because there was no artificial filling process, the barrier dam structure was loose and messy, sometimes overhead phenomenon occurred, and the substructure was relatively dense. In terms of the materials of barrier dam, the particle size distribution was wide, ranging from extremely fine particles to boulders, with various thickness of particle diameters and more complicated material composition.

5.3 Scouring Experiment for Wide Graded Mixture

The composition of Tangjiashan barrier dam was quite complex, which is composed of wide graded mixture ranging from fine particles to solitary boulders. During the discharge process of barrier lake at June 2008, the experts at the emergency site were quite concerned about the scouring characteristics for such kind of natural, loose and wide graded mixture. How large the block stone would be scoured and moved away in the corresponding flow condition? What was the differences of the scouring characteristics between the wide graded mixture and the uniform material? What was the differences between the continuous and discontinuous wide graded mixture? Thus, the experiment for wide graded mixture was particularly arranged to investigate the scouring characteristics.

Different kinds of wide graded mixture were arranged in the glass flume to figure out the movement characteristics in the condition of different water depth and velocity. Based on the incipient motion velocity of uniform granular (Shamov Formula), the incipient motion velocity of mixture was particularly investigated to figure out the critical velocity coefficient. Compared to the critical velocity of uniform

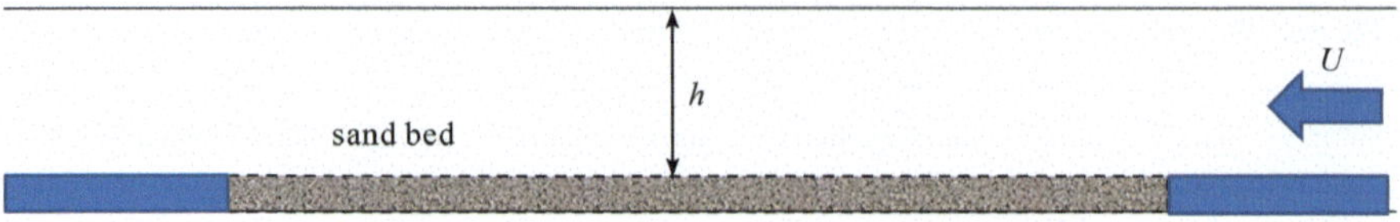

Fig. 5.2 Schematic diagram of flume arraignment for wide graded mixture

sands, the continuous and discontinuous wide graded mixture were comparatively studied to propose the practical expression of incipient motion velocity for wide graded mixture.

5.3.1 Experimental Design

5.3.1.1 Flume Arrangement

The whole experiment was arranged in a rectangular, flat glass flume. The geometrical size of the glass flume was 30 × 0.5 × 0.8 m (length × width × height). The horizontal moveable bed was 1.8 × 0.1 m (length × depth). The water level was determined by the point gauge arranged both at the upstream and the downstream. The average velocity was measured by the propeller current meter. The whole flume arrangement is shown in Fig. 5.2.

5.3.1.2 Sand Characteristics

Considering the wide graded mixture of the Tangjiashan barrier dam and the actual experimental flow conditions, the white alum sand (volumetric weight 2.8 kN/m^3, grain size 0.1–20 mm) was adopted to build the barrier dam. There successively arranged six kinds of sands, two kinds of uniform sands with different grain sizes, two kinds of continuous wide graded non-uniform sands and two kinds discontinuous wide graded mixture. The experimental mixtures were prepared in proportion with various grain size white alum sands. Six sand patterns are shown in Fig. 5.3, and the detail grain size for mixture is given in Table 5.3. Both the 1$^{\#}$ and 2$^{\#}$ were uniform sands with average grain size to be 3.5 and 10 mm. And the 3$^{\#}$, 4$^{\#}$, 5$^{\#}$ and 6$^{\#}$ were all wide graded mixture.

The non-uniformity coefficient C_u and the curvature coefficient C_c are defined as

$$C_u = \frac{d_{60}}{d_{10}} C_c = \frac{(d_{30})^2}{d_{60}d_{10}}.$$

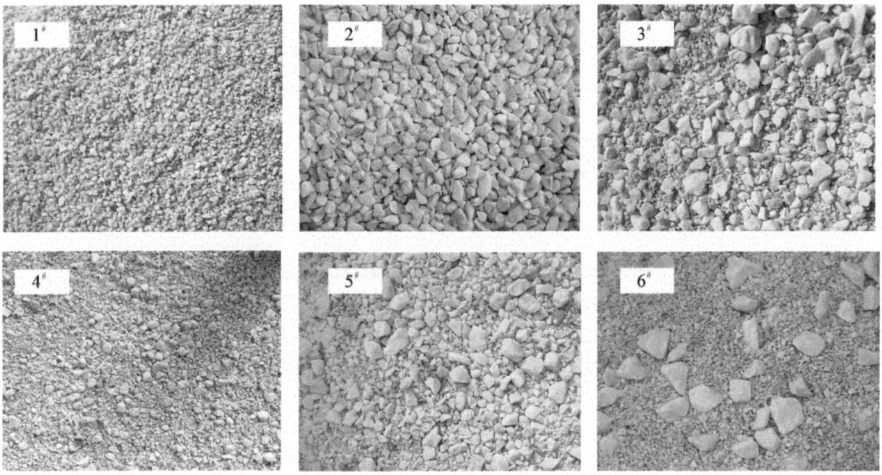

Fig. 5.3 Diagram of six kinds of mixture

Table 5.3 Mixture characteristics

No.	1#	2#	3#	4#	5#	6#
Pattern	Uniform sand	Uniform sand	Continuous non-uniform sand	Continuous non-uniform sand	Discontinuous non-uniform sand	Discontinuous non-uniform sand
Size (mm)	2–5	8–11	2–20	0.1–20	0.1–20, 6–9 missing	0.1–20, 5–12 missing
d_{50} (mm)	3.5	10	8.7	7.2	5.8	3.7
d_{60} (mm)	–	–	10.0	8.3	7.2	10.8
d_{30} (mm)	–	–	5.0	4.2	3.0	1.6
d_{10} (mm)	–	–	1.4	1.5	1.3	1.1
C_u	–	–	7.14	5.53	5.54	9.82
C_c	–	–	1.79	1.52	0.96	0.22

It could be found from Table 5.3 that the non-uniformity coefficient C_u for $3^{\#}$–$6^{\#}$ was all higher than 5 and belonged to the non-uniform mixture. And $3^{\#}$ and $4^{\#}$ were well-graded mixture with the curvature coefficient C_c higher than 1. $5^{\#}$ and $6^{\#}$ were the discontinuous wide graded mixture with the curvature coefficient C_c lower than 1 and missed some certain grain size.

5.3.1.3 Experimental Condition

In general, there prepared 50 kinds of experiment conditions with water depth to be 8–40 cm, velocity to be 40–120 cm/s and flow rate to be 40–120 L/s. The $1^{\#}$–$6^{\#}$

sands were successively laid as bed sand to figure out the corresponding movement characteristics in the condition of different water depth and flow rate.

5.3.2 *Criterion of Incipient Motion*

In view of the grain size, shape and location, the sands would be randomly moved away be the energetic flow, which caused uncertainty to the incipient motion criterion. At present, there widely used a qualitative concretion and a small amount of sediment movement would be set as the incipient motion criterion. Referring to the movement intensity at critical condition by H. Kramer, the detailed description about weak movement, medium movement and the general movement was as follows:

(1) Weak movement

Occasionally individual particles would intermittently move in local areas of the sand bed but would then gradually stop after rolling forward for a short distance. Such discontinuous movement would only make one or two sand pass through the downstream section during the long observation.

(2) Medium movement

There were a small amount of sands intermittently rolling above the bed and some tiny or flat sands were even moved to the downstream by the energetic flow. There presented no great change or deformation above the bed, and there were some particles intermittently passing through the downstream section.

(3) General movement

It presented obvious displacement movement around the bed, and all kinds of particles would be moved toward the downstream by the energetic flow. It could even observe some deformation above the bed within short time, and there were some sands continuously passing through the downstream section.

In addition to the sand movement characteristics, the subsequent development trend and bed morphology deformation were also particularly observed. The results showed that the sand bed for $1^{\#}$ and $2^{\#}$ would gradually stabilize after medium movement and the movement intensity would revert to weak movement, or even unmoved. But once the movement intensity increased to the general movement, the whole bed would be moved away layer by layer by the energetic flow, which presented flat surface at the end.

Once the $3^{\#}$ and $4^{\#}$ continuous wide graded mixture moved, it would form a shallow groove above the bed and all kinds of sands near the groove would easily move away by the energetic flow, which would longitudinally extend the groove. In such circulation, it would soon form a long strip gully above the bed with more and more drastic movement intensity. Consequently, the medium movement would transform into the general movement with great bed morphology deformation.

In view of some missing medium grain size particles at $5^{\#}$ and $6^{\#}$, the larger particles would still stabilize and protect the fine particles, making the fine particles rotate and move with little displacement movement near the large particles. Only the velocity increasing to some certain, the fine particles would seriously rotate and run away the nearby large particles. Consequently, as the fine particles moved away by the energetic flow, the large particles would then be exposed above the bed. Only the surrounding fine particles hollowed out, the exposed large particles could then start to move and leave a pit at the original location. Thus, it would present some pits at somewhere of the bed.

5.3.3 Test Results and Analysis

During the experiment, the movement characteristics of sand bed particles under different flow velocity and water depths were observed for the bed sand of each mixture; the velocity distribution was measured. It was observed in the experiment that given a fixed water depth, all the particles on the bed surface were in a static state with small flow rate, a few particles could be seen to start moving on the bed surface with increase of the flow rate, and the particles move at large scale, which caused sharp change of the bed surface and fast sediment discharge, with constant increase of the flow rate. The state in which a small amount of sand particles move is called incipient movement. Under incipient movement, the average flow velocity of the cross section is called incipient motion velocity.

For the wide graded mixture, it was observed that the small- and medium-diameter particles embedded in the large particles made incipient motion at first which then drove the larger particles to move. After all the small particles slid and rolled away, the particles with large particle diameter moved due to instability. At this time, the particles with various diameters of the mixture made incipient motion. This shows that the large particles in the mixture were easier to move than the homogeneous particles. This was due to the mixing of large particles with small- and middle-sized particles that increased the smoothness and reduced the roughness of bed surface relatively. It also showed that the smaller the roughness, the easier the bed particles move. Whether it was homogeneous or mixed, the initiation of particle motion was related to the fluctuation of water velocity of the bottom, the position of particles on the bed surface and other factors. Such factors were random, so it was necessary to repeat the test for necessary times.

Critical water depth h_c and critical velocity U_c corresponding to the weak and medium movement of sand beds of different mixtures based on statistical analysis are given in Tables 5.4 and 5.5.

It can be concluded from Table 5.4 that generally the critical velocity increases as the critical water depth becomes larger for each mixture. For different mixtures, under a fixed flow velocity, the critical water depth and critical flow velocity corresponding to weak and medium movement of sand bed varied in case of different average

Table 5.4 Critical water depth and critical velocity corresponding to weak movement of mixture

Mixture number	Median diameter d_{50} (m)	Flow Q (m^3/s)	h_{c1} (m) Weak movement critical depth	U_{c1} (m/s) Weak movement critical velocity	c_1 Weak movement velocity coefficient
1#	0.0035	0.0203	0.079	0.512	1.23
	0.0035	0.0413	0.140	0.588	1.28
	0.0035	0.0503	0.170	0.589	1.24
	0.0035	0.0622	0.201	0.616	1.26
	0.0035	0.0810	0.250	0.645	1.28
2#	0.0100	0.0645	0.151	0.851	1.29
	0.0100	0.0813	0.190	0.852	1.24
	0.0100	0.1003	0.217	0.921	1.31
	0.0100	0.1210	0.250	0.964	1.34
3#	0.0087	0.0650	0.160	0.809	1.27
	0.0087	0.0804	0.207	0.774	1.16
	0.0087	0.1010	0.227	0.886	1.31
	0.0087	0.1219	0.277	0.877	1.26
4#	0.0072	0.0652	0.180	0.722	1.18
	0.0072	0.0805	0.220	0.729	1.16
	0.0072	0.1030	0.249	0.824	1.28
	0.0072	0.1192	0.290	0.819	1.24
5#	0.0058	0.0658	0.188	0.697	1.22
	0.0058	0.0811	0.220	0.734	1.25
	0.0058	0.1012	0.260	0.775	1.24
	0.0058	0.1213	0.300	0.805	1.24
6#	0.0037	0.0402	0.149	0.537	1.14
	0.0037	0.0654	0.240	0.543	1.06
	0.0037	0.0811	0.291	0.555	1.05
	0.0037	0.1012	0.388	0.520	0.94

particle diameter, non-homogeneous coefficient and curvature coefficient grading characteristics of the mixture.

For the critical velocity of homogeneous sediment, predecessors proposed a series of empirical formulas on the basis of research. Among them, Shamov Formula, simple in form and widely used, aimed at homogeneous sand incipient movement of granular materials:

$$U_c = c\sqrt{\frac{\rho_s - \rho}{\rho} gd(\frac{h}{d})^{1/6}},$$

Table 5.5 Critical water depth and critical velocity corresponding to the movement in the mixture

Mixture number	Median diameter d_{50} (m)	Flow Q (m^3/s)	h_{c2} (m) Medium movement critical depth	U_{c2} (m/s) Medium movement critical velocity	c_2 Medium movement velocity coefficient
1#	0.0035	0.0203	0.070	0.578	1.41
	0.0035	0.0413	0.126	0.653	1.45
	0.0035	0.0503	0.152	0.659	1.42
	0.0035	0.0622	0.176	0.704	1.47
	0.0035	0.0810	0.237	0.681	1.36
2#	0.0100	0.0645	0.135	0.952	1.47
	0.0100	0.0813	0.164	0.988	1.48
	0.0100	0.1003	0.178	1.122	1.65
	0.0100	0.1210	0.216	1.116	1.59
3#	0.0087	0.0650	0.146	0.887	1.41
	0.0087	0.0804	0.163	0.983	1.54
	0.0087	0.1010	0.206	0.977	1.47
	0.0087	0.1219	0.242	1.003	1.47
4#	0.0072	0.0652	0.163	0.797	1.33
	0.0072	0.0805	0.200	0.802	1.29
	0.0072	0.1030	0.225	0.912	1.44
	0.0072	0.1192	0.262	0.906	1.40
5#	0.0058	0.0658	0.161	0.814	1.36
	0.0058	0.0811	0.190	0.850	1.39
	0.0058	0.1012	0.237	0.851	1.33
	0.0058	0.1213	0.246	0.982	1.34
6#	0.0037	0.0402	0.126	0.636	1.38
	0.0037	0.0654	0.197	0.661	1.33
	0.0037	0.0811	0.246	0.657	1.28
	0.0037	0.1012	0.326	0.618	1.15

where

U_C is the critical flow rate corresponding to starting;
ρ_s and ρ specific gravity of sand and water;
g gravity acceleration;
d average particle diameter of sand;
h water depth;
c critical velocity coefficient, taking 1.14 for homogeneous sand.

Table 5.6 Critical velocity coefficient

Mixture number	1#	2#	3#	4#	5#	6#
c_1	1.26	1.30	1.25	1.22	1.24	1.05
c_2	1.41	1.47	1.41	1.33	1.36	1.38

Here, in a similar form, the corresponding critical flow velocity upon incipient motion of the mixture is expressed as

$$U_c = c(C_u, C_c)\sqrt{\frac{\rho_s - \rho}{\rho} g d_{50}} (\frac{h}{d_{50}})^{1/6}.$$

In the formula, d_{50} is used to characterize the median particle diameter, and the influence of different gradation distribution of the mixture is attributed to coefficient c, that is, the critical velocity coefficient c is regarded as a function of the non-uniformity coefficient C_u and curvature coefficient C_c of the wide gradation mixture.

According to the experiment results, the critical velocity coefficients corresponding to weak movement and medium movement of homogeneous sand and non-homogeneous mixture can be obtained by fitting. In the experiment, the critical velocity coefficients corresponding to weak movement and medium movement are recorded as c_1 and c_2, respectively, and the statistical average of critical velocity coefficients obtained from the analysis of test results is given in Table 5.6.

It can be seen from Table 5.6 that the critical velocity coefficient obtained through experiment was close to the result of Shamov Formula, slightly larger than 1.14. The gradation of 6# mixture was wide with large range of missing particle diameter. The measured critical velocity coefficient was 1.05, which indicates that it was relatively easy to move.

For the discharge scouring of Tangjiashan barrier lake, it was more practical to use the critical velocity coefficient c_2 of particle motion to characterize the incipient movement scouring characteristics of wide graded mixture. The motion critical flow velocity coefficients of various mixtures are summarized in Table 5.7.

As can be seen from Table 5.7, the median critical velocity coefficient of continuous wide graded mixture and discontinuous wide graded mixture was 1.37, which was smaller than that of homogeneous mixture (1.44), indicating that scouring of graded mixture was more likely to occur under the same water flow condition. The size of the materials comprising Tangjiashan barrier dam varied considerably, ranging

Table 5.7 Motion critical velocity coefficient in mixture

Mixture characteristics	Homogeneous material	Continuous wide graded mixture	Discontinuous wide graded mixture
Critical velocity coefficient c_2 of particle motion	1.44	1.37	1.37

Table 5.8 Critical motion velocity in wide graded mixture

The median particle diameter d (mm)	$h = 1$ m	$h = 2$ m	$h = 4$ m
20	1.5	1.7	1.9
50	2.0	2.3	2.6
100	2.6	2.9	3.2
200	3.2	3.6	4.0
400	4.1	4.6	5.1

from fine silty loam to boulders with a diameter of up to 8 m. It was a wide graded mixture the same as that of 3#–6# mixture. It was more practical to use the "medium movement" in the experiment to start the scour particles discharged from the barrier. In view of this, according to the characteristics of the wide graded mixture in Tangjiashan barrier dam and its discharge and scouring based on the Shamov Formula for the incipient movement velocity of granular sediment, a practical expression for the incipient velocity of the wide graded mixture was put forward as below.

$$U_c = 1.37\sqrt{\frac{\rho_s - \rho}{\rho} g d_{50}} (\frac{h}{d_{50}})^{1/6},$$

where c—the critical motion velocity of the mixture, which was set as 1.37.

Therefore, the critical motion velocity of wide graded mixture under different water depths and particle diameters was listed in Table 5.8. The results showed that, under water depth h = 1–4 m, the corresponding median critical velocity was 1.5–1.9 m/s for the wide graded mixture with median particle diameter of 20 mm, 3.2–4.0 m/s for the wide graded mixture with median particle diameter of 200 mm and 4.1–5.1 m/s for the wide graded mixture with median particle diameter of 200 mm.

The experiment results were in good compliance with the field observation of the Tangjiashan barrier lake. The results and the practical expression of critical motion velocity of wide graded mixture can provide reference for scouring process in similar barrier dam in the future.

It can be seen that for the continuous wide graded non-homogeneous sand (3#, 4#), the critical coefficients c_1 and c_2 of incipient movement and scouring were obviously larger than those of homogeneous sand, which indicated that the continuous wide graded non-homogeneous sand in the test was more difficult to make incipient movement than the homogeneous sand of the same diameter. The c_2/c_1 value of these two kinds of non-homogeneous sand was smaller than that of homogeneous sand, which indicated that once the continuous and wide graded non-homogeneous sand begins to move, the sand movement is easy to intensify, and it is easier to form scouring and lead to the change of bed surface morphology. This had also been verified by the phenomenon observed in the experiment. The situation of discontinuous non-homogeneous sand (5#, 6#) was quite different. For the 5# mixture with short particle diameter-lacking section and high content of large particles, the critical coefficients c_l and c_2 corresponding to bed sediment incipient motion and scouring were obviously larger than those corresponding to homogeneous sediment. However, the coefficient of $6^{\#}$ mixture with longer particle diameter-lacking and high content of small particles was smaller than that of homogeneous sand. This is because that the incipient movement standard of $3^{\#}$–$5^{\#}$ non-homogeneous sand was based on the movement of medium-diameter particles. However, the $6^{\#}$ mixture lacked medium-diameter particles, while large-diameter particles were difficult to start. As a result, the movement standard of the $6^{\#}$ mixture was based on the movement of small particles, which led to the small critical velocity coefficient. In addition, c_2/c_1 values of $5^{\#}$ and $6^{\#}$ discontinuous wide graded non-homogeneous sand were larger than those of $1^{\#}$ and $2^{\#}$ homogeneous sand and $3^{\#}$ and $4^{\#}$ continuous wide graded non-homogeneous sand. This showed that the sand bed with discontinuous and non-homogeneous sand need more increase of flow velocity for the development form weak movement to medium movement. Under such circumstances, the scouring is difficult to form. This was because that the hidden blocking effect of large particles in the non-homogeneous sand in the particle diameter-lacking section was more obvious. Therefore, small-diameter particles could vibrate, rotate and move in a short distance in a small range, but it was difficult to move beyond large particles and make large displacement with the incoming flow.

5.4 Experimental Study on Barrier Dam-Break of Wide Graded Mixture

Tangjiashan barrier dam was formed by the landslides of bank slope, and its geological composition was mostly loose accumulation of earth and rock. Compared with the artificial earth-rock dam, the barrier dam was irregular in shape and rough on the crest, with overtopping flow most likely to occur in the weak area of the dam pass. In terms of the dam structure, since there was no artificial filling process, the dam structure was loose and disorderly, partly overhead, and tight in the lower layer. As for the dam materials, the accumulated particles of the barrier dam were wider in diameter, ranging from extremely fine particles to solitary boulders, with mixed thickness and more complex material composition.

During the formulation and implementation of the emergency handling scheme on site from May to June 2008, the on-site experts were most worried about whether the irregular loose accumulation of earth and rock formed by the natural force of the earthquake would induce the dam-break? Whether it would be sudden dam-break or gradual dam-break? What would happen in the dam-break process? How much would the peak discharge of the dam-break be? Whether the downstream cities, traffic bridges and important infrastructures could withstand the break flood safely? Whether or not to excavate a discharge channel in the emergency handling and disaster relief project? And how about the design scheme of the discharge channel? Since there was limited knowledge about the dam-break process of the artificially filled homogeneous earth-rock dam, especially the loose accumulation of earth-rock suddenly formed by natural forces, the experimental study on the barrier dam-break of wide graded mixture was specially conducted.

Tangjiashan barrier lake was huge in scale, with the dam body of 803 m long along the river and 612 m long across the river, the dam height of 82–124 m, the volume of 20.37 million m^3, the upstream catchment area of 3550 km^2 and the maximum water storage capacity of 316 million m^3, which made indoor simulation difficult. In addition, the model similarity theory was still immature for the experimental study of dam-break model, which needed to meet not only the similarity in geometry and gravity, but also the similarity in dam materials. Since the dam-break process was affected by many similar problems, such as hydraulics, soil mechanics, sediment dynamics, soil erosion and structural mechanics, the similarity of dam materials could not be completely satisfied in the model design. Moreover, in terms of the test means of dam-break experiment, the test of the breach expansion process was not perfect. In view of this, the design of indoor dam-break experimental model could only grasp the main contradiction and relax some secondary constraints to simplify and generalize the model appropriately. The experimental results of exploring the regularity of the dam-break process could provide not only the basic data for the study of the techniques for handling Tangjiashan barrier lake, but also the reference for the dam-break process of similar barrier dams, so as to provide scientific basis for the measures of emergency handling and disaster relief project.

5.4.1 *Experimental Design for Barrier Dam-Break Model*

Upstream of the model, the storage capacity of a certain capacity was simulated, with the barrier dam-building height of 60–100 cm and the downstream dam slope of 1:2–1:3. The layout of indoor experimental water tank is shown in Fig. 5.4. The upstream reservoir was 25 m long, 6 m wide and 1.2 m high, with the maximum storage capacity of about 80 m^3. The indoor experimental simulation took into account such conditions as different dam section shapes, different dam-building materials, different inflow and whether or not to open a discharge channel on the dam crest, etc., and made a comparative study of the barrier dam-break mechanism and the breach expansion process.

5.4.1.1 Measurement System of the Experiment

The measurement system of the experiment adopted omnidirectional (upstream and downstream, inside and outside the barrier dam, multi-means) measurement technology, with the arrangement of measuring points and testing means as follows:

(1) Timely water level: Four measuring points were arranged upstream and downstream of the barrier dam, and the dynamic hydrographs of water level were recorded by sensors and the DASP acquisition system.

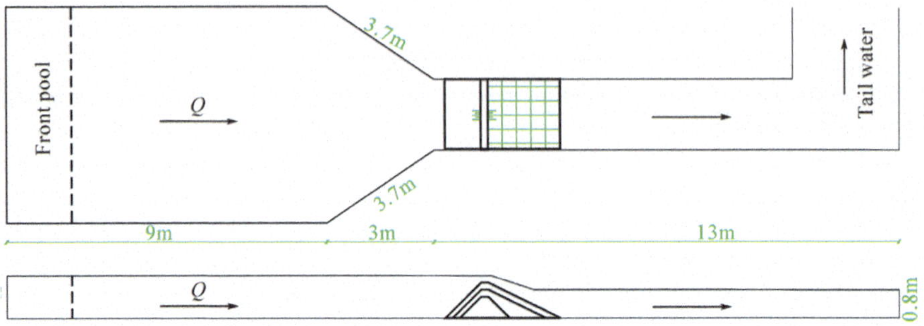

Fig. 5.4 Layout diagram of test water tank

(2) Discharge of breach: Calculated from the upstream stage-storage capacity curve.
(3) Flow velocity of breach: The measuring points were arranged at the breach of the discharge channel, and the data were dynamically measured by imported electromagnetic current meter.
(4) Shape of breach: Vertical scouring and down-cutting were recorded dynamically by embedded light erosion catchers, and horizontal expansion was recorded by grid technology, with 20cmx40cm grid drawn on the downstream slope of the barrier dam. Meanwhile, image analysis technology was used to record the breach process of in due course by two high-definition cameras arranged upstream and downstream of the barrier dam.

Working principle of erosion catchers: Before the experiment, several erosion catchers were embedded in the barrier dam, numbered according to the embedded elevation and section, and marked with different colors to make the distinction. The erosion catcher was a light ball with moderate volume and a diameter of 3–5 cm. It was easy to be carried out by flow and captured by high-speed camera with no effects on the erosion process of the surrounding barrier dam. During the experiment, once the dam erosion expanded to this position, the erosion catchers were carried out by flow and recorded and captured by high-speed camera, indicating that the instantaneous dam erosion had expanded to this position. By the analysis of the time when the erosion catchers embedded in different sections and elevations of the barrier dam were carried out by flow, the vertical and horizontal expansion of the barrier dam in the process of discharge could be obtained.

5.4.1.2 Material of Experimental Barrier Dam

On the basis of full collection of the data of material composition of Tangjiashan barrier dam, the wide graded mixture was selected as the dam-building material for the experimental simulation, with the particle diameter range of $d = 0.075$–40 mm, which could ensure the wide similarity of particle size distribution. The gradation curves were divided into continuous and discontinuous wide gradation, close to

Table 5.9 Gradation contents of 1# continuous wide graded non-cohesive earth-rock mixture

Particle diameter (mm)	20–40	10–20	5–10	2–5	1–2	0.5–1	0.25–0.5	0.1–0.25	Less than 0.1
Percentage content (%)	5.41	30.76	24.42	12.60	4.32	7.73	6.15	5.23	3.38

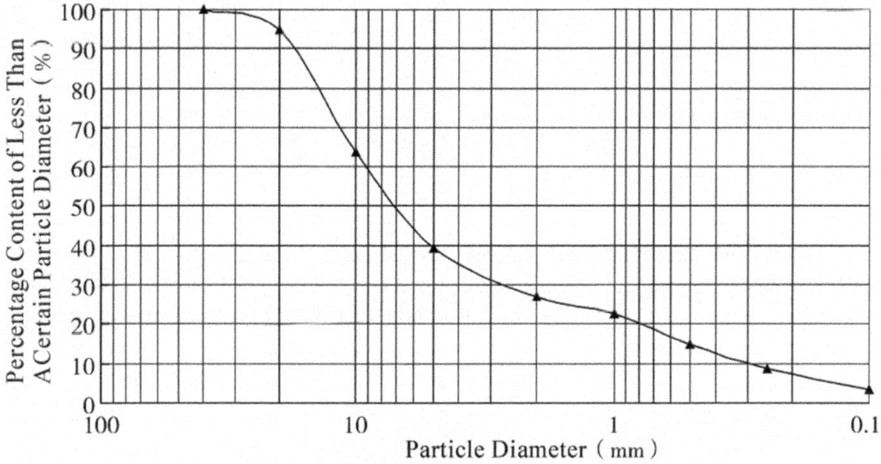

Fig. 5.5 Particle size distribution curve of 1# continuous wide graded earth-rock mixture

Table 5.10 Gradation contents of 2# discontinuous wide graded non-cohesive earth-rock mixture

Particle diameter (mm)	20–40	10–20	5–10	2–5	1–2	0.5–1	0.25–0.5	0.1–0.25	Less than 0.1
Percentage content (%)	8.93	23.86	1.65	15.68	8.09	14.07	12.52	10.71	4.49

the particle size distribution of Tangjiashan barrier dam. Of which, 1# dam-building material was continuous wide graded earth and rock, with the particle size distribution shown in Table 5.9 and Fig. 5.5. 2# dam-building material was discontinuous wide graded earth and rock, with the particle size distribution shown in Table 5.10 and Fig. 5.6.

5.4.1.3 Sections of Experimental Barrier Dam

Various dam sections were simulated, with the dam heights of 60, 80 and 100 cm, and the downstream dam slopes of 1:2 and 1: 3. The effects of the geometries of different

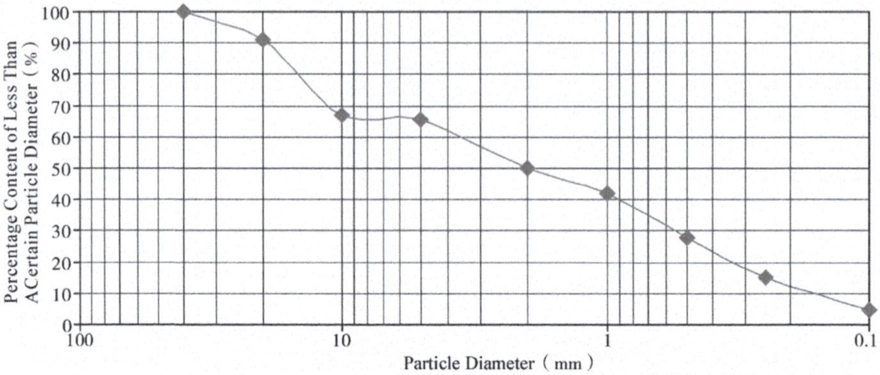

Fig. 5.6 Particle size distribution curve of 2# discontinuous wide graded earth-rock mixture

dam sections (including dam height and downstream dam slope) on the dam-break process, shape of breach and hydraulic elements of flood (water level, discharge and velocity) were studied. See Fig. 5.7 for the dam sections of the experiment.

During the dam-break, the upstream reservoir kept steady inflow. Three different scenarios ($Q = 10$ L/s, 30 L/s and 60 L/s) were experimented to study the effects of the upstream inflow conditions on the dam-break process, shape of breach and hydraulic elements of flood (water level, discharge and velocity).

In addition, the dam-break processes with or without discharge channel opened on the dam crest were studied, respectively (Fig. 5.8), aiming to compare the effects of the discharge channel on the dam-break process, shape of breach and hydraulic elements of flood (water level, discharge and velocity).

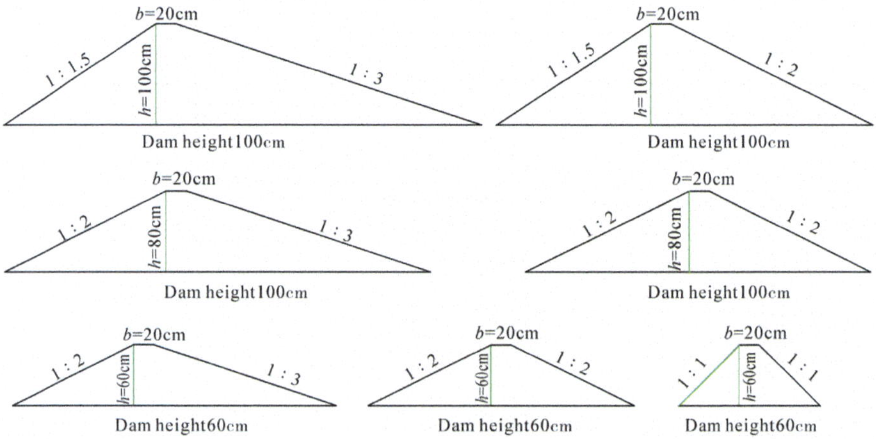

Fig. 5.7 Dam section forms of the experiment

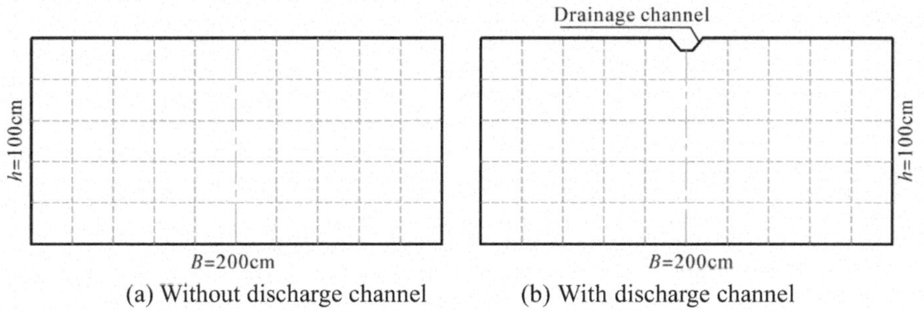

Fig. 5.8 Comparison of sections with or without discharge channel opened on dam crest

5.4.1.4 Conditions for the Experiment

The conditions for the experiment are given in Table 5.11. A total of 24 group of experiments were conducted, with the barrier dams for the dam-break scouring experiments under typical conditions shown in Fig. 5.9.

5.4.2 Experimental Study on Dam-Break Mode

The dam-break mode is the key and core of dam-break study, which determines the dam-break duration, the maximum dam-break peak discharge and the damage degree to the downstream. It is closely related to dam type and height and generally divided into sudden complete dam-break (also known as instantaneous dam-break) and gradual dam-break. Furthermore, the gradual dam-break is divided into gradual complete dam-break and gradual partial dam-break, which is divided into vertical partial dam-break and horizontal partial dam-break. The sudden complete dam-break usually occurs in gravity dam or arch dam, whereas the gradual dam-break generally occurs in earth-rock dam. Even so, the dam-break mode of the earth-rock dam is obviously different due to different conditions such as the overtopping discharge mode, dam-building materials, particle size distribution and compactness of materials.

According to the analysis of dam-break process and hydraulic elements, under the experimental conditions, the gradual dam-break could be roughly divided into three modes, namely homogeneous scouring layer by layer, full overtopping scouring and headcut scouring. Three conditions with the same section and inflow were selected for analysis. See Fig. 5.10 for the typical dam longitudinal section, with the dam height of 100 cm, the width of dam crest of 20 cm, the width of dam bottom of 470 cm, the upstream dam slope of 1:1.5, and the downstream dam slope of 1:3. During the experiment, the upstream reservoir had steady inflow of 30L/s.

Table 5.11 Conditions for each group experiment

Group	Dam material	Dam section				Discharge channel		Inflow (L/s)
		Dam height (cm)	Width of dam crest (cm)	Width of dam bottom (cm)	Downstream slope	Bottom width (cm)	Depth (cm)	
1	–	60	20	230	1:2	6	8	30
2	–	80	20	300	1:2	3	8	60
3	–	80	20	420	1:3	6	4	30
4	–	80	20	340	1:2	6	4	30
5	1#	80	20	380	1:3	6	4	30
6	1#	80	20	420	1:3	6	5	30
7	1#	80	20	420	1:3	6	8	60
8	1#	80	20	420	1:3	6	8	23
9	1#	80	20	420	1:3	6	8	12
10	1#	80	20	420	1:3	0	0	30
11	1#	100	20	420	1:3	6	8	31
12	1#	100	20	420	1:3	6	8	31
13	1#	100	20	470	1:3	6	8	62
14	1#	100	20	470	1:3	0	0	31
15	1#	60	20	290	1:3	6	8	30
16	1#	60	20	290	1:3	0	0	31
17	2#	100	20	470	1:3	6	8	30
18	2#	100	20	370	1:2	6	8	30
19	2#	100	20	380	1:2	0	0	30
20	2#	100	20	380	1:2	0	0	30
21	2#	80	20	380	1:3	0	0	30
22	2#	80	20	380	1:3	6	4	30
23	2#	80	20	380	1:3	6	4	30
24	2#	80	20	220	1:1	6	4	30

5.4.2.1 Homogeneous Scouring Layer by Layer

This mode was represented by condition 12. The dam-building material was 1# earth-rock mixture, and a discharge channel with trapezoidal cross section was opened on the dam crest, with the bottom width of 6 cm and the depth of 8 cm. The total observation recording time t_0 of the experiment was 420 s.

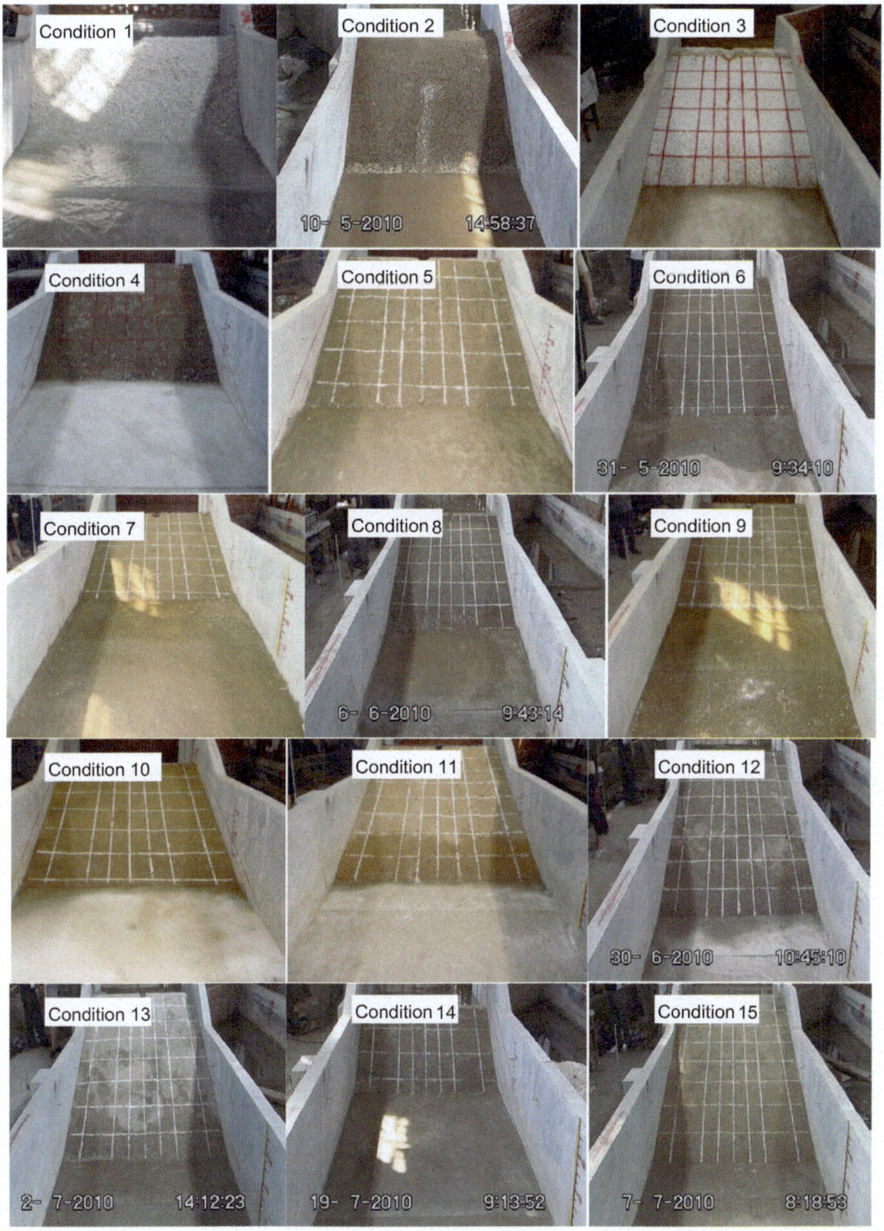

Fig. 5.9 Barrier dams of dam-break scouring experiment under typical conditions

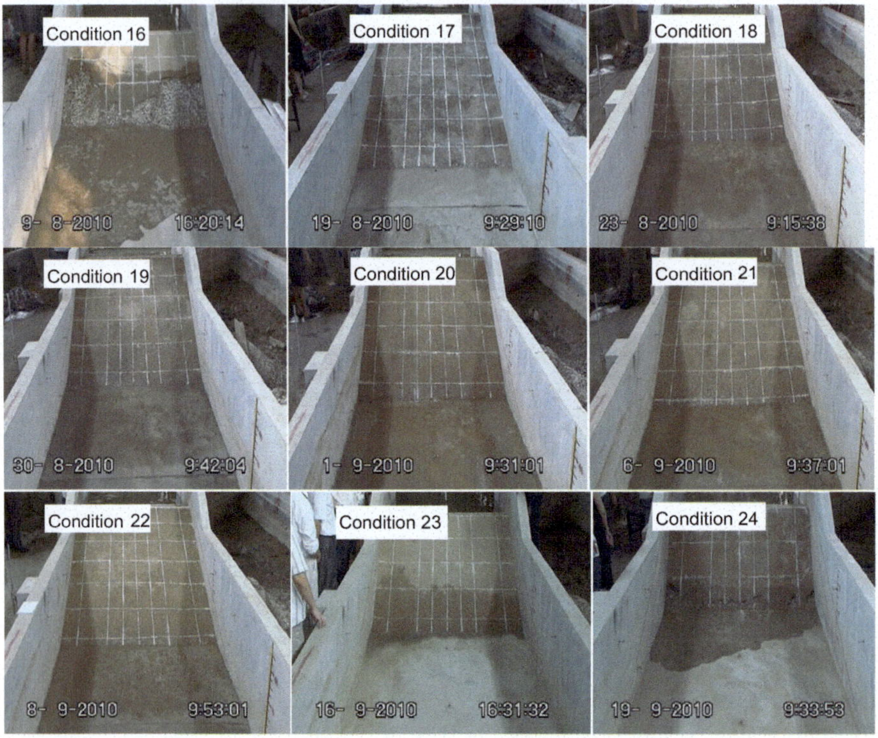

Fig. 5.9 (continued)

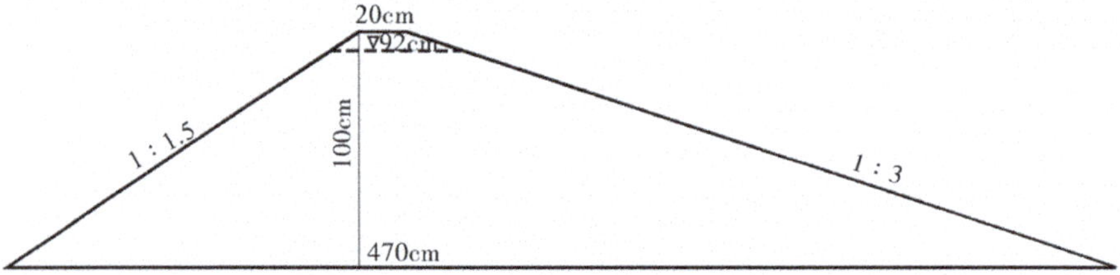

Fig. 5.10 Schematic diagram of typical dam longitudinal section

(1) Change process of stage-discharge

Under the condition of steady inflow to the reservoir of 30 L/s, the change of the upstream water level of the barrier dam could be roughly divided into four stages: ① slow rise ($t_0 = 0$–156 s) → ② slow fall ($t_0 = 156$–180 s) → ③ rapid fall ($t_0 = 180$–270 s) → ④ slow fall ($t_0 = 270$–420 s).

The discharge of the barrier dam could be roughly divided into five stages: ① no discharge (0–70 s) → ② slow rise (70–160 s) → ③ rapid rise (160–240 s) → ④ rapid fall (240–300 s) → ⑤ slow fall (300–420 s).

At the initial observation recording time ($t_0 = 0$ s), the water level of the upstream reservoir was 91.3 cm, lower than the dam crest, the elevation of the bottom of the discharge channel was 92 cm, with no discharge occurred, and all the upstream water inflow was stored in the reservoir, resulting in the increase of the reservoir capacity and the slow rise of the water level of the reservoir.

When $t_0 = 70$ s, the discharge channel began to discharge, but the discharge was little, less than the upstream inflow of 30 L/s. On the one hand, the upstream inflow overflowed from the discharge channel; on the other hand, it was stored in the reservoir, causing the water level of the reservoir to rise continuously.

When $t_0 = 156$ s, the instantaneous discharge of breach just increased to the upstream inflow of 30 L/s, and the upstream water level reached the peak of 97.7 cm. Then the discharge of breach on the dam crest gradually enlarged, causing the water level of the upstream reservoir slowly dropped.

When $t_0 = 180$ s, the water level of the reservoir dropped to 95.5 cm, and then the discharge of breach increased rapidly.

When $t_0 = 240$ s, the maximum discharge reached 459 L/s, indicating that the water level of the reservoir dropped the most sharply at the moment. Then the downstream water depth reached the maximum value, and the water level of the reservoir dropped rapidly.

When $t_0 = 270$ s, the upstream water level dropped to 51.3 m. Then the water level of the reservoir dropped slowly.

When $t_0 = 420$ s, the observation recording ended, the water level of the reservoir dropped to 35.6 cm, and the dam discharge was 41 L/s, slightly larger than the upstream inflow, indicating that the dam scouring gradually became stable.

See Fig. 5.11 for the hydrographs of stage-discharge.

(2) Change Process of Flow Velocity of Breach

As shown in Fig. 5.12, the change of the flow velocity of breach during the observation recording period (420 s) could be roughly divided into four stages: ① no discharge

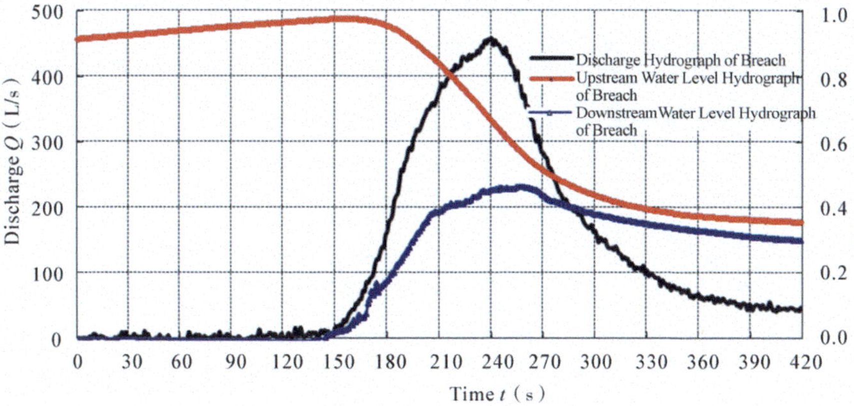

Fig. 5.11 Hydrographs of stage-discharge of breach under condition 12

(0–70 s) → ② gradual rise (70–218 s) → ③ gradual decrease (218–300 s) → ④ slow decrease (300–420 s). At the initial recording time ($t_0 = 0$ s), the water level was low with no overflow on the dam crest, so the velocity $V = 0$ m/s. When $t_0 = 70$ s, the dam crest began to discharge, but the flow was gentle and the velocity was slow. Afterward, the velocity rose gradually and reached the maximum value $V_{max} = 197$ cm/s when $t_0 = 218$ s, indicating the instantaneous water level up to the maximum. Then, the velocity of breach rapidly decreased to $v = 96$ cm/s when $t_0 = 300$ s. After that, the velocity slowly decreased and basically tended to a stable value $V = 76$ cm/s when the observation recording ended at $t_0 = 420$ s.

(3) Scouring Process of Barrier Dam

In order to dynamically monitor the scouring process of barrier dam, seven light erosion catchers were embedded in the barrier dam before the experiment (Fig. 5.13). According to the embedded elevation and section, they were numbered $1^\#$, $2^\#$, 3#, $4^\#$, $5^\#$, $6^\#$ and $7^\#$ in turn. During the experiment, once the dam scouring expanded to this position, the light erosion catchers were carried out by flow, indicating that the instantaneous dam scouring had expanded to this position.

The experiment under this condition showed that the sequence of the embedded light erosion catchers being carried out by flow was $1^\# \rightarrow 5^\# \rightarrow 6^\# \rightarrow 2^\# \rightarrow 7^\# \rightarrow 3^\#$

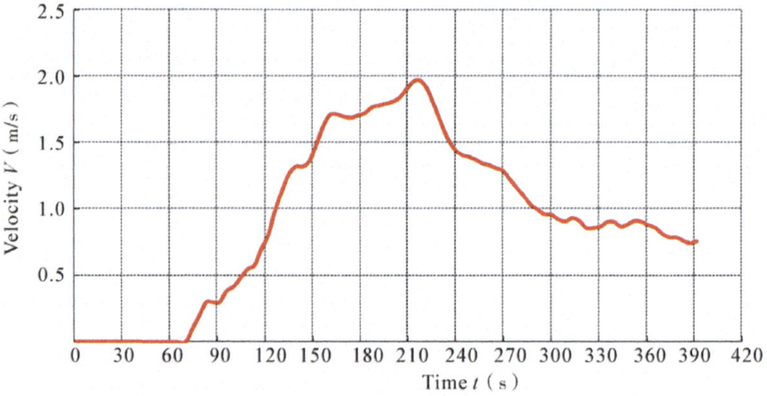

Fig. 5.12 Hydrograph of flow velocity of breach under condition 12

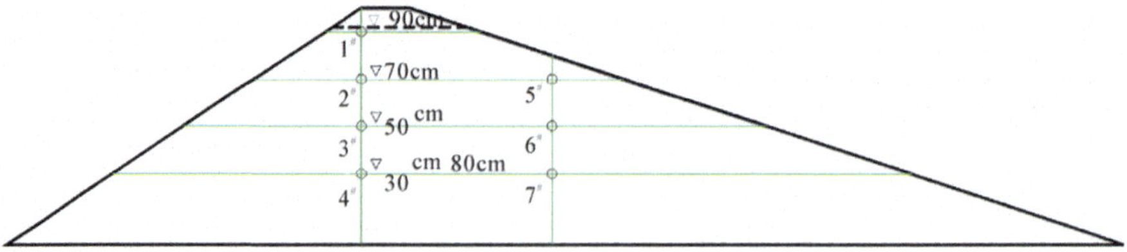

Fig. 5.13 Schematic diagram of embedded positions of dam erosion catchers under condition 12

Table 5.12 Time of embedded light erosion catchers being carried out

Catcher number	Embedded elevation (cm)	Embedded section	Capture time t_0 (s)
1#	90	Upstream barrier dam	136
5#	70	Downstream dam slope	143
6#	50	Downstream dam slope	172
2#	70	Upstream barrier dam	180
7#	30	Downstream dam slope	200
3#	50	Upstream barrier dam	210
4#	30	Upstream barrier dam	243

→ 4#. The dynamic capture moment was the observation recording time $t_0 = 136$ s, 243 s, 172 s, 180 s, 200 s, 210 s and 243 s in turn (Table 5.12).

According to the sequence of field test camera shooting and erosion catchers being carried out, the dam scouring process under condition 12 was drawn as shown in Fig. 5.14, and the dam scouring area gradually expanded from I → II → III → IV. See Table 5.13 for the dam scouring process and the characteristic values of hydraulic parameters. In order to facilitate the comparison of each group experiment, the moment when the downstream slope shoulder of the discharge channel began to discharge was defined as the dam-break reference zero time ($t = 0$), which was 70 s behind the observation recording time t_0. See Fig. 5.15 for the dam scouring process under condition 12.

It was found that the dam scouring process was relatively homogeneous under this condition. The overtopping flow was discharged from the discharge channel on the barrier dam crest, initially as thin sheet small flow, which sped up along the downstream slope and gradually increased its scouring capacity. At first, it scoured the fine particles on the barrier dam surface, forming rill surface erosion and then obvious erosion gully. The sediment carried by the slope flow was deposited in a fan shape around the groove of the downstream slope, forming an alluvial fan. After that, the erosion gully scoured deep and expanded continuously, making the slope of

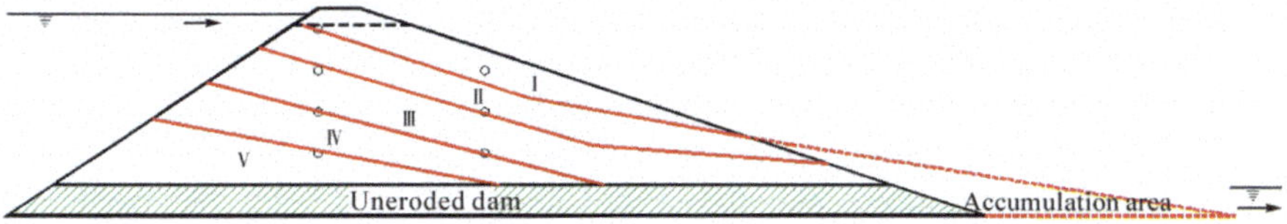

Fig. 5.14 Schematic diagram of dam scouring process under condition 12

Table 5.13 Dam scouring process and characteristic values of hydraulic parameters under condition 12

Recording time T (s)	Crest height of breach H (m)	Crest width of breach B (cm)	Upstream water level of breach (cm)	Downstream water depth of breach (cm)	Discharge of breach Q (L/s)	Flow velocity of breach V (cm/s)
0	92	16	94.4	0	0	2
30	91.5	16	95.9	0	2	42
60	91	20	96.9	0	7	114
90	85	50	97.6	4	44	172
120	66	180	92.7	25.6	267	178
150	48	200	78.4	41.3	412	197
180	28	200	60.6	45.7	432	138
210	27	200	48.2	41.3	220	110
360	26	200	35.6	29	41	75
Peak value	–	–	97.7 (87 s)	45.7 (181 s)	459 (170 s)	197 (147 s)

Note Based on the moment when the downstream slope shoulder of the barrier dam began to overflow, the reference zero time t ($t = 0$) was 70 s behind the initial observation recording time t_0

the downstream dam gradually become gentle. The scouring expanded to the main dam layer by layer, and the slope gradually inclined until the main barrier dam was almost completely scoured.

Based on the abovementioned change processes of hydraulic parameters of breach and scouring patterns, it could be found that:

(1) After the initial dam overtopping discharge, t = 90–180 s, the barrier dam scouring patterns changed the most violently. The vertical scouring of breach was downcut from the height of 85–28 cm, and the width of breach was extended from 50 to 200 cm. Both the velocity and peak discharge of breach were in this interval.
(2) When the barrier dam scouring time $t = 86$ s, the upstream water level of breach reached peak value $Z_{max} = 97.7$ cm. When the barrier dam scouring time t = 147 s, the velocity of breach reached peak value $V_{max} = 197$ cm/s. When the barrier dam scouring time $t = 170$ s, the discharge of breach reached peak value $Q_{max} = 459$ L/s. The water level of the upstream reservoir reached the highest value at first, and after a period of time, the velocity and discharge of breach reached the peak value one after another.
(3) The upstream water level of breach reached the highest value, indicating that the instantaneous discharge of breach just increased to the upstream inflow of 60 L/s. The upstream water level finally tended to a stable value, indicating that the barrier dam scouring expansion of breach was over, the breach was stable, and the upstream inflow was equal to the discharge of breach, i.e., "all the inflow is discharged". The discharge of breach reached the peak value, indicating that

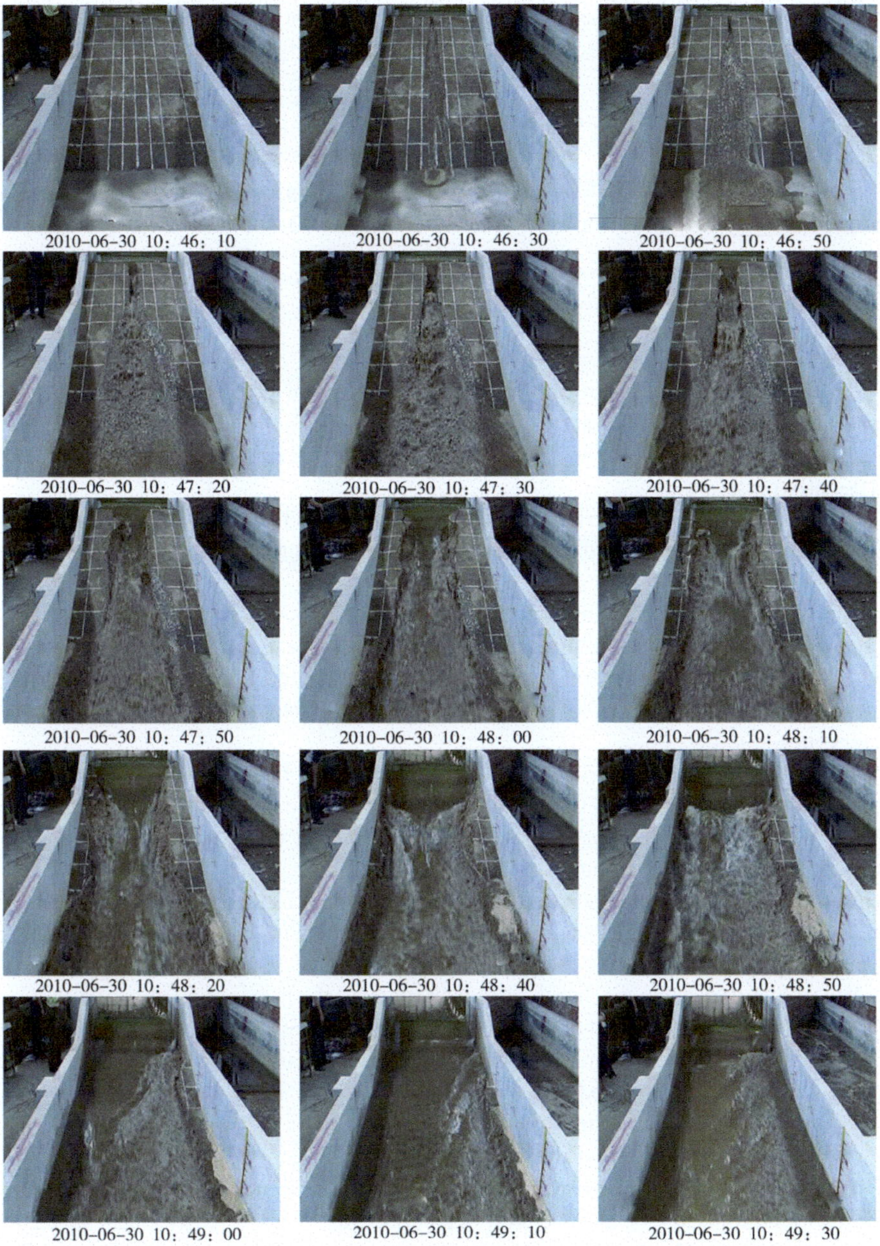

Fig. 5.15 Process diagram of dam scouring and break under condition 12

the water level of the reservoir dropped most sharply at the moment, and later on, the downstream water depth reached the maximum value.

5.4.2.2 Full Overtopping Scouring

This mode was represented by condition 14. The dam-building material was 1# earth-rock mixture, and compared with the condition 12, the only difference under this condition was that no discharge channel was opened on the dam crest. The total experimental observation recording time t_0 was 360 s.

(1) Change Process of Stage-Discharge

Under the condition of steady inflow of the reservoir of 30 L/s, the change of the upstream water level of the barrier dam could be roughly divided into three stages: ① slow rise (t_0 = 0–100 s) → ② rapid fall (t_0 = 100–210 s) → ③ slow fall (t_0 = 210–360 s).

The discharge of the barrier dam could be roughly divided into four stages: ① slow rise (0–10 s) → ② rapid rise (10–165 s) → ③ rapid fall (165–240 s) → ④ slow fall (240–360 s).

At the initial observation recording time (t_0 = 0 s), the water level of the upstream reservoir was 97.4 cm, which did not reach the dam crest, with no discharge from the barrier dam. All the upstream water inflow was stored in the reservoir, resulting in the increase of the reservoir capacity and the slow rise of the water level of the reservoir.

When t_0 = 70 s, the barrier dam began to overflow, but the dam discharge was little, less than the upstream inflow of 30 L/s. On the one hand, the upstream inflow overflowed from the dam crest, and on the other hand, it was stored in the reservoir, causing the water level of the reservoir to rise continuously.

When t_0 = 100 s, the discharge of breach just increased to the upstream inflow of 30 L/s, and the upstream water level reached the peak of 102.1 cm. Then the dam overflow quickly increased.

When t_0 = 165 s, the discharge of breach reached the maximum value of 526 L/s, indicating that the water level of the reservoir dropped the most sharply at the moment. Subsequently, the downstream water depth reached the maximum value, and the water level of the reservoir dropped rapidly.

When t_0 = 210 s, the upstream water level dropped to 54.5 m, and then the water level of the reservoir dropped slowly.

When t_0 = 360 s, the observation recording ended, the water level of the reservoir dropped to 41.5 cm, and the dam discharge was 39L/s, slightly larger than the upstream inflow, indicating that the dam scouring gradually became stable.

See Fig. 5.16 for the hydrographs of stage-discharge.

(2) Change Process of Flow Velocity of Breach

As shown in Fig. 5.17, the changes of the flow velocity of breach in the observation recording period (360 s) could be roughly divided into four stages: ① no discharge

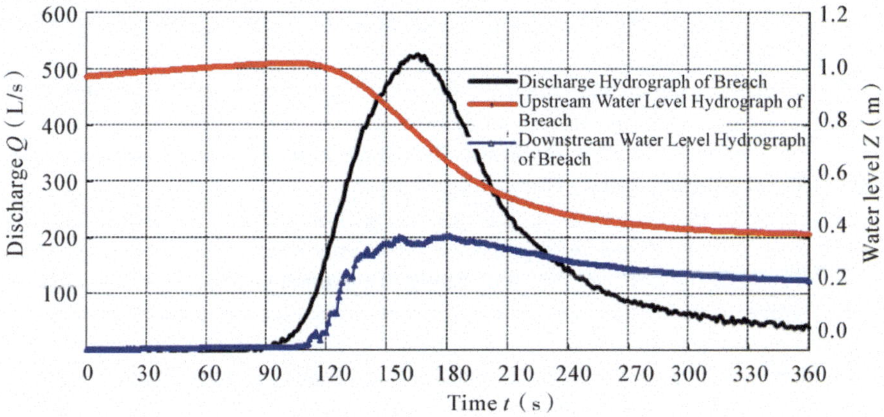

Fig. 5.16 Hydrographs of stage-discharge of breach under condition 14

(0–70 s) → ② rapid rise (70–160 s) → ③ rapid decline (160–270 s) → ④ slow decline (270–360 s).

When the dam crest began to discharge, the flow was gentle and the velocity was small. Afterward, the velocity gradually rose, reaching the maximum value $V_{max} = 220$ cm/s when $t_0 = 160$ s, indicating the instantaneous water level up to the maximum at this moment. Then the velocity decreased slowly and basically tended to a stable value of $V = 48$ cm/s when the observation recording ended at $t_0 = 360$ s.

(2) Scouring Process of Barrier Dam

In order to dynamically monitor the scouring process of barrier dam, seven light erosion catchers were embedded in the barrier dam before the experiment (Fig. 5.18).

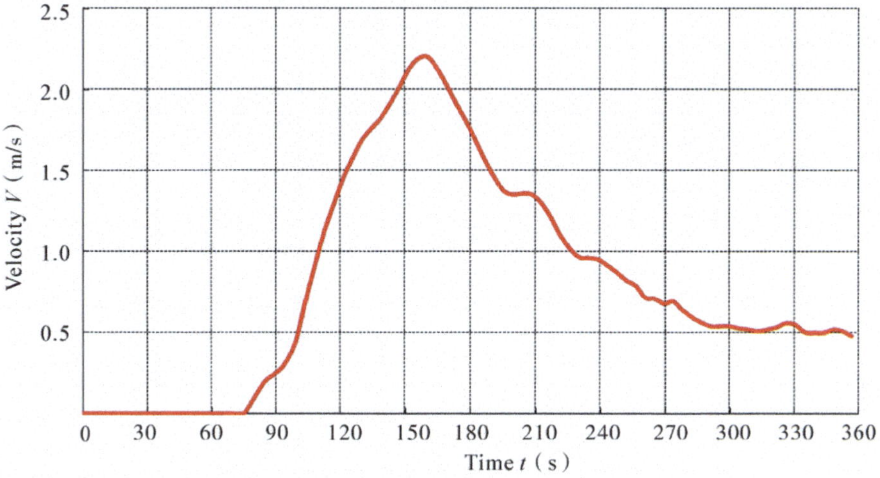

Fig. 5.17 Flow velocity of breach under condition 14

According to the embedded elevation and section, they were numbered 1#, 2#, 3#, 4#, 5#, 6# and 7# in turn.

The experiment under this condition showed that the sequence of the embedded light erosion catchers being carried out by flow was 1# → 5# → 2# → 6# → 3# → 7# → 4#. The dynamic capture moment was the observation recording time $t_0 = 110$ s, 130 s, 135 s, 154 s and 159 s in turn (Table 5.14).

According to the sequence of field test camera shooting and erosion catchers being carried out, the dam scouring process under condition 14 was drawn as shown in Fig. 5.19, and the dam erosion area gradually expanded from I → II → III → IV. See Table 5.15 for the dam scouring process and the characteristic values of hydraulic parameters. In order to facilitate the comparison of each group experiment, the moment when the downstream slope shoulder of the discharge channel began to overflow was defined as the dam-break reference zero time ($t = 0$), which was 70 s behind the observation recording time t_0. See Fig. 5.20 for the dam scouring process under condition 14.

It was found that the dam scouring process under this condition was as follows: The overtopping flow was discharged from the dam crest, initially as thin sheet small

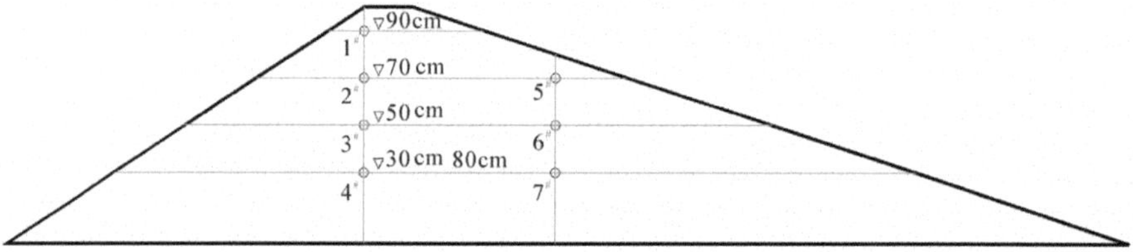

Fig. 5.18 Schematic diagram of embedded positions of dam erosion catchers under condition 14

Table 5.14 Time of embedded light erosion catchers being carried out under condition 14

Catcher number	Embedded elevation (cm)	Embedded section	Capture time t_0 (s)
1#	90	Upstream barrier dam	110
5#	70	Downstream dam slope	130
2#	70	Upstream barrier dam	135
6#	50	Downstream dam slope	154
3#	50	Upstream barrier dam	159
7#	30	Downstream dam slope	–
4#	30	Upstream barrier dam	–

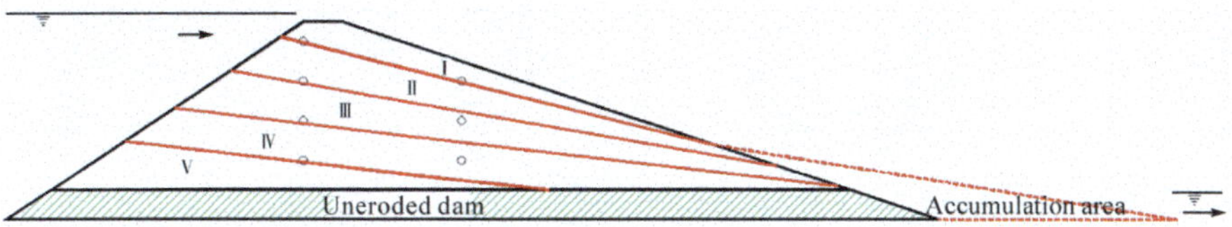

Fig. 5.19 Schematic diagram of dam scouring process under condition 14

Table 5.15 Dam scouring process and characteristic values of hydraulic parameters under condition 14

Recording time t (s)	Crest height of breach h (cm)	Crest width of breach B (cm)	Upstream water level of breach (cm)	Downstream water depth of breach (cm)	Discharge of breach Q (L/s)	Flow velocity of breach V (cm/s)
0	100	200	101	0	0	0
30	99	200	102.1	1	17	44
60	73	200	97.6	28	299	169
90	50	200	80.3	39	515	220
120	40	200	61.9	38.6	381	146
150	37	200	52.1	34.7	201	110
180	36	200	47	30.7	117	82
210	34	200	44.8	28.2	79	60
300	30	200	41.5	24.9	39	48
Peak value	–	–	102.1 (30 s)	40.6 (111 s)	526 (95 s)	207 (166 s)

Note Based on the moment when the downstream slope shoulder of the barrier dam began to overflow, the reference zero time t ($t = 0$) was 70 s behind the initial observation recording time t_0

flow, which sped up along the downstream slope and gradually increased its scouring capacity. At first, it scoured the fine particles on the dam surface, making the slope of the downstream dam gradually become gentle. The scouring expanded to the main dam layer by layer, and the slope gradually inclined until the main dam was almost completely scoured.

Based on the abovementioned change processes of hydraulic parameters and scouring patterns, it could be found that:

(1) After the initial dam overtopping discharge, t = 60–150 s, the barrier dam scouring patterns changed most violently. The vertical scouring of breach was downcut from the elevation of 73–37 cm. Both the velocity and peak value of breach were in this interval.
(2) When the dam scouring time $t = 30$ s, the upstream water level reached peak value $Z_{\max} = 102.1$ cm. When $t = 90$ s, the flow velocity of breach reached peak value $V_{\max} = 220$ cm/s. When $t = 95$ s, the discharge of breach reached peak value $Q_{\max} = 526$ L/s. The water level of the upstream reservoir reached

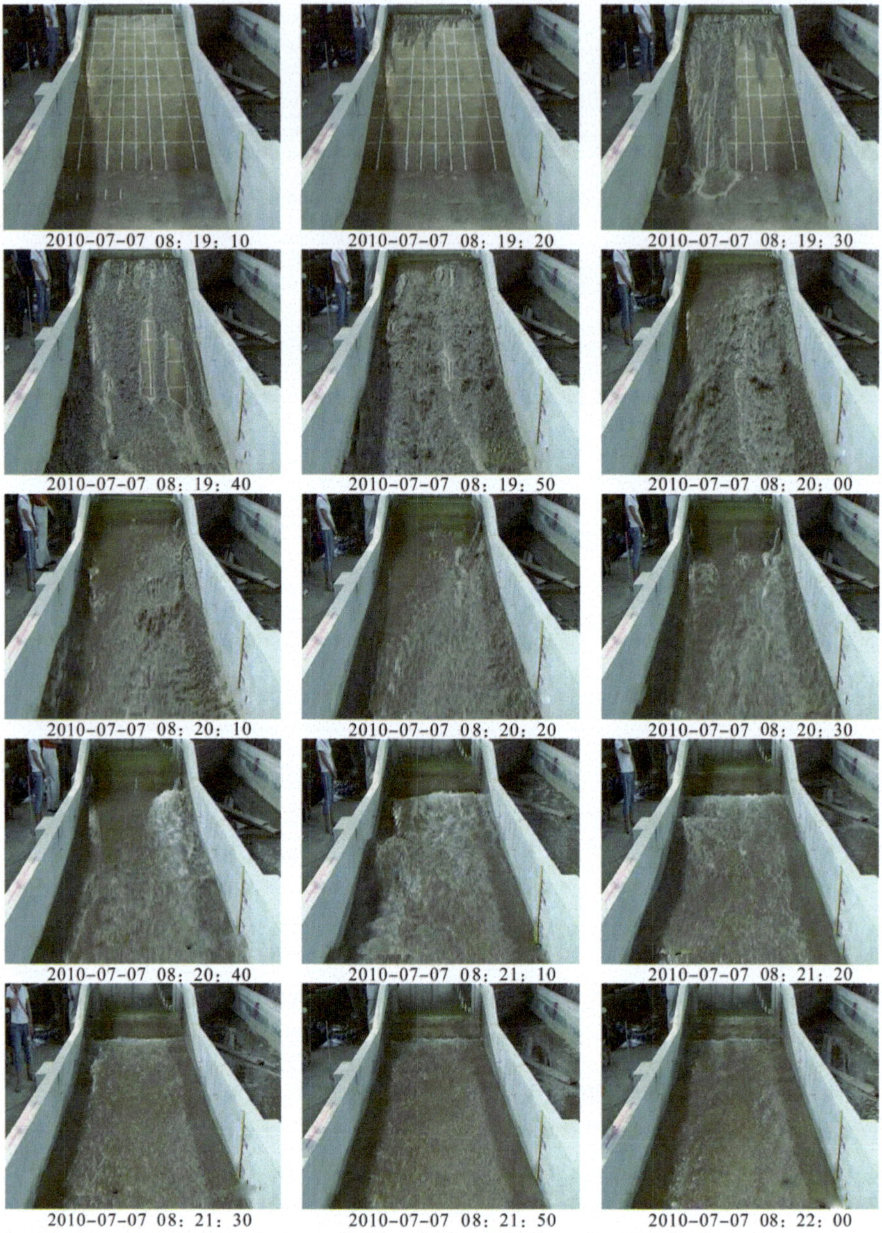

Fig. 5.20 Process diagram of dam erosion and break under condition 14

the highest value at first, and after a period of time, the velocity and discharge of breach reached the peak value one after another.

5.4.2.3 Headcut Scouring

This mode was represented by condition 17. The dam-building material was $1^{\#}$ earth-rock mixture, and a discharge channel with trapezoidal cross section was opened on the dam crest, with the bottom width of 6 cm and the depth of 8 cm. The total observation recording time (t_0) of the experiment was 480 s.

(1) Change Process of Stage-Discharge

Under the condition of steady inflow of the reservoir of 30L/s, the change of the upstream water level of the barrier dam could be roughly divided into four stages during the observation recording period: ① slow rise ($t_0 = 0$–150 s) → ② slow fall ($t_0 = 150$–250 s) → ③ rapid fall ($t_0 = 250$–330 s) → ④ slow fall ($t_0 = 330$–480 s).

The discharge of the barrier dam could be roughly divided into five stages: ① no discharge (0–50 s) → ② slow rise (50–250 s) → ③ rapid rise (250–290 s) → ④ rapid fall (290–30 s) → ⑤ slow fall (330–480 s).

At the initial observation recording time ($t_0 = 0$ s), the water level of the upstream reservoir was 91.7 cm, which did not reach the barrier dam crest, and the elevation of the bottom of the discharge channel was 92 cm, with no discharge of breach. All the upstream water inflow was stored in the reservoir, resulting in the increase of the reservoir capacity and the slow rise of the water level of the reservoir.

When $t_0 = 50$ s, the discharge channel began to discharge, but the discharge was little, less than the upstream inflow of 30 L/s. On the one hand, the upstream water inflow was discharged from the discharge channel, and on the other hand, it was stored in the reservoir, causing the water level of the reservoir to rise continuously.

When $t_0 = 154$ s, the instantaneous discharge of breach just increased to the upstream inflow of 30 L/s, and the upstream water level reached the peak of 97.7 cm. Then the discharge of breach gradually enlarged, and the water level of the upstream reservoir slowly dropped.

When $t_0 = 250$ s, the water level of the reservoir dropped to 89.5 cm. Afterward, the discharge of breach increased rapidly, up to the maximum discharge of 654 L/s at 289 s, indicating that the water level of the reservoir dropped the most sharply at this moment. Subsequently, the downstream water depth reached the maximum value, and the water level of the reservoir dropped rapidly.

When $t_0 = 330$ s, the upstream water level dropped to 43.2 cm, and then the water level of the reservoir dropped slowly.

When $t_0 = 480$ s, the observation recording ended, the water level of the reservoir dropped to 29.4 cm, and the dam discharge was 40 L/s, slightly larger than the upstream inflow, indicating that the dam scouring gradually became stable.

See Fig. 5.21 for the hydrographs of stage-discharge.

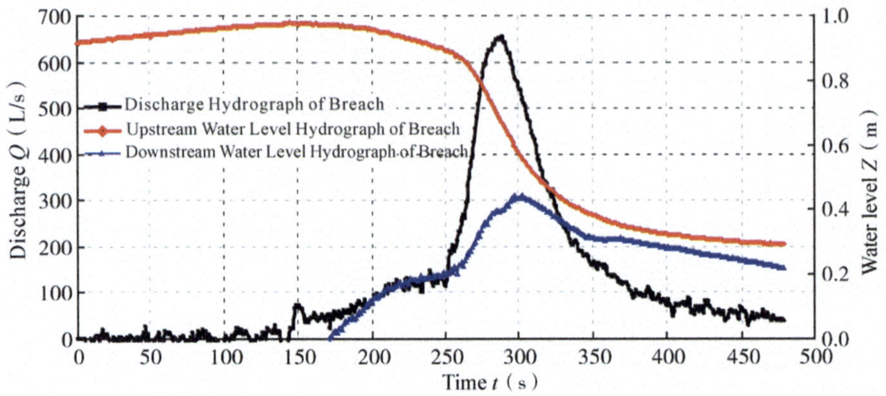

Fig. 5.21 Hydrographs of stage-discharge of breach under condition 17

(2) Change Process of Flow Velocity of Breach

As shown in Fig. 5.22, the change of the velocity of breach during the observation recording period (480 s) could be roughly divided into five stages: ① no discharge (0–50 s) → ② slow rise (50–250 s) → ③ rapid rise (250–280 s) → ④ rapid fall (280–320 s) → ⑤ slow fall (320–480 s).

At the initial recording time ($t_0 = 0$ s), the water level was low, with no discharge from the discharge channel, so the velocity of breach $V = 0$ m/s. When $t_0 = 50$ s, the discharge channel began to discharge, but the flow was gentle and the velocity was slow. Afterward, the velocity gradually rose. When $t_0 = 250$ s, $V = 148$ cm/s, the velocity of breach rose rapidly, reaching the maximum value $V_{max} = 242$ cm/s when $t_0 = 278$ s, indicating the instantaneous water level up to the maximum at this moment. Next, the velocity of breach rapidly decreased, reaching $V = 110$ cm/s

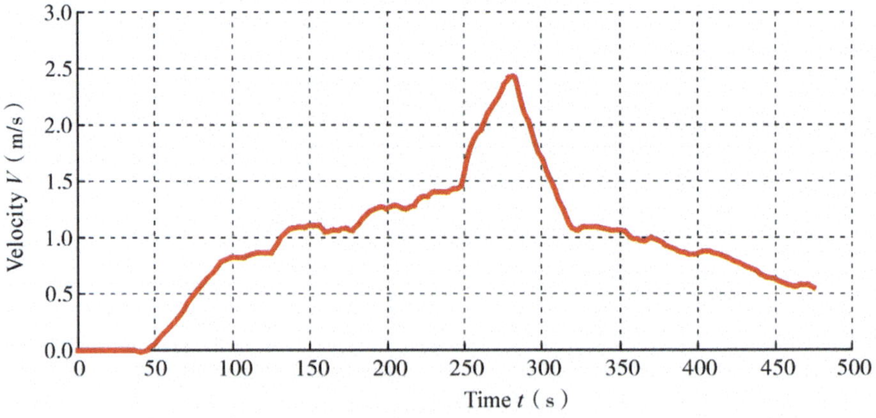

Fig. 5.22 Flow velocity of breach under condition 17

when $t_0 = 320$ s. Then the velocity decreased slowly and basically tended to a stable value $V = 59$ cm/s when the observation recording ended at $t_0 = 480$ s.

(3) Scouring Process of Barrier Dam

In order to dynamically monitor the scouring process of barrier dam, seven light erosion catchers were embedded in the barrier dam before the experiment (Fig. 5.23). According to the embedded elevation and section, they were numbered 1#, 2#, 3#, 4#, 5#, 6# and 7# in turn.

The experiment under this condition showed that the sequence of the embedded light erosion catchers being carried out by flow was 1# → 5# → 6# → 7# → 2# → 3# → 4#. The dynamic capture moment was the observation recording time $t_0 = 111$ s, 156 s, 159 s, 189 s, 256 s, 265 s and 268 s in turn (Table 5.16).

According to the sequence of field test camera shooting and erosion catcher being carried out, the dam scouring process under condition 17 was drawn as shown in Fig. 5.24, and the dam souring area gradually expanded from I → II → III → IV. See Table 5.17 for the dam souring process and the characteristic values of hydraulic parameters. In order to facilitate the comparison of each group experiment,

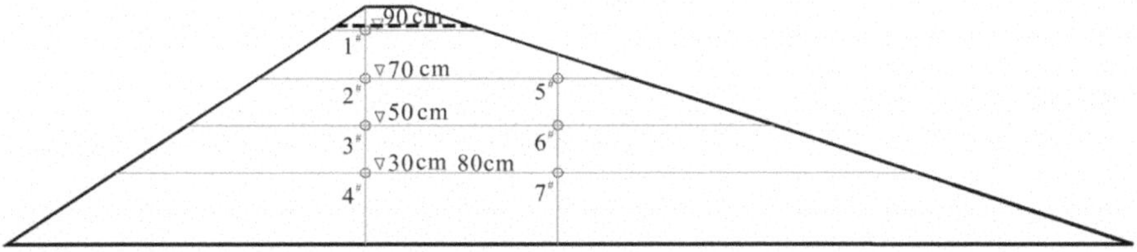

Fig. 5.23 Schematic diagram of embedded positions of dam erosion catchers under condition 17

Table 5.16 Time of embedded light erosion catchers being carried out under condition 17

Catcher number	Embedded elevation (cm)	Embedded section	Capture time t_0 (s)
1#	90	Upstream barrier dam	111
5#	70	Downstream dam slope	156
6#	50	Downstream dam slope	159
7#	30	Downstream dam slope	189
2#	70	Upstream barrier dam	256
3#	50	Upstream barrier dam	265
4#	30	Upstream barrier dam	268

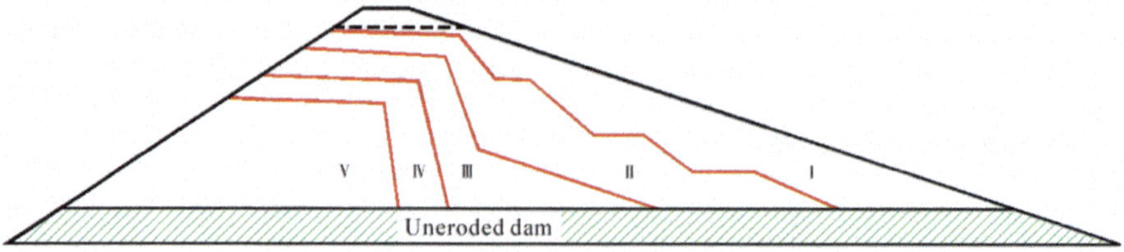

Fig. 5.24 Schematic diagram of dam scouring process under condition 17

the moment when the downstream slope shoulder of the discharge channel began to discharge was defined as the dam-break reference zero time ($t = 0$), which was 50 s behind the observation recording time t_0. See Fig. 5.25 for the dam scouring process under condition 17.

It was found that the barrier dam scouring process under this condition could be divided into the following four stages.

The first stage ($t = 0$–50 s) was the formation and development stage of scouring gully on the downstream slope. The overtopping flow was discharged from the

Table 5.17 Dam scouring process and characteristic values of hydraulic parameters under condition 17

Recording time t (s)	Crest height of breach h (cm)	Crest width of breach B (cm)	Upstream water level of breach (cm)	Downstream water depth of breach (cm)	Discharge of breach Q (L/s)	Flow velocity of breach V (cm/s)
0	92	16	94.3	0	0	0
30	91.5	16	95.5	0	0.6	5
60	91	16	96.5	0.4	1.7	86
90	90	28	97.5	0.7	2.2	109
120	88	44	97.2	5.4	52.7	108
150	84	67	95.7	11.3	74.2	126
180	78	89	92.4	18.3	115.1	140
210	69	144	87.1	22.7	233.2	196
240	34	196	65.8	40.1	640.2	203
270	31	200	46.8	39.0	320.7	110
300	30	200	38.5	31.3	159.7	106
330	29	200	34.2	30.0	98.4	92
360	28	200	31.9	27.5	74.4	86
390	27	200	30.4	25.2	42	67
420	27	200	29.7	22.7	40.5	59
Peak value	–	–	97.7 (104 s)	44.3 (248 s)	654.3 (239 s)	242 (228 s)

Note Based on the moment when the downstream slope shoulder of the barrier dam began to overflow, the reference zero time t ($t = 0$) was 50 s behind the initial observation recording time t_0

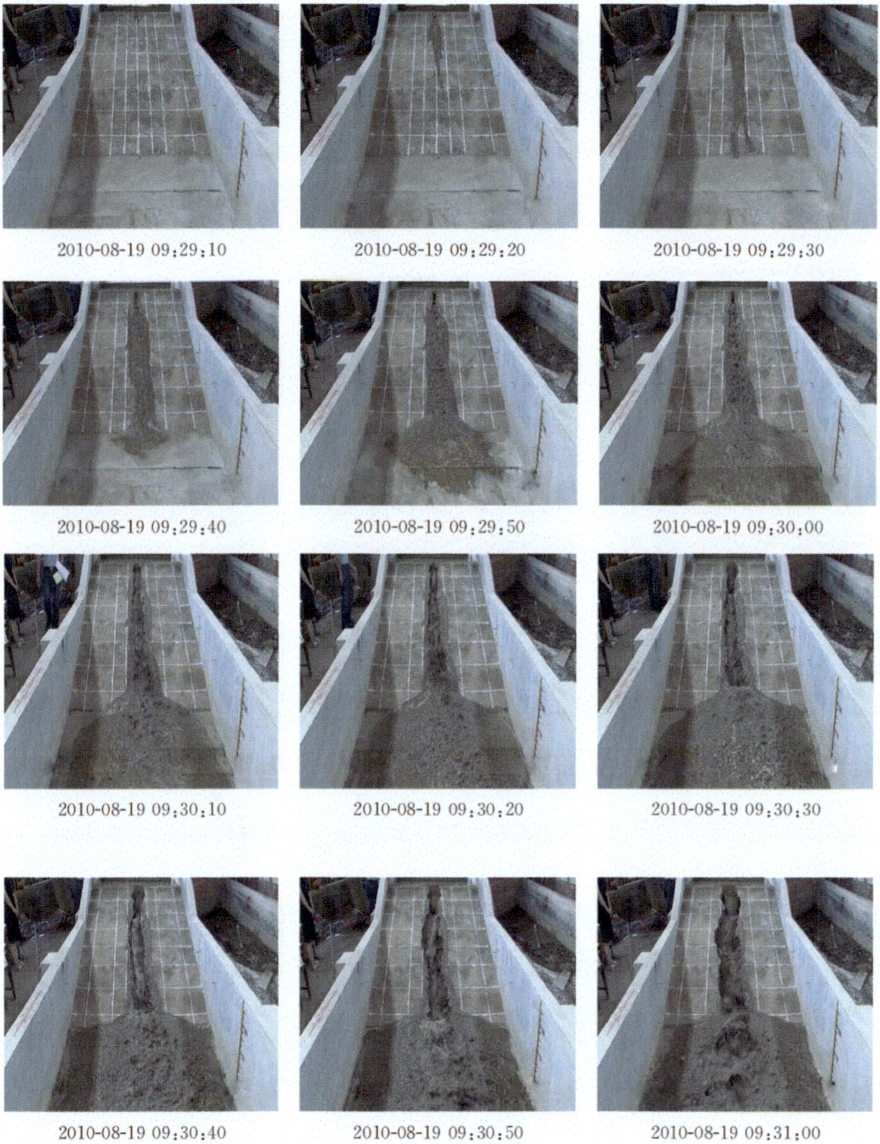

Fig. 5.25 Process diagram of dam scouring under condition 17

discharge channel, initially as thin sheet small flow, which sped up along the slope and gradually increased its scouring capacity. At first, it scoured the fine particles on the dam surface, forming rill surface erosion. The rill surface erosion further expanded, forming flow with fixed path, and the concentrated stream scoured and downcut the dam surface, forming obvious gullies. The erosion gullies were basically linear, with

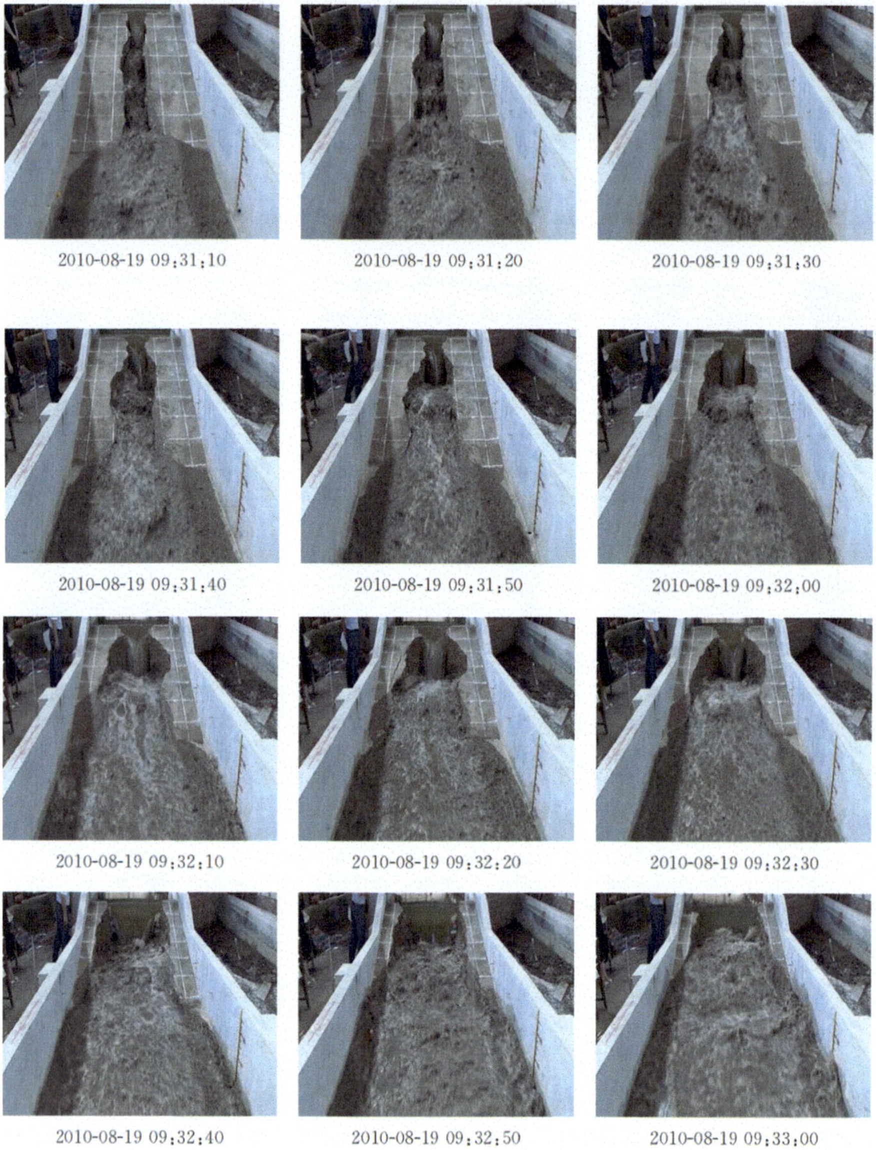

Fig. 5.25 (continued)

obvious gully edge, gully slope and gully bottom. Due to the erosion, transport and accumulation of the overtopping flow, the erosion gullies continuously scoured and downcut longitudinally, and expanded the undercutting of gully bank and gravity collapse horizontally. The sediment carried by the slope flow was deposited in a fan shape around the groove of the downstream slope, forming an alluvial fan. At this

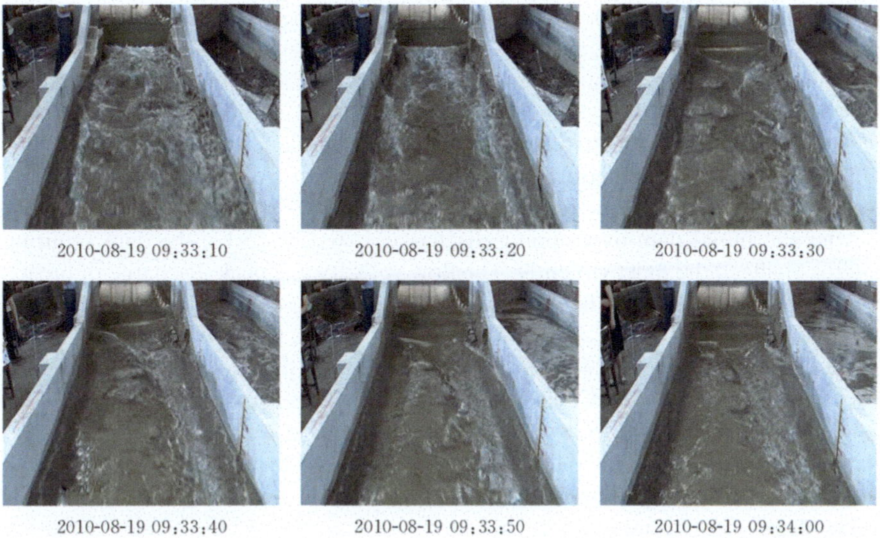

Fig. 5.25 (continued)

stage, as the dam scouring occurred on the downstream slope, the scouring process was slow and gradual. With little discharge and velocity of breach, the water level of the upstream reservoir was still in a slow rising stage.

The second stage (t = 50–150 s) was the multi-level drop-step scouring stage on the downstream slope. The erosion gullies further developed and continuously scoured deep and expanded. Due to the difference of the dam compactness and the inhomogeneity of the building material particles, the scouring resistance velocity of the barrier dam was different. In weak area, the incipient motion was easier to occur, with erosion first to form scouring holes. Under the continuous scouring and erosion of flow, the downstream dam slope formed multi-level small drop-steps. At this stage, as the barrier dam scouring still mainly occurred on the downstream slope, the scouring ability of the overtopping flow gradually increased. With the slow rise of the discharge and flow velocity of breach, the water level of the upstream reservoir began to decline slowly after reaching the highest value.

The third stage (t = 150–250 s) was the rapid scouring and collapse stage of the waterfall flow of the barrier dam. Retrogressively scouring occurred to the multi-level drop-steps, which continued to retreat upstream with time and finally merged into a big headcut, with waterfall flow at the breach. Upstream of the barrier dam, the height, width and geometry of the breach controlled the discharge of breach, and the dam crest formed the control section. At this stage, as the dam-break had expanded to the upstream edge of the dam crest, the scouring ability of flow was the strongest, and the dam-break and collapse process was very violent. With the rapid rise of the discharge and flow velocity of breach, up to the maximum value at this stage, the water level of the upstream reservoir dropped rapidly.

The fourth stage ($t = 250$–430 s) was the stabilization stage of dam scouring. The index of hydraulic elements, and the upstream water level, discharge and velocity of breach fell slowly after the maximum values. As the scouring ability of flow gradually weakened, the dam scouring process slowed down and eventually stabilized, with the upstream "inflow equal to discharge" of the barrier dam.

Based on the abovementioned change processes of hydraulic parameters and scouring patterns, it could be found that:

After the initial dam overtopping discharge, $t = 150$–250 s, the dam scouring patterns changed the most violently. The flow velocity and peak discharge of breach were both in this interval. When the dam scouring time $t = 104$ s, the upstream water level reached peak value $Z_{max} = 97.7$ cm; when $t = 28$ s, the flow velocity of breach reached peak value $V_{max} = 242$ cm/s; when $t = 239$ s, the discharge of breach reached peak value $Q_{max} = 654$L/s. Consistent with the previous conditions, the water level of the upstream reservoir reached the highest value at first, and after a period of time, the velocity of breach reached the peak value, followed by the discharge of breach, with an interval of about 10 s.

The dam scouring mode under this condition was close to the headcut scouring mode proposed by American scholars.

5.4.2.4 Analysis of Dam-Break Modes

See Table 5.18 for the summary of typical dam-break modes and scouring characteristics, and Table 5.19 for the results of hydraulic parameters of three dam-break modes for analysis and comparison.

According to the experimental results, the gradual break of barrier dam with wide graded mixture was roughly divided into three modes.

(1) Homogeneous Scouring Layer by Layer

The barrier dam scouring process was relatively homogeneous, and the erosion gullies on the downstream slope continuously scoured deep and expanded, which made the slope of the downstream dam gradually become gentle, the plane of breach expand synchronously, and the scouring expand to the main dam layer by layer. The scouring of the main dam lasted for a long time, about 90 s, and the water level of the upstream reservoir did not drop violently, with a distance of 17.3 cm within the maximum drop interval of 30 s. The peak type of the hydrograph of dam-break discharge was slightly fat, and the dam-break peak discharge was little.

(2) Full Overtopping Scouring

Similar to the homogeneous scouring layer by layer, the overtopping flow was discharged from the dam crest, and the scouring process was gradual. The erosion of the main dam lasted for a long time, about 90 s, and the water level of the upstream reservoir did not drop sharply, with a distance of 19.9 cm within the maximum drop interval of 30 s. With no discharge channel opened on the dam crest, the water level

Table 5.18 Different dam-break modes and scouring characteristics of barrier dam with wide graded mixture

No.	Dam-break mode	Scouring characteristics
1	Homogeneous scouring layer by layer	The scouring process was homogeneous, expanding layer by layer; the main dam scouring lasted for a long time; the peak type of the hydrograph of dam-break discharge was slightly fat, and the dam-break peak discharge was little
2	Full overtopping scouring	The scouring process was more homogeneous, expanding layer by layer; the main dam scouring lasted for a long time; the peak type of the hydrograph of dam-break discharge was slightly fat, and the dam-break peak discharge was larger
3	Headcut scouring	The scouring process was non-homogeneous, with slow scouring on the downstream slope → gradual expansion of multi-level drop-step scouring on the downstream slope → rapid scouring and collapse of the headcut → scouring tended to be stable; the main dam scouring lasted for a short time; the peak type of the hydrograph of dam-break discharge was thin, and the dam-break peak discharge was the largest

Table 5.19 Comparison of peak values of hydraulic parameters of different dam-break modes of barrier dam with wide graded mixture

Dam-break mode	Condition	Peak of upstream water level Z_{max} (cm)	Peak of discharge velocity of breach V_{max} (cm/s)	Peak of discharge of breach Q_{max}(L/s)	The main dam-break duration t (s)
Dam-break by homogeneous scouring layer by layer	12	97.7 (86 s)	197 (147 s)	459 (170 s)	90
Dam-break by full overtopping scouring	14	102.1 (30 s)	220 (90 s)	526 (195 s)	90
Dam-break by headcut scouring	17	97.7 (104 s)	242 (228 s)	654 (239 s)	60

of the upstream reservoir was high, and the maximum dam-break peak discharge was larger.

(3) Headcut Scouring

The scouring process could be divided into four stages: slow scouring on the downstream slope → gradual expansion of multi-level drop-step scouring on the downstream slope → rapid scouring and collapse of the headcut → scouring tended to be stable. The scouring of the main dam lasted for a short time, about 60 s. When the headcut expanded to a certain extent, the slopes of both banks became steeper. With the sudden collapse of breach horizontally and vertically, the upstream water level dropped abruptly by 24.8 cm within the maximum drop interval of 30 s. The peak type of the hydrograph of dam-break discharge was thin, and the dam-break peak discharge was the largest. The waterfall flow on the dam crest continuously scoured and eroded the bottom of the headcut, resulting in the instability of the bank slopes on both sides and gravity collapse, which continued to expand upstream and further increased the width and discharge of the upstream dam crest, and accelerate the rapid collapse of the barrier dam.

Under the same experiment conditions, different dam-break modes led to great differences in hydraulic elements of breach. For the homogeneous scouring layer by layer, the highest water level of the upstream reservoir was 97.7 cm, the peak of velocity of breach was 2.0 m/s, and the dam-break peak discharge was 459 L/s. For the full overtopping scouring, the highest water level of the upstream reservoir was 102.1 cm, the peak flow velocity of breach was 2.2 m/s, and the dam-break peak discharge was 526 L/s; for the headcut scouring, the highest water level of the upstream reservoir was 97.7 cm, the peak of velocity of breach was 2.4 m/s, and the dam-break peak discharge was 654 L/s, the maximum value of the three modes. For the homogeneous scouring layer by layer, due to the discharge channel on the dam crest, the water level of the reservoir dropped, and the dam-break peak discharge decreased by about 15%.

The discharge scouring process of Tangjiashan barrier lake was close to the first half of the condition of homogeneous scouring layer by layer. In the emergency handling and disaster relief project scheme for Tangjiashan barrier lake, since the discharge channel was excavated in advance, the water level of the upstream reservoir was lowered, achieving the aim of effectively reducing the dam-break peak discharge, and the adverse impact on the downstream cities, traffic bridges and important infrastructures, and ensuring the "safe and controllable" emergency handling and disaster relief. This could also be strongly proved in the experiment.

5.4.3 Analysis of Dam-Break Peak Discharge and Mechanism

5.4.3.1 Analysis of Dam-Break Peak Discharge of Different Dam Heights

For the convenience of comparative analysis, this discussion was represented by conditions 15, 7 and 12. Under the above three conditions, the discharge channel was opened on the dam crest, and the dam-building material was $1^{\#}$ material. The dam-break mode was homogeneous scouring layer by layer, but the dam heights were 60 cm, 80 cm and 100 cm, respectively. That was, with the same dam-break mode, overtopping discharge mode and dam-building material, the differences of hydraulic elements caused by different dam heights were analyzed.

As shown in Table 5.20, under condition 15, the dam height was 60 cm, with a discharge channel opened on the dam crest. In the experiment, the highest water level of the upstream reservoir reached 57.8 cm, the measured maximum dam-break peak discharge was 277 L/s, and the maximum flow velocity of breach was 169 cm/s. Under condition 7, the dam height was 80 cm, with a discharge channel opened on the dam crest. In the experiment, the highest water level of the upstream reservoir reached 77.9 cm, the measured maximum dam-break peak discharge was 384 L/s, and the maximum velocity of breach was 180 cm/s. Under condition 12, the dam height was 100 cm, with a discharge channel opened on the dam crest. In the experiment, the highest water level of the upstream reservoir reached 97.7 cm, the measured maximum dam-break peak discharge was 459L/s, and the maximum velocity of breach was 197 cm/s.

Therefore, with the increase of dam height, the maximum peak discharge caused by dam-break also increased.

Table 5.20 Comparison of maximum dam-break peak discharges of different dam heights

No.	Condition	Dam height *H* (cm)	Dam slope	Peak of upstream water level Z_{max} (cm)	Peak flow velocity of breach V_{max} (cm/s)	Peak discharge of breach Q_{max} (L/s)	Dam-break mode
1	15	60	1:3	57.8 (20 s)	169 (103 s)	277 (77 s)	Homogeneous scouring layer by layer
2	7	80	1:3	77.9 (70 s)	180 (119 s)	384 (126 s)	Homogeneous scouring layer by layer
3	12	100	1:3	97.7 (86 s)	197 (147 s)	459 (170 s)	Homogeneous scouring layer by layer

Table 5.21 Comparison of maximum peak discharge of different dam longitudinal sections

No.	Condition	Dam slope	Inflow Q (L/s)	Peak of upstream water level Z_{max} (cm)	Peak flow velocity of breach V_{max} (cm/s)	Peak discharge of breach Q_{max} (L/s)	Dam-break mode
1	17	1:3	30	97.7 (104 s)	242 (228 s)	654 (239 s)	Headcut waterfall flow
2	18	1:2	30	96.6 (57 s)	230 (154 s)	943 (113 s)	Headcut waterfall flow

5.4.3.2 Dam-Break Analysis of Different Dam Sections

This analysis was represented by conditions 17 and 18, with the only difference in the downstream dam slope. The downstream dam slope under condition 17 was 1:3, whereas the downstream dam slope under condition 18 was 1:2, which was slightly thin.

Under condition 17, the dam-building material was $2^{\#}$ material, with the dam height of 10 cm and the downstream dam slope of 1:3, the discharge channel was opened on the dam crest, and the upstream steady inflow was 30 L/s. In the experiment, the highest water level of the upstream reservoir reached 97.7 cm, and the measured maximum dam-break peak discharge was 654 L/s. Under condition 18, the dam-building material was $2^{\#}$ material, with the dam height of 10 cm and the downstream dam slope of 1:2, the discharge channel was opened on the dam crest, and the upstream steady inflow was 30 L/s. In the experiment, the highest water level of the upstream reservoir reached 96.6 cm, and the measured maximum dam-break peak discharge was 943 L/s (Table 5.21).

In addition, through the comparative analysis of different upstream inflow, it was found in the experiment that, as the upstream inflow accounted for less than 10% of the dam-break peak discharge, the effects of different inflow on the dam-break peak discharge were not obvious.

Under the experiment conditions, due to the little difference in particle diameter gradation between the two dam-building materials $1^{\#}$ and $2^{\#}$, their effects on the results of dam-break experiment were also not significant.

5.4.3.3 Analysis of Dam-Break Mechanism of Barrier Dam

For the barrier dam with wide graded mixture, through the flood overtopping dam-break experiment, the following preliminary understanding of the scouring expansion of breach and the dam-break mechanism of barrier dam was obtained.

(1) The dam-break mechanism of barrier dam and the expansion process of breach were complicated due to many factors involved, such as hydraulics, soil

mechanics and sediment dynamics. The expansion process of breach of barrier dam was closely related to dam type, geometric section shape, dam-building materials, gradation and compactness. The scouring and dam-break of barrier dam were the process of interaction between water and sand (rock). The vertical direction was dominated by flow scouring and down-cutting, whereas the horizontal expansion was dominated by flow scouring and then gravity collapse. The shape of breach changed violently and expanded nonlinearly with strong three-dimensional characteristics. It was trapezoidal or rectangular, with steep side wall and almost upright.

(2) For the dam-break mode of homogeneous scouring layer by layer, the dam scouring process was relatively homogeneous and gradual. The overtopping flow was discharged from the dam crest and formed obvious erosion gullies. The erosion gullies continuously scoured deep and expanded, making the slope of downstream dam become gentle gradually. Then the scouring expanded layer by layer to the main dam area, the dam slope became gentle, and the plane of breach expanded synchronously until the main dam was completely eroded.

(3) For the dam-break mode of headcut waterfall flow, the rapid collapse of the dam occurred in the formation stage of the big headcut waterfall flow. Firstly, the dam at the drop point of water jet was eroded, resulting in the increase of waterfall flow drop and the improvement of flow scouring ability. The turbulent flow at the drop point of water jet was violent, which continuously eroded the bottom of the headcut, causing the instability of the bank slopes on both sides and gravity collapse. The collapsed body intermittently rolled to the bottom of the gully and was carried to the downstream by the flow. The collapse continued to expand to the upstream, which led to the enlargement of the width of the upstream dam crest, the increase of the discharge and the rapid collapse of the dam. In the experiment, the erosion volume accounted for about 50% of the dam within 100 s. The process of barrier dam-break was obviously non-equilibrium, namely nonlinear. In the initial stage, the scouring of the downstream slope expanded slowly and gradually, and then the scouring ability of the multi-level drop-steps gradually increased. Once the scouring expanded to the inside of the dam near the upstream edge of the dam crest, the waterfall drop flow was formed, accelerating the collapse of the dam. In the process, the index of hydraulic elements of breach showed strong unsteady flow characteristics. As the discharge and flow velocity of breach rose and fell sharply, the upstream water level dropped rapidly, and the measured discharge of breach increased more than 5 times in 40 s, from 120 to 654 L/s.

5.4.4 Conclusions

5.4.4.1 Main Study Results

(1) Through field survey and data collection, the geometric and physical characteristics of the material composition of Tangjiashan barrier dam were analyzed.

Tangjiashan barrier dam was formed by landslides, and its geological composition was mostly loose accumulation of earth and rock, with strong material heterogeneity. The barrier dam was divided into four layers from top to bottom: gravel earth layer, block gravel layer, solitary block gravel layer and gray–black silty gravel layer. Compared with the artificial earth–rock dam, the barrier dam was irregular in shape and rough on the crest, and the overtopping flow was most likely to occur in the weak area of the dam pass. In terms of the dam structure, since there was no artificial building process, the dam structure was loose and disorderly, partly overhead, and compact in the lower part. Most of the dam materials were silty loam, coarse particles, gravel, crushed stone, and even mixed with solitary boulders. The coarse and fine particles were mixed, forming a wide range of particle diameters. From extremely fine particles to isolated boulders, the composition of the barrier dam was more complex.

On the basis of full analysis of the material composition of Tangjiashan barrier dam, the simulation design of particle diameter range and unit weight for indoor experiment and study was conducted. In the experiment, building alunite and natural sand were selected to prepare the wide graded mixture with different particle diameters. The particle diameters ranged from 0.075 to 40 mm, which could ensure the wide similarity between the particle diameter gradation and the material composition of Tangjiashan barrier dam.

(2) The scouring and incipient motion characteristics of homogeneous material, continuous wide graded mixture and discontinuous wide graded mixture were studied by using glass sink.

In the experiment, the changes of the motion and bed surface patterns of various mixture bed sands under different water depth and velocity conditions were observed, and the relation between the critical velocity and the characteristics of bed sand particle diameter was measured and analyzed. The comparative analysis showed that, compared with the homogeneous material, once incipient motion occurred with the continuous wide graded mixture, and the particle motion was more likely to intensify and form scouring to cause the change of bed surface pattern. However, for the discontinuous wide graded mixture, due to the obvious hidden blocking effect of large particles, it was still difficult for the particles to develop into large displacement motion after incipient motion, and there was little change in bed surface pattern under the same flow conditions. According to the experiment results, a practical expression of erosion incipient velocity of wide graded mixture was proposed as follows:

$$U_c = 1.37\sqrt{\frac{\rho_s - \rho}{\rho} g d_{50} (\frac{h}{d_{50}})^{1/6}}.$$

The experiment results were consistent with the discharge scouring by field observation of Tangjiashan barrier lake. The formula could provide reference for the overflow scouring of similar barrier dams in the future.

(3) With the indoor experiment as the technological means and the barrier dam with discontinuous wide graded mixture as the study object, the dam−break process of barrier dam was recorded dynamically by using the embedded light erosion catchers for the first time.

Under the prerequisite of upstream steady inflow, the scouring process of barrier dam and the change process of hydraulic elements of breach under the flood overtopping condition were studied. With the consideration of the conditions of different dam section shape, different dam materials, different inflow and whether the discharge channel was opened on the dam crest, a series of dam scouring and dam−break experiments were conducted. The scouring process of barrier dam and the change process of hydraulic elements under each experiment condition were recorded and observed in detail, which could provide technical support for the decision−making of disaster relief and emergency handling.

(4) The barrier dam−break mode determined the dam−break duration, the maximum dam−break peak discharge and the damage degree to the downstream.

Due to the different conditions of overtopping discharge mode, dam-building material, particle diameter gradation and material compactness, there were obvious differences in the barrier dam-break modes. According to the experiment results, the gradual dam-break of the barrier dam with wide graded mixture was roughly divided into three modes: homogeneous scouring layer by layer, full overtopping scouring and headcut scouring.

For the dam-break mode of homogeneous scouring layer by layer, the dam scouring process was relatively homogeneous, and the erosion gullies on the downstream slope continuously scoured deep and expanded, which made the slope of the downstream dam gradually become gentle, the plane of the breach expand synchronously, and the scouring expand to the main dam layer by layer. The scouring of the main dam lasted for a long time. The peak type of the hydrograph of dam-break discharge was slightly fat, and the dam-break peak discharge was little.

The full overtopping scouring mode was similar to the process of homogeneous scouring layer by layer. The overtopping water flow was discharged from the dam crest, and the scouring process was gradual. With no discharge channel opened on the dam crest, the water level of the upstream reservoir was high, and the maximum dam-break peak discharge was larger.

For the dam-break mode of headcut scouring, the scouring process could be divided into four stages: slow scouring on the downstream slope → gradual expansion of multi-level drop-step scouring on the downstream slope → rapid scouring and collapse of the headcut → scouring tended to be stable. The scouring of the main dam lasted for a short time. The peak type of the hydrograph of dam-break discharge was thin, and the dam-break peak discharge was the largest. The waterfall flow on the dam crest continuously scoured and eroded the bottom of the headcut, resulting in the instability of the bank slopes on both sides and gravity collapse, which continued to expand upstream and further increased the width and discharge of the upstream dam crest, and accelerate the collapse of the barrier dam.

Under the same experiment conditions, different dam-break modes led to great differences in hydraulic elements of breach. For the homogeneous scouring layer by layer, the highest water level of the upstream reservoir was 97.7 cm, the peak flow velocity of breach was 2.0 m/s, and the dam-break peak discharge was 459 L/s; for the full overtopping scouring, the highest water level of the upstream reservoir was 102.1 cm, the peak of velocity of breach was 2.2 m/s, and the dam-break peak discharge was 526 L/s; for the headcut scouring, the highest water level of the upstream reservoir was 97.7 cm, the peak flow velocity of breach was 2.4 m/s, and the dam-break peak discharge of dam-break was 654 L/s, the maximum value of the three modes. For the homogeneous scouring layer by layer, due to the discharge channel opened on the dam crest, the water level of the reservoir dropped, and the dam-break peak discharge decreased by about 15%.

The discharge scouring process of Tangjiashan barrier lake was close to the first half of the condition of homogeneous scouring layer by layer. In the emergency handling and disaster relief project scheme for Tangjiashan barrier lake, since the discharge channel was excavated in advance, the water level of the upstream reservoir was lowered, achieving the aim of effectively reducing the dam-break peak discharge, and the adverse impact on the downstream cities, traffic bridges and important infrastructures, and ensuring the "safe and controllable" emergency handling and disaster relief. This could also be strongly proved in the experiment.

(5) The dam-break peak discharge was closely related to the dam height, dam longitudinal section, dam-break mode, overtopping mode, dam material characteristics and compactness.

Under the experiment conditions, with the increase of dam height, the maximum dam-break peak discharge also increased. With the experiment dam heights of 60 cm, 80 cm and 100 cm, respectively, the measured maximum dam-break peak discharge was 277 L/s, 384 L/s and 459 L/s, respectively. The thinner the dam longitudinal section, the shorter the dam-break duration and the greater the dam-break peak discharge. However, the effects of upstream steady inflow on the dam-break peak discharge were not obvious.

(6) The special experimental study and the practice of discharge and scouring process of Tangjiashan barrier lake showed that when the hazard of barrier lake occurred, the construction technology of excavating discharge channel could reduce the water level of the reservoir in front of the dam, effectively cut the peak value of dam-break discharge, and mitigate the hazard to the downstream. During the emergency handling, various measures should be taken to achieve

the dam-break mode of homogeneous scouring layer by layer, aiming to prolong the dam-break duration and decrease the dam-break peak discharge.

5.4.4.2 Suggestions for Future Dam-Break Study

(1) It is of great significance to conduct systematic and targeted dam-break experiments, especially large-scale experiments, to observe the dam-break process, master the dam-break mechanism, check the dam-break model and improve the prediction accuracy of the model. Full attention should be paid to the effects of the model scale and the scale in the experiments.
(2) In terms of the dam-break simulation, the most critical problem is to simulate the scouring process and velocity of flow on the dam. This is especially important for cohesive earth-rock dam. As the scouring of cohesive earth by flow is a very complicated issue, the current study is still far from perfect. Strengthening this study in the future will help to better simulate the dam-break of earth-rock dam.
(3) Headcut scouring plays an important role in the dam-break process of earth-rock dam, but the study on this issue is still in the initial stage. It is suggested to strengthen the study on the formation, expansion and scouring mechanism of headcut.
(4) Due to the current limited understanding of the dam-break process and mechanism, the effects of such factors as the composite structure of the dam itself and the slope protection on the dam surface are not considered in most of the existing dam-break models. With the improvement of study level, the reasonable consideration of these factors will be an inevitable trend of model development.
(5) Natural earth-rock dam formed by landslides and debris flow has many similarities with artificial earth-rock dam in dam structure and dam-break characteristics, but at the same time they both have their own characteristics. In the future, according to the characteristics of natural earth-rock dam, we shall thoroughly study its dam-break modes and mechanisms and make full use of the existing study results and experience of dam-break simulation of artificial earth-rock dam to conduct the study of dam-break simulation of natural earth-rock dam.

References

1. Rongyu Q. Starting rule of uneven sand. J Sedim Res. 1980;1:83–91.
2. Yuaner C, Jianheng X. Preliminary research on the laws of threshold motion for non-uniform sediments. J Wuhan Univ Hydr Elec Eng, 1988;3:28–37.
3. Han Q, He M. Incipient mechanism and incipient velocity of nonuniform sediment. J Yangtze Riv Sci Res Inst. 1996;3:15–20.
4. He W, Fang D, Yang S, et al. Study on laws of threshold motion for non-uniform sediment. J Hydro Eng. 2002;2:72–77.
5. Tang Z, Fang D. Problems related to non-uniform sediment incipient motion. Sichuan Hydro Power. 1997;1:69–72, 81.

6. Wang T, Liu X, Nie R, et al. Comparative study on the incipient motion conditions of non-uniform non-cohesive sediment. J Hydro Eng. 2010;29(6):126–131.
7. Mai J-Z, Wang L-P. Summarization on non-uniform sediment transport specialty and law. J North China Inst Water Conserv Hydro Power. 2008;4:29–32.
8. Xu HT, Lu JY, Liu XB. Non-uniform Sediment incipient velocity. Int J Sedim Res

Chapter 6
Study on Emergency Handing Scheme of Tangjiashan Barrier Lake

6.1 Domestic and International Handing Technologies of Barrier Lake

During the rescue action of the Wenchuan earthquake barrier lake on May 12, a number of organizations collected and compiled domestic and international cases of barrier lake emergency handing and analyzed the handing techniques. Table 6.1 listed ten typical barrier lake cases at home and abroad. It could be seen that there were many differences in the handing methods due to the different formation conditions, barrier lake characteristics and traffic conditions [1–5]. But in general, they all followed the following three principles.

(1) When making the handing plan, it was proposed to avoid severe working conditions and finish the handing task before the arrival of the flood, thus avoiding the risk of barrier dam collapse caused by the superposition of the flood and the high water level of the barrier lake.
(2) According to the topographic and geological conditions of the barrier dam, it was supposed to take advantage of the situation and combine organically engineering measures with the non-engineering measures.
(3) The engineering measures should facilitate the rapid construction. It was supposed to achieve the minimum engineering handing objectives by guaranteeing preparatory measures. In order to cope with the impact of unforeseen factors, the preparatory program should start quickly to ensure the achievement of the goal of removing the risk when the implementation of the recommended program meets difficulties.

Q. Yang and W. Yang, *Emergency Handling of the "5.12" Tangjiashan Barrier Lake*,
https://doi.org/10.1007/978-981-99-1706-8_6

Table 6.1 Comparative analysis of typical barrier lake emergency handling technologies

Formation conditions of typical barrier lake	Tangjiashan Earthquake landslide	Egong Rainfall after snow melting	Yinjiang River Excavation landslide	Canning, Madison, USA Earthquake landslide	Tsao–Lin, Taiwan Earthquake landslide
Barrier dam material composition	Fractured rock, gravelly soil and gray-black mud-bearing fine sand	Fine-grained soil accounting for more than 70%	Gray rock, slope soil, etc	Boulder stones	Weathered rock and crumbling deposits
Volume of barrier (million m^3)	20.37	300	1.80	–	126
Barrier capacity (billion m^3)	0.316	4	0.039	0.002	0.046
Reservoir capacity when discharge or collapse (billion m^3)	0.2466	3.2			
Affecting population	1.3 million	Thousands	60,000		Tens of thousands
Stability evaluation	It was basically judged that collapse could be avoided by taking engineering measures	The collapse was sure to occur, and engineering measures were taken to reduce collapse water head and damage	Collapse	–	Collapse could be avoided by taking engineering measures
Traffic conditions	It is likely to rely entirely on air transportation as in the absence of land or water transportation	Transportation was extremely difficult at altitude of more than 3000 m in mountainous terrain	Convenient transportation	Convenient transportation	Convenient transportation
Treatment options		Drainage and predischarge: excavation of open discharge channels	Excavation of flood relief tunnels	Drainage open channel	Discharge channel

(continued)

Table 6.1 (continued)

Formation conditions of typical barrier lake	Tangjiashan Earthquake landslide	Egong Rainfall after snow melting	Yinjiang River Excavation landslide	Canning, Madison, USA Earthquake landslide	Tsao–Lin, Taiwan Earthquake landslide
Construction method	Large excavators, bulldozers or blasting construction	Combination of bulldozer drainage scouring and hand wind auger drilling or blasting for rock removal	Underwater rock foundation blasting construction	Blasting	Mechanical excavation
Construction time (d)		33	8	1	–
Volume of earthwork (million m^3)		1.355	0.035	–	–
Engineering treatment effect		Flood head reduced by 18.16 m and potential water storage decreased by 1 billion m^3	All emptied	Maintained water levels within safe limits	Water head reduced by 1.5 m and potential water storage decreased by nearly 14 million m^3
Peak flood flow (m^3/s)		124,000	–	–	–
Formation conditions of typical barrier lake	Ershan, Jiufen,Taiwan Earthquake landslide	Longquan stream, Taiwan Rainfall and earthquake collapse	Terano, Japan Earthquake landslide	Higashi Takesawa, Japan Earthquake landslide	SpiritLake, USA Volcanoes accompanied by earthquake landslides
Barrier dam material composition	Weathered rock and crumbling deposits	Boulder stones	Crumbling material	Crumbling material	Volcanic ash and boulders
Volume of barrier (million m^3)	70	0.60	1	1.30	
Barrier capacity (million m^3)	1.94	0.8	0.43	0.47	618

(continued)

Table 6.1 (continued)

Formation conditions of typical barrier lake	Ershan, Jiufen,Taiwan Earthquake landslide	Longquan stream, Taiwan Rainfall and earthquake collapse	Terano, Japan Earthquake landslide	Higashi Takesawa, Japan Earthquake landslide	SpiritLake, USA Volcanoes accompanied by earthquake landslides
Reservoir capacity when discharge or collapse (billion m^3)	–	–	–	–	–
Affecting population	–	–	–	–	–
Stability evaluation	Collapse could be avoided by taking engineering measures	Collapse	Collapse could be avoided by taking engineering measures	Collapse could be avoided by taking engineering measures	High water levels would cause a collapse due to a piping phenomenon in the volcanic ash bottom layer
Traffic conditions	–	–	–	–	–
Treatment options	Excavation of spillway	Excavation of spillway	Construction of discharge channels and spillways	Excavation of diversion channels	Drainage by 3.4 m diameter culvert pipe
Construction method	Mechanical excavation	Mechanical excavation	Mechanical excavation	Mechanical excavation	Mechanical excavation
Construction time (d)	–	–	–	–	–
Volume of earthwork (10^4 m^3)	8.88	–	–	–	–
Engineering treatment effect	1.1 million m^3 of potential water storage was reduced	Water level lowered by 10 m	Capable to carry a 100-year flood	Water level lowered by 6.5 m	Water level kept within safe limits
Peak flood flow(m^3/s)	–	–	–	–	–

6.2 Overall Emergency Handing Plan of Tangjiashan Barrier Lake

6.2.1 Necessity and Overall Objective for Emergency Handling

The Tangjiashan barrier dam was a secondary disaster caused by the strong earthquake and blocked the Tongkou River to cause inundation damage to the upstream towns of Yuli and Xuanping. With the increasing water level and greater inflow, the barrier would directly threaten more than 1.3 million people in the downstream cities of Mianyang and Suining once it collapsed.

There were a number of cases in the history that the casualties and property losses caused by barrier lake collapse were greater than those caused by the earthquake. On June 1, 1786, a 7.5 magnitude earthquake in Kangding caused a huge landslide in Luding and Hanyuanalong, the Dadu River, blocked the Dadu River and cut off its flow for 10 days. On June 11, the barrier lake collapsed, causing huge casualties of cities along the lower river reach in Leshan, Yibin and Luzhou as the lake surged from tens of meters high. On August 25, 1933, a major earthquake occurred in the Diexi Town, and 45d after the earthquake, the Diexi barrier lake collapsed, causing a serious secondary disaster and a large number of casualties.

The threat of the Tangjiashan barrier lake was mainly presented in two aspects. The first one was the continuously upwelling water head at the barrier lake. Taking the lowest notch of 752 m for calculation, the water head would reach 87 m once the water overtopped the barrier dam top. The second one was the water volume of the barrier lake. The total water volume of the barrier lake would reach 316 million m^3 once the water overtopped the barrier dam top from the lowest notch.

With the approaching of the major flood season from June to September, the probability of experiencing a large flood was increasing as the water level of the barrier lake kept in a high level. According to the results of hydrological analysis of the Tongkou River, the maximum multi-year average flow at the dam site was 1600 m^3/s, and the 5-year flow could reach 2190 m^3/s. If the flood came down from the barrier dam top, the high water level and the large flow would increase the possibility of rapid collapse. With the reservoir water level increasing, the water head difference between the front and back of the barrier dam kept increasing. And the possibility of piping and leakage rose with the collapse risk increasing.

About 3 km upstream of the barrier dam, there was a giant Malingyan paleolandslide with a total volume of about 16.5 million m^3. As the water level of the barrier lake rose to soak the slope foot, the stability of the paleolandslide body was gradually reduced. Coupled with aftershocks and rainfall, it was much possible for the paleolandslide body to activate. Once it slid into the reservoir, the induced surge would approach to the front of the barrier dam and the consequences were unpredictable.

Once the barrier lake was fully collapsed, the huge water level difference and the huge water volume would pose a serious threat to the downstream. According to the breach calculation, the higher water head would result into higher peak discharge and higher water level lifting in the downstream in the condition of same collapse bottom elevation and duration time, causing greater inundation damage in the downstream. Therefore, the Tangjiashan barrier lake must be handled as soon as possible.

In order to ensure the goal set by the Party Central Committee and the State Council (no single death should be caused by the overflow at the Tangjiashan barrier lake), the overall goal of the Tangjiashan barrier lake emergency response was to adhere to the people-oriented policy and prioritize the safety of people's lives and implement simultaneously engineering measures against risks and personnel transfer for safety. It was supposed to actively take engineering measures against risks, try the utmost to reduce the water level and barrier lake volume as soon as possible under the premise of controllable engineering measures and strive to eliminate the threat posed by the barrier lake before the main flood season to the maximum extent.

6.2.2 Basic Principles of Emergency Handling

6.2.2.1 Active Handling and Avoid Passivity

Proactive handling was conducive to master an advantage position for controlling the risk. The proactive engineering measures against risk were to induce the discharging path by artificially arranging a discharge channel, then deepen and widen the discharge channel along the pass by using the huge kinetic energy transformed from the potential energy of water flow and gradually improve the discharging capacity under the controllable principle. On the one hand, the water level of the barrier lake was greatly lowered and the water volume of the barrier lake was reduced to avoid abrupt collapse and active Malingyan paleolandslide body. On the other hand, through the scouring of the discharging lake water, a stable new river channel with certain flooding capacity was formed, which fundamentally eliminated the threat of the barrier to the upstream and downstream.

6.2.2.2 Early Handling to Reduce Difficulty and Risk

The water level of the barrier lake rose at a rate of about 2 m per day. The elevation difference from the water level to the dam notch was only 39 m in the first site survey, leaving only ten days to make plan, prepare construction and implement engineering. If taking measures one day earlier, it was supposed to decrease more than 6 million m^3 of water into the barrier lake and dominate more in the race against water level and suffer less risk.

Meanwhile, the main flood season was also around the corner. It would greatly decrease down the emergency handling difficulty and risk if finishing the arrangement before the main flood season as soon as possible.

Once there was a heavy rainfall in the reservoir area, it would seriously threat the emergency handling scheme. Since there was no land or water transportation at the construction site, the fuel, equipment and support materials were completely dependent on air transportation, thus there was always a risk of interruption due to clouds, rain, fog, wind and other weather conditions. Early handling before the arrival of the flood season would allow the scheme to be achieved more fluently.

6.2.2.3 Dynamical Balance Between the Assured and Striving Goals, Realization of Goals Against Risks Supported by Preparatory Scheme

As there were a lot of uncertain factors to affect the implementation of the risk eliminating scheme, personnel, machinery, weather, logistics, etc., might all become the control link for the achievement of the target. Different risk eliminating targets should be set for various levels of construction efficiency and the amount of water coming from the lake to fit the change of the external conditions and be able to flexibly swift in time. To cope with the impact of unforeseen factors, there should be preparatory scheme for the proposed scheme in the case of encountering difficulties, so as to ensure the goals of eliminating risks. The construction methods should also be adapted to the level of air transport guarantee.

6.2.2.4 Technical Scheme Featuring Trend Following

It was impossible to remove the barrier dam within a short time by relying merely on manual force. The water potential energy should be made full use to achieve twice the result with half the effort by capitalizing on the trend on a controllable basis. The section of the artificial discharge channel should be narrow and deep in order to pass the flow as early as possible, limit the rise of the reservoir water level and widen the deep cross section by utilizing the scouring power of water flow at the same time. It was supposed to bring forward the working period, which was beneficial for achieving the goal of risk elimination and avoiding the large amount of water from flood season. It would be appropriate to arrange water flow in the early June. According to data of years, the incoming water was relatively small during the early June, and the gradual collapse by scouring was expected to fall into an acceptable range.

6.2.2.5 Engineering Measurement Facilitating Rapid Construction

The implementation plan should be sound and reliable with simple process, and it should meet the requirement of large-scale construction. Meanwhile, there should be less crossover and interference, and the situation of multiple working faces could be formed.

6.2.2.6 Simultaneous Formulation of Engineering Risk Elimination Plan and Personnel Evacuation Risk Avoidance Plan, Keeping Risk Elimination and Avoidance at the Same Step

The risk elimination and risk avoidance measures were a pair of mutually protective and complementary measures. Many changing factors in the emergency engineering required the risk elimination and risk avoidance measures to have the necessary safety margins. In order to achieve no casualty, the engineering measures should endeavor to do the best, and the risk avoidance measures should consider the worst situation.

6.2.3 The Overall Emergency Handling Plan

Regarding the requirements of basic emergency handling principle and no casualty target, the overall plan for emergency response was consisted of an emergency treatment engineering plan, a personnel risk avoidance plan, and emergency protection measures. These three plans should support and complement each other, and none could be neglected.

6.2.3.1 Emergency Handling Engineering Plan

The recommended emergency engineering plan was to excavate a 5–10 m deep discharge channel, and the alternate plan was to form a 5 m deep discharge channel by blasting. Under the controllable principle, it was supposed to gradually improve the discharging capacity. On the one hand, it was able to gradually decrease the water level and water volume of the barrier lake to avoid the concurrence of the abrupt collapse. On the other hand, a stable new river with a certain flooding capacity would be formed by the scouring of the discharging lake water to eliminate the threat of barriers to the upstream and downstream.

6.2.3.2 Personnel Evacuation Risk Avoidance Plan

According to the analysis of the breach possibility, three different kinds of personnel transfer plan under the modes of 1/3 dam-break, 1/2 dam-break and full dam-break to arrange the personnel evacuation risk. By analysis, the possibility of 1/3 dam-break mode was most likely to happen, and it was the recommended implementation plan. For the 1/2 dam-break and full dam-break plans, contingency plans were made in advance and necessary trainings were conducted.

6.2.3.3 Emergency Security Measures

In order to ensure the safe implementation of engineering measures and personnel evacuation measures, it was necessary to establish a reliable security system. By analysis, the emergency security systems which are needed to be established included rapid decision-making and response mechanism, water and precipitation prediction system covering the possible impact area of the Tangjiashan barrier lake, remote real-time video monitoring system to facilitate real-time visual understanding of the barrier status, safety monitoring system in the barrier dam area, communication security system on the barrier dam and to the outside world, as well as emergency security measures such as expert consultation and decision-making mechanism for preventing dam-break.

6.3 Emergency Engineering Plans of Tangjiashan Barrier Lake

6.3.1 Draining and Scouring Principle and Realization Way of Discharge Channel

6.3.1.1 Draining and Scouring Principle of Drainage Channel

The principle of discharging and scouring was that the artificially formed discharge channel with small section induced the overflow path, and the energetically overtopping water scoured the channel with mud and stones moved away along the channel. With the increase of the flow speed and flow rate, the carrying capacity of the water became larger, which gradually deepened and widened the discharge channel and keeps expanding the cross section of the discharge channel. With the expansion of the discharge channel cross section, the discharging volume also increased and the carrying capacity of the water was further enhanced, which further deepened and widened the discharge channel cross section until it scoured to the rock and soil mass with corresponding scouring resistance capacity, and then a stable new channel was formed.

The transport capacity of water flow increased with the rise of flow velocity. According to the experimental studies of Grote and Ishbash, the diameter of a single nudging mass on the riverbed was proportional to the square of the water velocity, i.e., $V = [4.89\text{–}6.82]\sqrt{d}$. Since the weight of sand and gravel was proportional to the third power of its particle size, the weight of the nudging mass was proportional to the sixth power of the flow velocity, i.e., $M = \mathrm{CV}^6$. It was the Airy's law of fluid dynamics. When the water velocity increased n times, the weight of the particles that could be taken away by the water was n^6 times, in other words, when the flow rate increased by 1 time, the weight of the particles that could be moved by was increased by 64 times.

6.3.1.2 Realization Path of Draining and Scouring

The barrier lakes threat was mainly reflected on the water head and volume. In terms of the energy, the higher water head would result into greater unit energy. Using the principle of discharging and scouring, it was able to form a discharge channel with a certain width and depth at the notch on barrier top, inducing the discharging path. The discharge channel was deepened and widened along the path with the kinetic energy transformed by water potential energy, so as to reduce the barrier lake level and eliminate the lake threat.

Tangjiashan barrier dam was long in plain form with a crest barrier width of 300 m and undulating topography and was higher in the left and lower in the right in the transverse direction. There was a trench at the right side of the barrier dam through the upstream and downstream with right bow-shaped surface, the bottom width was 20–40 m, and the elevation of the highest point of the notch was 752.2 m. The trench through the upstream and downstream provides more favorable topographic conditions for the implementation of the draining and scouring.

On the one hand, most of the 20–40-m-thick loose deposits in the upper barrier dam were softened by water and its viscosity increases. As its scouring resistance capacity was very poor, it was easy to form dilute debris flow. On the other hand, due to the extreme nature of the barrier material gradation, when the content of grain size material gradation in the starting state exceeded a certain amount, the large stone of barrier dam migrated under movable bed status.

Cataclastic rock underlays along the discharge channel, and fractured rock lays in the lower part. The favorable media structure provided a material basis for a controllable discharge process. By using the longitudinal slope of the drainage channel with a gentle front slope and a steep back channel, the critical water depth appeared at the junction of the gentle slope and the steep channel during the initial flow, and the water surface line of the discharge channel section was a precipitation curve on the steep slope, and the flow velocity increased rapidly along the course. The steep slope section was a fractured rock, which had a strong ability to resist water scouring and transport, and was not easily hollowed out unless it came to a high-speed water flow. Therefore, there would not appear abrupt failure at the beginning of water flow. The water surface line of the gentle slope section was the precipitation curve on the gentle slope, the flow velocity increased gradually along the course, and the water carrying capacity increases with the rise of flow velocity. If the material composition was the same, the debris and soil deposited on the residual gentle slope section would be scoured and transported by the water, the scouring would be gradually increased along the course, and the scouring process would be controlled in stages. After entering the underlying fractured rock layer, the initial erosion and headcut scouring coexist because of the strong scouring resistance capacity of the fractured rock, and the scouring resistance was gradually strengthened as the completeness of the fractured rock increases and finally entered a stable state.

6.3.2 Ideas and Principles of Design

(1) Taking manual intervention measures to control the water flow in the manageable range. It was necessary to not only avoid the rapid collapse caused by strong hydraulic scouring, but also to make full use of the hydraulic transport capacity to widen and deepen the channel.
(2) Taking advantage of the situation, it was necessary to make full use of the natural conditions and selected the part of geological weaknesses on the barrier dam where were most suitable for rapid construction and arrange engineering measures to achieve the purpose of rapid risk removal.
(3) Within the utmost range achieved by construction strength, the initial section of the discharge channel should meet the water flow capacity requirements under a certain standard during the flood season as far as possible.
(4) The design section should match the construction equipment with simple structure to benefit easy and fast construction.
(5) The temporary stability of the design section should be guaranteed during the process of the construction and the initial practice.
(6) Due to the limitation of air transportation capacity, only a single unit of construction machinery with a weight less than 15t could be used.

6.3.3 Route Selection and Arrangement

(1) The discharge channel should be arranged in the place where the original topography was low and the particle composition was quite fine, in order to reduce the difficulty and load of excavation work, speed up the excavation progress, and at the same time, it was convenient to give full play to the water carrying capacity.
(2) The canal line should be as straight as possible, the corner of the turning section should not be too large, to ensure smooth flow.
(3) The outlet of the discharge channel should be set in a place where it was easy to be scoured, in order to accelerate the formation of the scouring air surface and to achieve the purpose of headcut erosion.

According to the topography and geological conditions at the barrier site, by comprehensive comparison, within the three available natural gullies, it was allowed to arrange the discharge channel following the terrain by choosing the middle one which was located on the right side of the barrier dam with the lowest natural notch elevation. The discharge channel was arc-shaped convex to the right bank in the plane. The layout plan of the discharge channel is shown in Fig. 6.1.

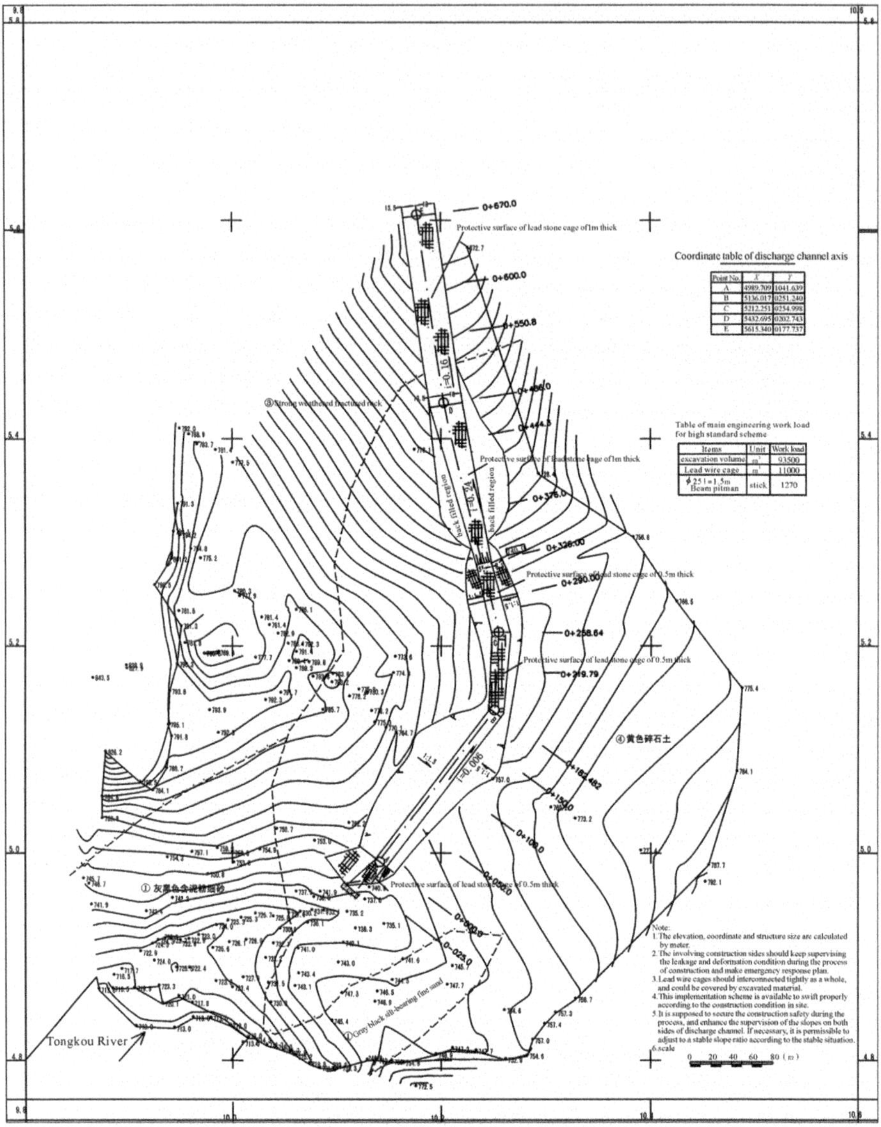

Fig. 6.1 Plain layout of Tangjiashan barrier discharge channel

6.3.4 Structure Design of Drainage Channel

6.3.4.1 Structural Design Principles

(1) The longitudinal slope of the channel should gradually get steeper from upstream to downstream in order to induce the headcut scouring.

(2) The slope of the weaker section at channel outlet should be protected partially to avoid the blockage to channel by slope collapse when the flow was small.
(3) Since the loose residual slope deposits were concentrated on the right side of the channel, in principle, channel slope protection targets at the right side rather than the left side.

6.3.4.2 Structural Form and Scheme

1. Structural form

The discharge channel adopted trapezoidal cross section. The side slopes on both sides were 1:1.5.

Since this project was a rescue project, the aftershocks were constantly occurring during the construction period, and rocks were constantly falling from the right bank landslide. There was no land transportation at the construction site, completely dependent on so air transportation. At the same time, due to the limitation of lifting capacity, only machinery and equipment with a single weight of less than 15t could be used, leading to a limitation of the construction means. Fuel, equipment and logistic support materials had to be transported by air, which meant that the risk of interruption was possible due to cloud, rain, fog, wind and other weather conditions. The uncertainty of the water conditions also leaded to the uncertainty of the actual construction period. All above restricted the construction capacity.

There existed the following preconditions: (1) construction conditions and capacity of the Tangjiashan barrier emergency response and the limited construction period estimated by the hydro-meteorological forecast, (2) the right side barrier dam material composition dominated by residual slope deposits in the surface layer and fractured bedrock landslide deposits in the middle and lower parts of the barrier, (3) the basic judgment that water entrainment scouring subsequent to water flow in the discharge channel will deepen and widen the flow section and increase the flow capacity and (4) the basic condition that the emergency project would be guaranteed by the airborne troops and the local people's government for air transportation and material logistics, thus three excavation plans were made during the designing with the same opening line, same slope ratio, various channel bottom elevations and widths. The total excavation volume of the three options was 50,000 m^3, 70,000 m^3 and 90,000 m^3, respectively, corresponding to the three targets of low standard (high bottom elevation), medium standard (medium bottom elevation) and high standard (low bottom elevation), which were dynamically adjusted according to the meteorological and transportation conditions during implementation. The implementation was dynamically adjusted according to the actual construction capacity, water conditions and other dangerous situations.

2. Three excavation options

(1) Low standard scheme with the highest bottom elevation and the least excavation work. For the discharge channel, the inlet elevation was 747 m, the bottom width

was 28 m, and the total length was 670 m. The gentle section of upstream was 305 m long with a longitudinal slope of 0.6%, and the steep slope section of downstream was 193 m and 172 m long with longitudinal slopes of 24% and 16%, respectively. The volume of earthwork was about 50,000 m^3. When the water level of the upstream barrier lake was heading up to the same level as the lowest notch at an elevation of 752.2 m, the allowable flow rate could be 700–900m^3/s, which corresponded to a roughness of 0.04–0.05 of the discharge channel.

(2) Medium standard scheme with the second-highest bottom elevation and the medium excavation work. For the discharge channel, the inlet elevation was 745 m, the bottom width was 22 m, and the total length was 680 m. The gentle section of upstream was 324 m long with a longitudinal slope of 0.6% and the steep slope section of downstream was 184 m and 172 m long with longitudinal slopes of 24% and 16%, respectively. The volume of earthwork was about 70,000 m^3. When the water level of the upstream barrier lake was heading up to the same level as the lowest notch at an elevation of 752.2 m, the allowable flow rate could be 1000–1300m^3/s, which corresponded to a roughness of 0.04–0.05 of the discharge channel.

(3) High standard scheme with the lowest bottom elevation and the largest excavation work. For the discharge channel, the inlet elevation was 742 m, the bottom width was 13 m, and the total length was 695 m. The gentle section of upstream was 351 m long with a longitudinal slope of 0.6%, and the steep slope section of downstream was 172 m and 172 m long with longitudinal slopes of 24% and 16%, respectively. The volume of earthwork was about 90,000 m^3. When the water level of the upstream barrier lake was heading up to the same level as the lowest notch at an elevation of 752.2 m, the allowable flow rate could be 1400–1800 m^3/s, which corresponded to a roughness of 0.04–0.05 of the discharge channel.

The longitudinal section of the Tangjiashan barrier dam drainage channel (inlet elevation 742 m) is shown in Fig. 6.2.

6.3.4.3 Discharge Channel Protection Program

The discharge channel was mainly for lowering the lake level and guiding the path of headcut erosion, so it did not require the long-term stability of the channel. In order to control the speed of headcut scouring, it was necessary to protect the easy siding section in the lower and middle parts of the channel, so as to avoid the time constrains for personnel evacuation resulted from failure status within short time subsequent to water flow and the channel collapse and blockage with small flow.

When the overall scheme was designed, only steel wire cages were designed for slope protection to simplify the construction. The protection areas were

(1) The steep slope section and its slopes on both sides were protected by double-layer steel wire cages with a thickness of 0.5 m per layer. The double-layer steel

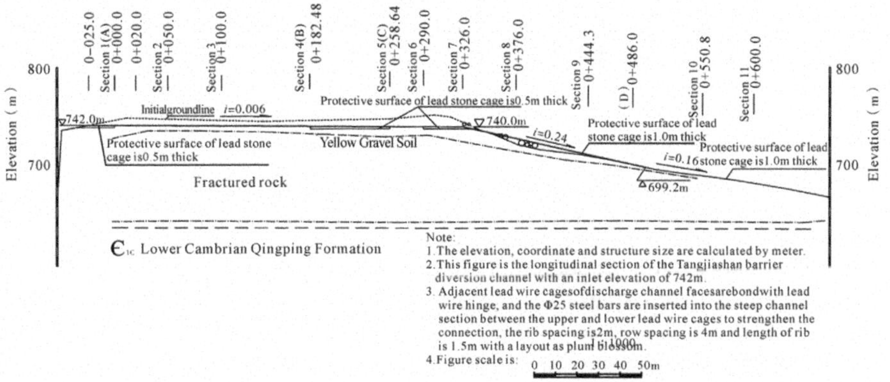

Fig. 6.2 Schematic diagram of the longitudinal section of the Tangjiashan barrier drainage channel (inlet elevation 742 m)

wire cages were connected by inserted steel, and the inserting depth into the barrier dam was not less than 1 m.

(2) The downstream slope falling into a range of 50 m along the river and the discharge channel concave slope were protected by a single layer of steel wire cages with a thickness of 0.5 m.

In the process of implementation, the protection area was optimized in line with the actual geological conditions.

6.4 Emergency Personnel Evacuation Plan of Tangjiashan Barrier Lake

6.4.1 Necessity of Personnel Evacuation to Avoid Danger

Once a large barrier collapses, the large flow generated by the collapse would seriously threaten the life and safety of people along the downstream. Statistics showed that 93% of the barrier lakes had suffered collapse, resulting in a large number of casualties.

There were cities of Downstream of the Jiangyou, Mianyang, Santai, Shehong and Suining in the downstream of the Tangjiashan barrier lake. All of these towns were with economic and population importance in central Sichuan. The analysis results of the dam-break flood showed that the population affected by the downstream flooding area would be 1.3 million in case of full dam-break within 1 h.

Although it was decided to take engineering measures to remove the risk for the Tangjiashan barrier dam, however, the following risks still existed before, during and after the implementation of the engineering treatment plan.

(1) Heavy rainfall caused the lake dam to be overtopped before or during the implementation of engineering measures.

The Tangjiashan barrier had a rainfall catchment area of 3550 km^2. Tongkou River has entered into the flood season in May. The main flood season was approaching, and the water level in front of the barrier continues to rise, and the damage energy was increasing. Calculated by the rainfall runoff and reservoir capacity curve, the water level in front of the dam was lifted by about 1 m resulted from the inflow water produced by every 2 mm of net rain. Even if it did not rain, the daily base flow into the lake could raise the water level in front of the dam by about 1 m. The measured data showed that the multiple year average rainfall of Tangjiashan barrier lake on June was 15.5 d and the maximum daily rainfall was 201.4 mm. Once there was heavy rainfall during the construction period, if overtopping flow appears, a certain scale of flooding would occur.

Due to the destroy of roads after the earthquake, land transportation was interrupted air traffic was constrained by terrain and weather conditions, and the construction schedule of the emergency handling project was difficult to settled accurately in advance. In fact, the initial proposed construction period for the excavation of the drainage channel was 10 d, i.e., from May 25 to June 3, 2008. Due to the absence of land transportation, the transportation of construction personnel, equipment, materials, supplies, etc., could only be done by air. The helicopter take-off and navigation were highly dependent on the weather. The uncertainty of microclimate forecast in the mountainous area made it very difficult to transport the construction equipment in time and quickly. Until 25th, due to weather conditions, the helicopter was unable to transport the equipment. The Tangjiashan barrier lake Emergency Rescue Project Command decided to activate the alternate construction plan of blasting into the channel. More than 1,000 officers and soldiers from Chengdu Military Region carried the explosives manually on their backs late at night on 25th, and the weather improved on 26 and the excavation plan was reintroduced. The uncertainty of the construction plan and construction period led to uncertain effect of danger removal and also increased the possibility of heavy rainfall.

(2) The material composition of the barrier dam could not be accurately predicted. The barrier dam instability might occur at a high water level or the discharge channel headcut scouring effect was not ideal to remove the danger before the main flood period. All these lead to a rapid collapse of the barrier after the main flood.

The upper part of the barrier dam was residual slope of gravel soil, and the lower part was fractured rock. Experts at the site made the evaluation without any exploration data that the upper and lower junction was at an elevation of 710–730 m. Experts mentioned that the possibility of full dam-break was small, but after the water level rose to 710 m, the loose soil layer started to block the water. Thus, the possibility of partial collapse increased. Once the water level rose to the elevation above 752 m of the barrier top notch, there was overflow within the loose layer in a wide range. The upper gravel soil layer was extremely weak in scouring resistance, and there was the possibility of rapid collapse.

The elevation of the channel bottom was above 740 m. The head difference between upstream and downstream was more than 77 m. There were two sections of fractured rocks exposed at the discharge channel 130–180 m away from the inlet, and at the exit of the flat section, the scouring resistance was relatively strong. The experts concluded that the scouring would start when the flow velocity was greater than 5 m/s. However, if the scouring continues at low flow rate, the discharge channel cannot be deepened and widened. In case of a 10-year or 20-year flood, the water level would rise to 752.88 m and 754.75 m which exceed the elevation of the lowest notch. The risk of collapse still existed when overflow of loose body occurred.

Through analysis and calculation, in case of a 10-year flood, the water level rose to 752.8 m and a collapse occurred after the excavation of the drainage channel was completed. In the case of 1/3 dam-break which meant collapse to the elevation of 720 m, assuming that the breach lasts 1–3 h, the flow rate at the breach would still as high as 1630–2640 m^3/s. The peak flow of Fujiang River Bridge was still 10,480–10,860 m^3/s, and the collapse flow exceeded the design flood control standard along the Tongkou River in the downstream of the barrier lake.

(3) Factors like strong aftershocks and paleolandslides activation, either alone or in combination, might cause sudden destabilization of the barrier dam. After the Wenchuan earthquake, aftershocks continued to occur in the affected areas. Calculations showed that the barrier dam would enter into a state of partial instability or extreme stability in the event of a strong aftershock of VII degree. If an earthquake of VIII degree occurred, the possibility of barrier destabilization would be greatly increased. In case of a combination of several unfavorable factors, the possibility of barrier destabilization was extremely high.

There were many uncertainties before and during the implementation of the emergency engineering treatment of Tangjiashan barrier lake, resulting in uncertainties in the duration and flow of the flooding. As long as the water energy in the barrier lake was not released to a safe level, there still existed risk of barrier collapse, and the threat to people's lives in downstream would not be relieved. In order to ensure the safety of people's lives, engineering measures and non-engineering measures must be implemented simultaneously.

6.4.2 Emergency Personnel Evacuation Plan

In order to avoid casualties in downstream, and in accordance with the basic principles of "safe, scientific and quick" and "engineering measures with best endeavor, risk avoidance measures in terms of the most unfavorable factors", experts proposed three generalized model plans for 1/3 dam-break, 1/2 dam-break and full dam-break according to the risk probability, among which the occurrence of 1/3 dam-break mode dominated more. The *Design and Construction Plan of Tangjiashan barrier lake emergency handling* pointed out that although the barrier dam-break flood deducing was still immature and there were uncertainties in the model and the parameters selection, the occurrence of 1/3 dam-break mode dominated more based on the actual field geological conditions and the material composition of the barrier dam. However, when a large flood occurred, the occurrence of 1/2 dam-break or full dam-break was still possible, which should be taken seriously. For the safety evacuation of the downstream people, the local people's government should make plans in three levels of high, medium and low as soon as possible according to the three working conditions of full, 1/2 and 1/3 dam-break, and start them at the right time in accordance with the actual situation.

By the decision of the Tangjiashan barrier lake Emergency Engineering Command Headquarters, advanced evacuation was made for the people within the inundation area of the 1/3 dam-break plan, evacuation contingency plans were prepared for the people within the inundation area of the 1/2 and full dam-break plans. Through analysis, the local people's government concluded that the difference was not significant in the number of people within the inundation area of the ½ and full dam-break plans, so the evacuation contingency plan was made according to the full dam-break plan.

The water conservancy group of Earthquake Relief Headquarters of the State Council proposed the evacuation criteria for the construction team in *Design and Construction Plan of Tangjiashan barrier lake emergency handling,* to specify it, when the difference was 2 m between the water level and the inlet elevation of the discharge channel in the approved construction plan, the construction team should be evacuated in time to ensure the safety of personnel. In case of abnormal deformation and leakage of the barrier dam, it was supposed to conduct timely analysis. The team should be evacuated in a timely manner after approval by the Emergency Response Command of Tangjiashan barrier lake if it was a major hazard that threatened the safety of the construction team. This standard was also for starting the downstream personnel evacuation plan. The actual situation was that due to the forecast of 50–70 mm rainfall on May 25th–27th, the water level of the barrier lake might rise rapidly and might not be able to start the implementation of engineering measures for the reason of the weather. Therefore, the local people's government started to implement the evacuation plan under 1/3 dam-break scenario on May 24. However, the actual rainfall did not occur as predicted, and the difference was large, which led to a severe complain from the people evacuated in Mianyang on June 4.

Through analysis and calculation, impact area of 1/3 dam-break flood included the cities of Mianyang, Suining, with 277,590 population. Before the water flow of discharge channel, these people had been safely transferred in group. Among them, 129,866 people were centralized resettled, and 147,724 people were relying on relatives and friends. The danger of the barrier dam was relieved after the water flow of the discharge channel, and all the people who had been transferred returned on June 11.

The local people's government had prepared an evacuation contingency plan for the 1.3 million people affected by the full dam-break flood while transferring the people affected by the 1/3 dam-break flood. A sound system had been established for monitoring, warning and command evacuation, as well as a plan for transferring people under the yellow, orange and red warning and forecasting mechanism, respectively. All the people in the affected cities, counties, districts, towns, villages and groups were given a timely evacuation plan, specifying the people to be evacuated, the paths, locations, early warning measures and people responsible for the evacuation, and necessary trainings were conducted. The local people's government sets up the danger observation stations around the Tangjiashan barrier lake and the riparian safety zone for 24 h real-time observation and report. The relevant counties, cities and districts also sent their staff to continuously observe the danger and set up a double backup 24 h on-duty hotline to notify every person who needed to be evacuated and transferred in case of a full dam-break, except for those who had already been transferred. Once the risk of collapse occurred, the upstream observation station would continuously send 30 signal flares in every 20 s as an alarm. When the alarm was issued, the Mianyang City Command would simultaneously use hotline, cell phone, text message, radio station, television station, tweeter, radio and other auxiliary methods to announce relevant information to the community.

In fact, the actual rapid discharging process of the discharge channel showed that the maximum flow rate of 6500 m^3/s exceeded the 100-year flood flow rate of 6040 m^3/s at the Tangjiashan barrier site and also far exceeded the measured maximum flow rate of 4270 m^3/s. The peak flood flow rate of the discharge channel exceeded 6000 m^3/s all along the Tongkou River, which far exceeded the flood control standards at every location along the river. If the personnel were not transferred to avoid danger in advance, it would certainly cause casualties. Although the actual flow rate exceeded the flood control standard of the lower reach of the Tongkou River, it was still below the controllable flow rate of the proposed 1/3 dam-break evacuation plan. After the water from the Tangjiashan barrier lake was released through the discharge channel, there was no casualty along the downstream. Therefore, the simultaneous implementation of the drainage channel emergency engineering plan and the downstream personnel transfer plan ensured the decisive victory of the Tangjiashan barrier lake rescue.

6.5 Emergency Response Security Measures of Tangjiashan Barrier Lake

Emergency response measures included rapid decision-making and response mechanism, water and rainfall forecasting system, remote real-time video monitoring system for barrier dam, safety monitoring system for dam area, information communication security system and expert consultation and decision-making mechanism for breach prevention.

6.5.1 Rapid Decision-Making and Response Mechanism

The response time of the Tangjiashan barrier lake was extremely urgent, the later the response, the greater the risk. At the same time, the barrier dam was in a huge scale, it was unable to accurately master the meteorological and hydrological conditions, together with lack of basic information, poor construction conditions and other potential effects such as post-earthquake secondary disasters. All these led to an extremely complex situation for which the response was extremely difficult and the elimination of disaster threat was impossible by conventional operation. For this reason, the priority of emergency response security was to establish a rapid decision-making and response mechanism.

6.5.1.1 Leaders Attached Great Importance and Made Forward Command

The Party Central Committee and the State Council attached great importance to the treatment of the Tangjiashan barrier lake. General Secretary Hu Jintao asked that the emergency work for Tangjiashan barrier lake must adhere to the people-oriented principle and give first priority to the safety of people's lives, conduct careful organization and thoughtful arrange, strengthen the inspection and make emergency planning in order to ensure no major problems in emergencies. Premier Wen Jiabao went to Mianyang City twice, climber up to Tangjiashan three times, studied the specific plan for the treatment of the barrier lake and established the Tangjiashan emergency dredging project command by People's Government of Sichuan Province, the Ministry of Water Resources (MWR), armed police hydropower troops, Chengdu military air force, etc.

MWR, as a core member of the Water Conservancy Group of the State Council's Earthquake Relief Headquarters, was responsible for the earthquake relief work in the disaster area and, at the same time, made all efforts and took priority to provide technical support and guidance for the emergency response strategy of the Tangjiashan barrier lake. The main leaders of the MWR had stood fast in Mianyang City since May 23, the other leaders of the MWR and the key personnel of CWRC, and the chief engineer of MWR had stuck to the posts in Mianyang City since May 19. They were directly responsible for decision-making.

People's Government of Sichuan Province and the people there also made the greatest efforts for it. The major leaders of Sichuan Province served as the commander of the Tangjiashan barrier Emergency Rescue Headquarter. Mianyang City guaranteed the material supply needed for rescue and relief.

6.5.1.2 Rapid Decision-Making and Response Mechanism

The earthquake relief headquarters of State Council set up the Tangjiashan Emergency Response Command, the risk avoidance command departments of Mianyang and Suining cities, and a frontline command at the Tangjiashan site. The leaders of the MWR directed at site in Mianyang or Tangjiashan, forming a frontline command center in Mianyang, a command center in Beijing and three technical support centers in Wuhan, Beijing and Chengdu. A mechanism was set up featuring continuous work by front and back side day and night and carried out a large amount of analysis and demonstration work around hydrological and water conditions, barrier dam stability, flood analysis and emergency handling plan in a very short time. The General Staff of the Chinese People's Liberation Army, the Armed Police Headquarters and the Armed Police Hydropower troops participated in the command at site. The frontline expert group was mainly composed of experts from the CWRC, together with experts from POWERCHINA Chengdu Engineering Corporation, Armed Police Hydropower troops and PLA University of Technology. Under the direct leadership of the CWRC and Chief Engineer of the MWR, by gathering the wisdom of the whole water industry, it was able to put forward a scientific technical program in short time, for which four large dynamic adjustments were made according to the actual situation in the field, so as to provide technical support for the successful implementation of the discharge channel engineering measurements. Armed police hydropower troops with PLA officers and soldiers fought continuously with a total of 1142 officers and soldiers, 108 sets of large equipment were gathered, and 136,000 m^3 of earth stones were excavated. Two Mi-26 helicopters and ten helicopters from the Air Force Army Aviation troops transported rescue personnel and materials, serving as strong support for rescue work of Tangjiashan barrier lake. Mianyang City sets up the emergency rescue headquarters, and a frontline command was set up at the Tangjiashan site, which made it possible to realize rapid mobilization, coordinate every force and conduct multi-sectoral cooperation. A core expert group was formed to provide technical support and dynamic optimization. This rapid response and decision-making mechanism ensured timely decision-making and implementation in the shortest time.

6.5.1.3 Work Mechanism of Top-Down Coherence, Military Coordination, Departmental Linkage, Regional Collaboration

Each department organized a group of experts consisting of a large number of professional engineers and technicians, to propose a scientific plan and provide strong technical support for on-site decision-making by division of responsibility, tight cooperation and danger judgment. The Armed Police Hydropower troops played a vital role in the risk treatment construction of the Tangjiashan barrier lake. Chengdu Military Air Force opened up an air channel to carry out super conventional air transportation. Land aviation regiments of Jinan Military Region and Chengdu Military Region flew with extra intensity to provide a strong support for the risk removal of the barrier lake. The International Cooperation and Scientific and Technical Department of the Ministry of Water Resources, Ministry of Foreign Affairs, the Ministry of Public Security, the PLA General Staff Department and the Civil Aviation Administration of China handled the entry procedures of the Mi-26 helicopter as quickly as possible. The Chinese Academy of Sciences installed a remote broadband video surveillance system on the Tangjiashan barrier in time. The Office of the National Flood Control and Drought Relief Headquarters, the Hydrology Bureau of the Ministry of Water Resources, the China Institute of Water Resources and Hydropower Research, the Chinese Academy of Sciences, the Sichuan Provincial Water Resources Department, the Sichuan Hydrology Bureau, the Sichuan Meteorological Bureau, the Mianyang Hydrology Bureau and many other units also participated directly in the emergency response to the Tangjiashan barrier lake.

It lasted one month from the formation of Tangjiashan barrier lake to the smooth discharge of the lake water. Under the direct leadership of the Party Central Committee and the State Council, all departments and units worked together selflessly, cooperated across ministries, specialties and fields and made full use of all favorable conditions to achieve the decisive victory of the Tangjiashan barrier lake risk removal project, creating a miracle of eliminating the danger of the barrier lake.

6.5.2 Prediction and Forecasting System of Water and Precipitation

In order to strengthen the monitoring of hydrological elements along the Tangjiashan barrier lake, on the basis of making full use of the existing stations, the number of the observation stations was increased appropriately to meet the monitoring requirements. In addition to the existing rainfall stations at Baishi, Piankou in the basin above the barrier, six additional rainfall stations were set up at Tangjiashan, Fushun, Zhicheng and Youfang to monitor the rainfall situation in the basin. Beichuan hydrology station had automatic measuring and reporting equipment, in order to ensure the reliability of upstream water level observation, an additional automatic water level monitoring station was set up. Tongkou temporary observation station was set up about 1.5 km downstream of Tongkou power station, and Xiangshui temporary observation station was also set up at downstream of Tongkou temporary observation station.

The existing automatic measurement equipment at Beichuan station could monitor rainfall and water level and can monitor normally under the condition that flooding or aftershocks did not cause damage to it. Through the observation of the water level and based on the stable water level and flow relationship, it was able to deduce the flow when the flow did not exceed 1500 m^3/s at Tongkou observation point. When the flow exceeded 1500 m^3/s, two 800 m prism-free total stations were used to simultaneously observe the upstream and downstream water surface elevation at the two pre-set observation points at the mid-hillside, and the flood flow was calculated according to the observed water level at the upper and lower reaches, using the ratio drop-area method and the midstream float method. At the Xiangshui observation point, a 120 m prism-free total station was set up on the left slope of the monitoring point to monitor the water level, and the flow was deduced by using the water level-flow relationship of the existing Jiangjunshi hydrological station 600 m in the upstream.

According to the communication conditions of each monitoring station, the communication plan was determined as follows: The information of each hydrological monitoring station at upstream of Tangjiashan was automatically collected and transmitted by Beidou satellite to Mianyang subcenter; the monitoring information of Beichuan hydrological station was automatically transmitted by Beidou satellite to Mianyang subcenter; the observation stations of Tongkou and Xiangshui were communicated by walkie-talkie, and the observation information was transmitted to Mianyang subcenter by satellite phone and voice. The above information was sent to the Hydrological Information Center of the Ministry of Water Resources and the Hydrology Bureau of the Changjiang Water Resources Commission through the existing water information network.

6.5.3 Remote Real-Time Video Monitoring System of the Barrier Dam

In order to ensure the safety of the personnel, it was required to evacuate them when the discharge channel formed by the engineering risk treatment measures was ready to the discharge. It was difficult for people to approach the discharge channel after its discharge, so a remote real-time video monitoring system was set up subsequent to the evacuation of personnel in order to monitor the dynamic situation of the barrier dam after the discharge in real time.

The remote real-time video monitoring system set up 14 monitoring points along the route from the front part of Tangjiashan barrier dam to Fujiang River Bridge in Mianyang. Among them, five points were on the barrier dam, and nine points were located in the downstream of Beichuan, Tongkou Power Station, Fujiang River Bridge, Xiangshui Town, Qinglian and other places. The monitoring points used unattended video cameras to collect data, including five infrared video cameras, two of which were placed on the barrier dam, so that monitoring could be done at night. The camera could be controlled to rotate 360° left and right, 90° up and down and 36 times zoom. The collected data could be transmitted instantly via satellite to various earthquake relief headquarters in Beijing, Shanghai, Mianyang and Chengdu. The clear monitoring images provided by the remote real-time video monitoring system could provide important support for decision-making. This technology provided greater transmission bandwidth than the current wireless network to transmit high-definition images with large data volumes. This was the first time that the system had been used in the field of emergency communications on a national scale.

6.5.4 Safety Monitoring System in the Dam Area

During the construction period, armed police hydropower troops formed a dam area safety monitoring inspection team to monitor the barrier cracks, leaks, abnormal sound and slope collapse on both banks regularly.

After the completion of the first phase of construction, a safety monitoring system was set up in the dam area in order to have real-time accurate information about the stability and deformation of the barrier dam. The system integrated geological disaster monitoring data collection, networked information transmission, information analysis and processing, information dissemination and emergency command. It could monitor and warn the stability and deformation of the barrier dam and grasp the occurrence, development and changes of the barrier dam deformation in real time.

After the survey on site, five measurement points were selected at the low step of downstream barrier dam, and six sets of surface displacement wireless monitoring equipment were installed. The wireless monitoring equipment could conduct dynamic observation to monitor the overall movement of the barrier dam, the impact of fissure seepage on the barrier dam and the precursor of deformation and collapse of the barrier dam in real time in an unattended situation.

6.5.5 Communication Security System

Electricity supply in Tangjiashan dam area was paralyzed due to the earthquake, and the external communication was interrupted. China Mobile overcame the difficulties to install a communication base station on the barrier dam, use gasoline generator to provide power for the base station, ensuring the smooth communication between the upstream and downstream dam. Handheld walkie-talkie was used for communication within the construction area. The Armed Police Hydropower troops also installed wireless radio stations on the barrier dam to ensure the communication in special circumstances.

6.5.6 *Expert Consultation and Decision-Making Mechanism of Collapse Prevention*

The Emergency Response Engineering Command of Tangjiashan barrier lake established an expert consultation group in Mianyang. The expert group was composed of a number of senior experts with rich professional experience. According to the meteorological, rainfall, water regime, on-site construction progress, logistics and material security and other comprehensive real-time situation, the expert group held regular meetings at 9:00, 15:00 and 21:00 every day for research, real-time dynamic analysis and timely proposal of the plan for the leadership decision. An expert group was also set up at the construction site to study and solve problems in construction together with the Armed Forces Hydropower Command. The group held a daily coordination meeting on site regularly to summarize the construction progress of the day, implement the decision of the earthquake relief front command, and research and deploy the construction tasks for the second day.

6.6 Summary

(1) By summarizing the cases of barrier treatment at home and abroad, it was able to obtain the following rules:

① Emergency handling should be completed before the arrival of the flood, so as to avoid the severe working conditions.
② For the emergency handling for large barrier of high risk, it was supposed to carry out an organic combination of engineering measures and non-engineering measures.
③ Engineering measures should take advantage of and follow the situation based on the topography and geological conditions of the barrier dam, facilitating rapid construction.
④ Achievement of the minimum engineering risk removal target should be supported by preparatory measures. When the implementation of the recommended plan encountered difficulties, the preparatory plan should be quickly activated.

(2) From the comprehensive analysis of Tangjiashan barrier dam formation characteristics, stability performance, barrier destabilization damage form and dam-break characteristics, downstream risk loss, etc., Tangjiashan barrier was in an extremely high risky situation and must be treated as soon as possible.
(3) The presence of heavy rainfall in the Tangjiashan barrier area could lead to an early overflow collapse, barrier dam destabilization at high water level, strong aftershocks and paleolandslides activation. The combination of various adverse factors may lead to uncertain factors as rapid barrier collapse and unsatisfactory effect of drainage deepening and widening, etc. In order to ensure the safety

of people's lives and property in the downstream areas, it was necessary to implement personnel transfer to avoid danger.

(4) In order to avoid the catastrophic damage to the downstream caused by the rapid collapse of the barrier dam, it was unrealistic to take only a single engineering measure to eliminate the threat of the dam-break flood to the safety of downstream people's lives and properties within a limited time. It was required to adopt comprehensive plan featuring simultaneous implementation of engineering and non-engineering measures. The comprehensive plan included three aspects: emergency engineering measures for risk removal, personnel transfer for risk avoidance and emergency security.

(5) In response to an emergency like the Tangjiashan barrier lake, which posed a great danger to life, leaders must take the lead and implement an emergency response mechanism with the cooperation of multiple parties.

References

1. Jinyang S. Hydrological emergency treatment of Jiweishan Dammed Lake in Wulong. Hydrology. 2009;19:55–7.
2. Yueping Y. Study on the characteristics and disaster reduction of the huge landslide of the Bomi Yigong Expressway in Tibet. Hydrogeol Eng Geol. 2000;4:8–11.
3. Cheng Z, Cui P, Li Y, et al. Characteristics and countermeasures of landslide and debris flow dammed lake disaster. J Mount. 2008;26(6):733–738.
4. Xiangqian L, Yihao C, Yingqi W. Treatments of landslide dam and its cases by blasting. Hydroelectr Power. 2009;35(10):88–90.
5. Hu X, Lu X, Huang R, et al. Developmental features and evaluation of blocking dangers of Dashui ditch debris flow in Tangjiashan dammed lake. Chin J Rock Mech Eng. 2009;28(4):850–858.

Chapter 7
Construction of Emergency Handling of the "5.12" Tangjiashan Barrier Lake

7.1 The Organization and Implementation of Construction

7.1.1 Construction Features of Emergency Handling of the "5.12" Tangjiashan Barrier Lake

7.1.1.1 Project Overview

The basic scheme for the Emergency Handling of the "5.12" Tangjiashan barrier lake was to excavate at fastest reasonable speed a discharge channel and build steel gabion lining at the right bank and part of the steep slope section of the channel. Based on the actual construction conditions, taken into account the limited construction time under hydrological and meteorological forecast, three schemes of lower standard, medium standard and high standard were proposed with the same opening line, same slope ratio but different channel bottom elevations. During construction, timely and dynamic adjustment was carried out according to construction capability, water regime conditions, meteorological conditions and transportation conditions.

Lower-standard scheme: Elevation of discharge channel inlet was 747 m, and earthwork quantity was 56,200 m^3.

Medium-standard scheme: Elevation of discharge channel inlet was 745 m, and earthwork quantity was 75,300 m^3.

High-standard scheme: Elevation of discharge channel inlet was 742 m, and earthwork quantity was 93,500 m^3.

The discharge channel was divided into upstream gentle slope section and downstream steep slope section. In high-standard scheme, the approximate length of gentle slope section was 326 m and that of steep slope was 344 m. The excavation section of the channel was trapezium-shaped, with design excavation slope ratio of 1:1.5. The gentle slope section passes through the upper part of the barrier dam. The original terrain was saddle-shaped, high in both ends and low in the mid. The highest position of axis was around pile number 0 + 300 with elevation of 752.2 m. The

Q. Yang and W. Yang, *Emergency Handling of the "5.12" Tangjiashan Barrier Lake*,
https://doi.org/10.1007/978-981-99-1706-8_7

Table 7.1 Bill of quantity of emergency handling discharge channel

Scheme	Quantity of Excavation (m^3)	Gabion (m^3)	Number of Connecting Rebar (deformed rebar, Φ25, length: 2 m)
Lower-standard scheme	56,200	12,000	1500
Medium-standard scheme	75,300	11,500	1380
High-standard scheme	93,500	11,000	1270
Actual construction	135,500	Approx. 3000	0

second highest position was around pile number 0 + 130 with elevation of 749.7 m. The lowest position was at pothole part with elevation of 741 m. The elevation of the gentle slope section inlet bottom was 742 m. The average longitudinal slope ratio was approx. 0.6%. The elevation of outlet was 740 m. The maximum excavation depth was approx. 12.2 m, and maximum excavation width was approx. 40 m. The width of channel bottom was 13 m. The steep slope section was basically build alongside the downstream slope of the barrier dam, and the average longitudinal slope ratio was approx. 21%. The elevation at the end of open channel was 669 m. The bill of quantity of emergency handling discharge channel was shown in Table 7.1.

During preparation of the scheme for emergency handling of Tangjiashan barrier lake, in order to safeguard the fulfillment of the basic target of emergency handling, an alternative scheme of blasting method was proposed for the situation that air transportation was obstructed, which was carrying explosives by manpower crossing mountains to the site. From May 23rd to 25th, due to continuous rainy weather, the alternative scheme was once launched by the Earthquake Relief Headquarters under the State Council. On May 26th, with the weather condition becoming positive, the original scheme of channel excavation by machines was readopted.

7.1.1.2 Construction Features of Stratum

Based on the analysis of the cause of formation of the barrier dam and analysis on topographical and geological conditions, the stratum alongside the discharge channel was divided into 4 sections.

(1) From 0 + 000 to 0 + 180, approx. 180 m long, formed by the thick slope deposits (stone chipped clay loam) and underlying weathered rock mass around the Tangjiashan villages, this section was featured by high terrain and dense vegetation, with soil layer thickness of over 10 m, which was suitable for ordinary earthwork machinery construction. But at the channel bottom and the slopes on both banks, especially on the left bank, there were large clasolite rock blocks which may possibly need to be handled by shallow hole blasting method.

(2) From 0 + 180 to 0 + 280, approx. 100 m long, this section was featured by low terrain and there was water in low-lying areas. It was estimated to be formed by slope soil and siltation of original riverbed. It was suitable for ordinary earthwork machinery construction.
(3) From 0 + 280 to 0 + 330, approx. 50 m long, this was the junction of gentle slope section and steep slope section, formed by the relatively complete landslide deposits on the right bank. This section was formed by high terrain, maximum excavation depth, hard rocks and large isolated boulders were widely distributed. It was difficult for ordinary earthwork machinery construction, which may possibly need to be handled with the assistance of blasting and other crushing methods.
(4) From 0 + 330 to 0 + 570, approx. 340 m long, this was the downstream slope of the barrier dam. The vertical elevation difference was approx. 70 m. There was mixed distribution of isolated boulders and detrital soil which made the slope surface uneven. It was designed for filling or cut and fill balance, keeping the channel bottom smooth as possible, and certain protection method was considered.

7.1.1.3 Construction Period Analysis

According to the analysis on the measured hydrological data in Tangjiashan area and its vicinity, the water regime during earthquake was relatively stable, which was about equal to the water regime in normal years. According to the annual average inflow data of Beichuan hydrological station, the annual average daily inflow volume was 7–7.5 million m^3 in Tangjiashan area from May to June. After the earthquake, numerous landslides occurred on both banks of Tongkou River, thus the stage-discharge curve data, which was calculated based on topographical data before the earthquake, was considered large than the present situation. After appropriate adjustment, calculated based on the upstream inflow and stage-discharge curve, under the condition that the upstream inflow was stable, it was estimated to raise the water level to elevation 742 m on the 15th day (June 7th), to elevation 745 m on the 17th day (June 9th), and to elevation 747 m on the 19th day (June 11th). These were the design elevations of the discharge channel bottom of the three different schemes. On the 23rd day (June 20th), the water level will be raised to elevation 752 m, water will flow over the barrier dam. If the estimation was based on the measured data of Zhicheng water level station, the date that the water level reaches such elevation will be 3 days in advance.

Under actual inflow condition, water begins to flow through the Tangjiashan discharge channel on June 7th. The water level of the lake at dam was elevation 740.39 m.

Fig. 7.1 Land transportation map of Tangjiashan

7.1.2 External Transportation Conditions

7.1.2.1 Analysis of Land Transportation Conditions

The distance from Tangjiashan to Beichuan Town is 3.2 km. Before the earthquake, the 302 provincial highway was available on the right bank of Tangjiashan, through which Beichuan, Jiangyou and Mianyang could be reached in the downstream and Xuanping, Zhicheng, Maoxian could be reached in the upstream. The transportation condition was good. After the earthquake, the whole 302 provincial highway was damaged, the Beichuan bridge collapsed, and the highway was buried under the barrier dam at downstream of Kuzhu dam. Due to the raise of water level caused by the barrier lake at upstream of Tangjiashan, the 302 provincial highway was submerged under water; thus, the external land transportation route of construction area was difficult to be made through in short term (Fig. 7.1).

7.1.2.2 Analysis of Water Transportation Conditions

Tongkou River was a mountainous river that had never been navigable and had no shipping facilities along the river. Although the water surface width and depth conditions of the barrier lake after earthquake was suitable for the sailing of motor boats and other small-sized boats, the barrier lake could not serve as the transportation route for the sailing of large ships and the landing of construction machinery, nor could it serve as the smooth passage to connect the supply base. Apparently, the water transportation in this region could not be achieved.

7.1.2.3 Analysis of Air Transportation Conditions

After the earthquake, the land transportation route to Tangjiashan was totally blocked and could not be made available in short term. Air transportation became the only possible option for the access of personnel, equipment and construction materials. On May 22nd, the Water Resource Division of Earthquake Relief Headquarters under the State Council stated in the meeting held in Mianyang, the transportation task will be undertaken by the air force by using helicopters. A special air transportation line from Mianyang to Tangjiashan was achieved. Two Mi-26 heavy-lift helicopters, one requisitioned from a Chinese company and one supported by the Russian Government, were used for the lifting and transporting of heavy equipment. Premier minister Wen Jiabao approved the plan in the evening on May 22nd.

The main models of helicopters participating in the Tangjiashan Earthquake Relief were Mi-17, S-70C (Black Hawk) and Mi-26. The main features of the helicopters were shown in Table 7.2.

From May 26 to June 11, 2008, there were in total 741 sorties of above mentioned helicopters flying to Tangjiashan barrier dam, among them 92 were Mi-26. The total transportation volume completed was: 92 sets of machinery equipment, containers and fuel tanks, 179.9 tons of other materials and 3142 person-time travels.

Mi-26 helicopters were mainly used in flying between Leigu Town and Tangjiashan, with airline distance of 12 km, approx. 30 min for round trip. Mi-17 and S-70C helicopters were mainly used in flying between Mianyang airport and Tangjiashan, with airline distance of 64 km, approx. 60 min for round trip.

As the main transportation scheme for the construction, air transportation had advantages of smooth connection with supply base, shorter transportation distance and high efficiency. The disadvantages were restriction by weather conditions, limited

Table 7.2 Main features of helicopters

Model	Mi-26	Mi-17	S-70C
Length (m)	33.727	18.42	19.76
Height (m)	8.145	4.755	5.13
Empty weight (kg)	28,200	7100	5118
Maximum take-off weight (kg)	56,000	13,000	9200
Diameter of main rotor (m)	32	21.29	16.36
Diameter of tail rotor (m)	7.61	3.9	3.35
Flying range (km)	800	495	660
Service ceiling (m)	4600	5000	5800
Cruise speed (km/h)	255	240	268
Maximum speed (km/h)	295	250	292

weight-carrying capacity of helicopters and restriction by landing conditions on site, etc.

The result from construction had shown that the air transportation scheme was the important guarantee to the success of the emergency handling of the "5.12" Tangjiashan barrier lake.

7.1.3 Construction Scheme and Optimization Measures During Implementation

7.1.3.1 Construction Scheme and Implementation

According to the features of this emergency handling project, through scheme comparison, machinery excavation was chosen as the main method, supported with small-scale blasting works (blasting isolated boulders, cleansing stone blockages, etc.), and manual installation of steel gabions for slope protection. The main construction machinery selected in the design were bulldozer, backhoe excavator, wheel loader, dump truck, pneumatic drill and gasoline rock drill. Restricted by the lifting capacity of Mi-26 helicopter, the maximum weight of single machinery was all under 15t. The main features of machinery actually lifted to the barrier dam were shown in Table 7.3.

During construction, restricted by the air transportation capacity, wheel loaders were not transported to the site. In the afternoon on May 28th, four dump trucks were transported to site for auxiliary construction work. Bulldozer and backhoe excavator were the major construction machinery used in this project.

Based on the topographic and geological features alongside the open channel, the construction site was divided into three operation zones: upstream high ground as

Table 7.3 Main features of machinery

Machinery	Model	Capacity of main motor (kW)	Bucket capacity/carrying capacity	Number
Bulldozer	Shantui SD13	99.5		13
	Dongfanghong T120N	88		8
	Dongfanghong TS100	73.5		5
Backhoe excavator	Guizhou Xiaosong PC130	66	0.28 m^3	3
	HITACHIZx120	66	0.52 m^3	6
	HYUNDAI110-7	70	0.30–0.59 m^3	5
Dump truck			10 t	4

Zone 1, sink area in the middle as Zone 2, downstream stone blockage and steep slope section as Zone 3.

Due to the very limited construction period and difficulty in acquiring the accurate condition on weather, aftershock, seepage and water regime, the key focus of excavation was on Zone 1 and Zone 3. The excavation slag discharge direction in Zone 1 was mainly toward the upstream. The slope height in Zone 2 was not high, and the space was relatively open; the slag discharge direction was mainly toward outside the slopes of the two sides (more toward right side). The slag discharge in Zone 3 was mainly toward downstream; the construction was combined with slope trimming of steep slope section and balanced cut and fill. The basic methods of earthwork excavation were bulldozer relay excavation, backhoe relay excavation, backhoe slag excavation, bulldozer slag pushing, backhoe slope trimming, etc. Steel gabion slope protection was constructed by workers manually.

7.1.3.2 Dynamic Optimization During Construction

Dynamic optimization was the principle that must be followed in emergency handling. During Tangjiashan emergency handling, in order to safeguard the fulfillment of the basic target, maximize construction efficiency and shorten construction period, water resource experts participated as guidance for site construction and carried out scheme optimization and adjustment.

(1) In Zone 1, trees were dense and thick which make sawing efficiency low. Through experiments, it was adopted to take blasting clearance method by burying explosives at the pits excavated near tree roots.
(2) In Zone 3, rocks were hard and there were a large number of large isolated boulders, which hinders excavation progress. The method of slag clearance after blasting in the original design was optimized, because each blasting quantity was small thus too much time was wasted during machinery shutdowns and sheering-offs. The CAPF Hydropower Force headquarter mobilized machinery with best performance and operators with most experience, overcame the 50 m long extremely difficult construction section and achieved complete clearance, using the progressed excavation method.
(3) In order to shorten the time for machinery shutdowns and sheering-offs during blasting work, hydraulic static rock ripper was adopted for the breaking of giant isolated boulders and showed good result.
(4) CAPF Hydropower Force overcame difficulties of rainstorms, aftershocks, landslides, and guaranteed the 24 h non-stop construction of each construction machinery, fully utilizing machinery efficiency. On May 30th, it had been excavated to the design elevation 742 m, and slope trimming and protection was completed. But at this time, the upstream lake water level was only 733.88 m, still 8.12 m distance to channel bottom. Front command in Tangjiashan decided to keep lowering the channel bottom elevation. On May 31st, the channel bottom

elevation was further lowered to 739–740 m, with bottom width of 7 m. From May 26th till then, the total earthwork excavation volume was 135,500 m^3.

(5) Considering that the cataclasite rocks in Zone 3 had blocked and delayed the downward water flow, after site study, the stone gabion bottom protection for steep slope section was canceled, and the stone gabion protection for right side slope was adjusted.

7.1.4 Construction Progress and Construction Period

In determining the reasonable construction period for handling the Tangjiashan barrier dam, principles of "proactive", "as early as possible" and "deep excavation" were followed, which means to lower as much as possible the water level of barrier lake within limited time and reduce the water volume in the lake. It was difficult to make accurate estimation on the lake water rising process due to effect of runoff and precipitation from upstream river basin. On May 21, 2008, the water level measured at dam front was 711 m, with 39 m height difference from the lowest pass point of weir crest. Continuous observations in the following days shown that the daily reservoir inflow water volume was around 7,500,000 m^3. Calculated based on stage-discharge curve, the daily rise of water level was estimated to be around 2 m, the daily rise at upper part was around 1 m. If the reservoir inflow volume remains unchanged, and not engineering methods taken, the barrier lake will be overtopping on around June 20th. If engineering method of discharge channel excavation was implemented, the lake water level will come to the designed channel bottom on around June 5th, at inlet elevation 742 m.

May 21st was determined in the scheme as the official launch time of the emergency handling project. Given consideration to time needed for construction preparation and necessary time for withdrawal before discharge of lake water into discharge channel, the headquarter had confirmed May 25, 2008, as the official commencement date, June 3rd as the completion date and total construction period of 10 days. The excavation period was 8 days and protection lining period was 2 days. The target was to make condition for the implementation of high-standard scheme, scheme 3. The average excavation volume was 13,000 m^3/d. The peak excavation volume was 16,000 m^3/d. Restricted by the lifting capacity of air transportation, the construction scheme could only select T130 below-standard bulldozers, bucket capacity of 0.6 m^3 below-standard backhoe excavators, bucket capacity of below 2 m^3 wheel loaders and carrying capacity of 8–10 t below-standard dump trucks as the main equipment for the open channel excavation. It was calculated to need a total number of 26 bulldozers, 18 backhoe excavators, 12 loaders and 10 dump trucks. Eventually, 24 bulldozers, 16 backhoe excavators and 4 dump trucks were put into operation. It was planned to allocate 200 construction personnel on the dam section. CAPF Hydropower Force had eventually allocated a total number of 1142 military personnel, with 637 persons for the site of Emergency Handling of Tangjiashan barrier lake, 94 persons for equipment transportation at Mianyang airport and guarantee of provisions, 404 persons for supply of provisions and guarantee of road transportation condition of Beichuan and 7 persons for commanding and telecommunication at front command in Mianyang.

Due to numerous uncertain conditions of the emergency handling of Tangjiashan barrier lake, the above estimation on construction period was rough. In case of encountering unfavorable circumstances such as delay in construction preparation, failure of organizing certain construction resources, obstruction of air transportation route, rapid rising of lake water level due to upstream rainstorm and occurrence of aftershock landslide, the available construction duration world be severely impacted. Therefore, dynamic command and establishment of a fast decision-making and fast-response mechanism was necessary during construction implementation.

After reaching Tangjiashan on May 21st for the first time, the bad weather from May 22nd to 25th had made it impossible for helicopters to operate, thus the command headquarter decided to launch the alternative scheme for blasting. From the evening of 25th to the early morning of 26th, around 1200 military personnel from the PLA troop and CAPF Hydropower Force had carried the explosives and detonators with them and climbed the mountains until. Later on 26th when the weather condition turned better, the helicopters re-started to transport and the alternative scheme was shutdown.

Coordination meetings were jointly held by the site expert group and CAPF Hydropower Force everyday on the construction site. Plans were studied thoroughly and implemented efficiently according to the latest situation on site. In regards to construction safety management, 24 h monitoring on precipitation, landslides, seepage, water level, etc. were carried out. Safety and risk-avoidance plan were prepared and counter-measures for accidents would be effectively adopted.

The target was to achieve high-standard scheme in the construction arrangement. The principle for the construction of discharge channel was "focusing on the two ends and then getting to the mid" and excavate step by step from high to low elevation, so that it was possible to quickly change to medium-standard or lower-standard scheme in case of encountering unusual rise of water level.

Regarding construction organization, the potential of machinery was fully utilized, the best operation personnel were inputted and effectively arranged. The principle was to "machines don't stop while personnel take shifts" and keep every machine working continuously and efficiently above quota level.

The dynamic command won valuable time for the project, increased construction efficiency and guaranteed personnel safety.

7.2 Dynamic Optimization and Adjustment of Discharge Channel

7.2.1 Dynamic Adjustment

The official commencement of the construction of discharge channel was on May 26th and completed at 24:00 on May 31st. During implementation, the structural design scheme of discharge channel was optimized and adjusted according to the actual exposed geological conditions and construction capacity.

(1) The excavation exposure showed that loose residual slope deposits were mainly distributed along the discharge channel, and the exposed cataclasite rocks at around 60 m from the outlet of gentle section could function as the lock in controlling rapid retrogressively scouring. For that, all the original protection measures at the outlets of steep slope section and gentle section were canceled. The original steep slope section could be used as waste disposal area, under the requirement that the slag stacking elevation shall not be higher than the outlet elevation of the gentle section. The slags would be carried away by water flow.
(2) In order to lead the water flow into the river as soon as possible, the channel route design scheme was slightly adjusted, and the outlet direction of the gentle section was slightly adjusted to the left compared with the original direction.
(3) In construction on gentle section upstream, combination of backhoe excavators and bulldozers was adopted for the excavation and transportation of slag. Adapting to this construction method, the 10.6% slope ratio section around 170 m at the gentle section inlet was adjusted to backward slope above pile number 0 + 170. The actual vertical section of the discharge channel bottom was humped-shape.
(4) The actual construction capacity was significantly enhanced than the original estimation. Close to achieving the high-standard scheme, the outlet elevation had been reduced from 742 to 741 m. Close to the completion of optimized construction scheme, the channel bottom elevation was reduced by 1 m. Eventually, the gentle section inlet elevation was 740 m, the gentle section outlet elevation was 739 m, and the channel bottom width was 7–10 m.

7.2.2 Scheme of Excavating a New Branch Channel

After the discharge channel was completed according to the original scheme, there was less water from the upstream; thus, the water level of the lake rises slowly and failed to reach the inlet elevation of drainage tunnel. In order to achieve the emergency handling principle of "safety, scientific, fast", to effectively enlarge the discharge section area, to increase the discharge capacity and scouring efficiency of discharge channel, after site consultation meeting on June 6th, it was decided to excavate a new discharge branch channel at the left side of the end of discharge channel gentle section and lower the branch channel elevation as much as possible to improve discharge and scouring efficiency. The design dimension of the branch channel was 110 m in length, 10 m in channel bottom width, 1:6–1:4 in longitudinal slope of channel bottom and 1:1.5 in slope ratio. For slope height above 10 m, bridle path of 3 m wide was arranged. On the right side, the inlet and the original discharge channel, rock ridges of width above 3 m were kept and would be blasted and demolished after the construction of branch channel. The construction requires that backhoe excavator and bulldozers shall be used for excavation as first option. In case of encountering cataclasite rocks, assistance with blasting methods could be adopted considering the actual situation in combination with slagging by backhoe excavators and bulldozers. During construction, if the cataclastic rocks exposed in the original discharge channel was found to have been pinched out in the branch channel, the branch channel bottom could be narrowed. In order to speed up the construction, on the premise of ensuring operation safety, the construction principle of "machines don't stop while personnel take shifts" was adopted.

During construction, the already completed part of discharge channel began to let water flow through on June 7th, a steep channel with a depth of 10 m had been formed at the outlet section on June 9th, retrogressively scouring was going rapidly, the scheme of excavating a new branch channel was stopped.

7.2.3 Scheme of Blasting Method for Channel Excavation

7.2.3.1 Alternative Scheme of Blasting Method for Channel Excavation

After the *Technical Scheme for Design and Construction of Tangjiashan Emergency Handling Project* was officially submitted to the Earthquake Relief Headquarters under the State Council on May 23rd, the expert group started to study the alternative scheme for situation when air transportation was not available under unfavorable weather condition.

(1) Manually dig pits for explosives along the discharge channel in upstream and downstream sections separately. Upstream section was 130 m in length and downstream section was 40 m in length. Every pit was 2.5 m in depth, in the shape of inverted cone with underpart diameter of 0.6 m and upper opening diameter of 2.0 m. Considering the restricted conditions of dam, no support methods could be taken, this might be the maximum depth of digging.
(2) Pits for explosives were arranged in a square shape on the plane with 4 m spacing distance of holes. Three lines and 42 rows were arranged along the discharge channel. Altogether 126 pits.
(3) The volume of explosives was designed according to the strengthened throwing blasting, and the blasting action index was taken as 1.5. Each pit was for emulsified ammonium nitrate explosive of 50 kg, in total 6300 kg explosives and 1000 detonators.

(4) In order to reduce the shock impact to the surrounding unstable mountain massif, non-electric millisecond relay network for initiation was adopted. Single section explosives was 150 kg.
(5) The explosion was foreseen to form a channel with 13 m in width and 2.3 m in depth. If the construction machinery could still not be brought to the dam afterward, the water flow level of barrier lake could only be reduced by around 2.3 m and the storage reduced by 20,000,000 m^3. If the construction machinery could be brought to the dam after blasting works, and on the premise of having enough time for operation, then further excavation could be continued on the loose layer after blasting, which could improve the excavation efficiency of construction machinery.

7.2.3.2 The Suspension of Alternative Scheme

From May 23rd, it had been rainy and foggy for several consecutive days. The visibility of Tongkou River valley was less than 500 m, and the helicopters were commanded to return to base after taking-off for several times. The personnel left on-site for observation after reaching the top on May 21st had no supplies and lost communication. On May 24th, one flight was made to reach the top under extreme adverse weather, but the flight was forced to stop shortly after. Under the situation that short-term weather forecast could not estimate on when the weather would clear up, the water level of barrier lake (taken from water level of upstream Zhicheng city) had risen from 711 to 721 m from May 21st to 25th, and kept rising at a speed of around 2 m/d. The construction scheme of excavation by machinery could not be implemented; the situation became extremely urgent.

In the morning on May 25th, the scheme of blasting method for channel excavation was officially submitted to the project headquarter for discussion. After the meeting, the headquarter invited blasting experts to Mianyang for consultation. It was agreed with consensus that the scheme of blasting method for channel excavation was technically viable and was the only possible alternative scheme under the adverse situation that air transportation was obstructed. But the drainage capacity by the alternative scheme will be far inferior to that of the machinery excavation scheme.

In the afternoon on May 25th, the weather forecast had estimated the rainy and foggy weather would continue in the coming days thus could not determine the time when flying could become available again. In addition, a 6.2 aftershock occurred on the same day, and the headquarter meeting decided not to wait passively for the weather getting favorable and immediately organized around 1200 military personnel from PLA and CAPF Hydropower Force to manually carry explosives, detonators and equipment to cross the mountains by walking for 5–8 h and reach the crest of barrier dam from early morning to the afternoon on May 26th.

In the morning on May 26th, the weather turned favorable, the scheme of blasting method for channel excavation was suspended, and the method of machinery excavation was re-implemented. The first Mi-26 helicopter carrying bulldozer successfully arrived on site at around 13:00. Through this whole operation, the military force had shown brave, unyielding and fast-response spirits, and the command headquarter had shown proactive capability. Also, due to numerous restriction factors, it had shown that the weather forecast played an important role for the decision-making during emergency handling of Tangjiashan barrier dam.

7.2.4 Scheme of Channel Bottom Blasting and Deep Excavation

After the official construction commencement of discharge channel on May 26th, 44 construction machinery were assembled from 26 to 28th, the construction work had entered the right track quickly and construction intensity was increasing every day. Considering the 1200 military forces had already overcame numerous difficulties and brought 10.16 t explosives and 1200 detonators to the dam on 26th, in order to make advantage of it, it was proposed to further reduce the elevation of channel bottom by excavation and blasting. The scheme of channel bottom blasting and deep excavation was as follows:

(1) When the excavation of discharge channel had reached design elevation, after the withdrawal of construction machinery, taking advantage of the 1–2 days before lake water level reaching the elevation of channel bottom, keep 2 backhoe excavators there at the channel bottom for digging pits or troughs for explosives and carry out blasting work in the pits or troughs to let the lake water discharge in advance.
(2) The design depth of pits or troughs were 2.5 m. The volume of explosives was designed according to the strengthened throwing blasting. One line was arranged along the channel direction, with 3 m spacing of pits. In total, 100 explosive charges were arranged, with single charge weight of 48 kg, and total weight of 4800 kg.
(3) In order to reduce the shock impact to the surrounding unstable mountain massif, non-electric millisecond relay network for initiation was adopted. Single-section explosives were controlled to be less than 96 kg.
(4) After the explosive charges were in place, the best initiation time was selected to make the lake water flow through the bottom of the blasting funnel, which was expected to reduce the water level of the barrier lake by about 3 m.

After this theme was submitted, the command headquarter considered that the project progress had been ahead of schedule, the water level rose more slowly than expected, and it was possible to further deepen the channel bottom with excavators, and some experts believed that the height of the channel slope had exceeded 12 m, and blasting operations might lead to slope collapse, which had certain risks, so the scheme was not adopted. Generally, the explosive materials once left warehouse cannot be brought back for further use, so the detonators were destroyed on site and the explosives were transported back to Mianyang for disposal.

7.3 Risks in the Implementation of the Drainage Scheme and Countermeasures

7.3.1 Risks in the Implementation of the Drainage Scheme

7.3.1.1 Extreme Hazard of Barrier Dam Accident

The size of barrier dam was enormous, with storage of 316,000,000 m^3, affecting downstream population of over 1,300,000. Once the barrier dam-break occurred in a sudden, the hazard will be extreme.

7.3.1.2 Unpredictable Meteorological and Hydrological Conditions

Limited by the current meteorological forecast level, reliable numerical forecast cannot be made for long-term weather conditions of more than 3 days. Under the tense situation that the water level of the lake was rising at the speed of 1–2 m every day, the time for formulating the scheme, preparing for construction and completing the construction was extremely urgent. Moreover, it was flood season, heavy rainfall may occur at any time in the reservoir area, seriously threatening the smooth implementation of the emergency handling scheme. There was no land transportation condition at the construction site, and the equipment, fuel and logistics support materials were all dependent on air transportation, which was greatly affected by meteorological conditions and was may be obstructed at any time due to cloud, rain, fog, wind, etc. The continuous precipitation from 24 to 25th had made it unable to be transport construction machinery to site; thus, alternative scheme of blasting method was forced to be implemented until the weather condition turned good later. The forecast of heavy rainfall in late May also had a great impact on the construction schedule.

7.3.1.3 Insufficient Basic Data

Barrier dam was formed in a short time due to landslide caused by earthquake. The landslide mass was composed of soft and hard materials, with different water permeability. Due to the urgency of emergency handling, there was no time nor condition for material tests. There were no survey data that could meet the accuracy requirements of engineering for the reservoir and dam area of the barrier lake; thus, reservoir capacity, lake water storage and other basic data were difficult to be determined accurately.

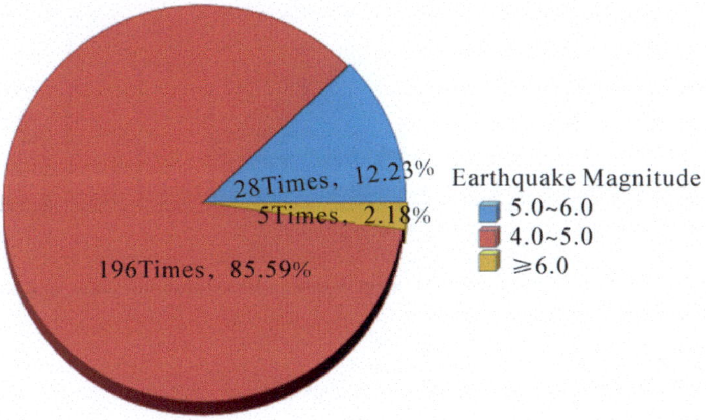

Fig. 7.2 Magnitude statistics of aftershocks of magnitude above 4.0 (Till 08:00 July 5th)

7.3.1.4 Impact of Aftershock on Construction

During construction, aftershocks continued. During the aftershocks, the number of rolling stones on the landslide surface on the right bank of the dam area increased significantly, which seriously affected the safety of equipment and personnel in the construction area. Till 18:00 June 12th 2008, a total number of 208 aftershocks above magnitude 4 occurred, among which 175 aftershocks of magnitude 4.0–4.9, 28 aftershock of magnitude 5.0–5.9 and 5 aftershocks of magnitude 6.0–6.9. The maximum aftershock magnitude was 6.4, occurred at 16:21 on May 25th, in Qingchuan county of Sichuan Province, about 110 km straight-line distance from Tangjiashan barrier dam. The aftershock closest to the epicenter was 15 km from Tangjiashan barrier dam, occurred at 18:51 on June 8th, with magnitude of 4.8.

Till 08:00 July 5th 2008, there had been a total number of 15,703 aftershocks occurred in the aftershock area of Wenchuan 8.0 Earthquake, among which 196 aftershocks of magnitude 4.0–4.9, 28 aftershock of magnitude 5.0–5.9 and 5 aftershocks of magnitude 6.0–6.9, the maximum aftershock magnitude was 6.4. The magnitude statistics of aftershocks of magnitude above 4.0 was shown in Fig. 7.2. The time series of aftershocks of magnitude above 5.0 was shown in Fig. 7.3.

7.3.1.5 Potential Impact of Secondary Disasters Such as Landslide in Reservoir on Construction Safety

The Malingyan ancient landslide mass was located at the upstream of the reservoir, within the valley area about 4.0–4.6 km on the left bank of Tangjiashan barrier dam, in between of Malingyan gully and Zhangjiashuitou gully. The mass was about 650 m long along the river, with rear edge elevation of about 1047 m and front edge elevation of about 665 m. The area was about 300,000 m^2, and the volume was about

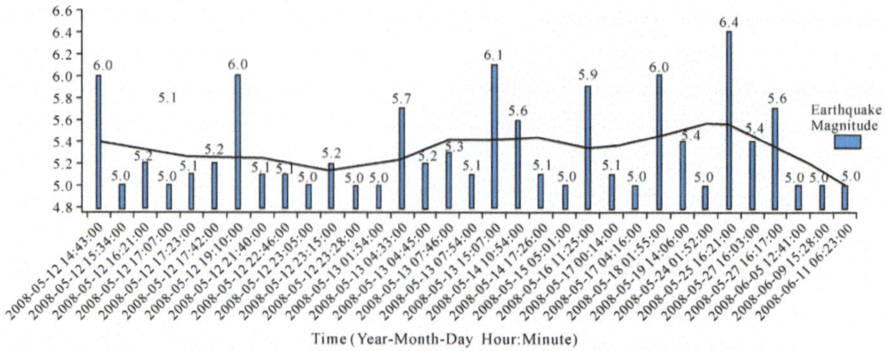

Fig. 7.3 Time series of aftershocks of magnitude above 5.0

16,500,000 m^3. It was difficult to assess the stability of the landslide due to limited study at that time on the geological structure and other basic data of the Malingyan ancient landslide mass.

7.3.1.6 Impact of Seepage Behind Dam and Piping Effect on the Safety of Barrier Dam

After reaching the dam on May 21st, at around elevation 669 m of the downstream slope toe of barrier dam, 15 seepage points were found, with seepage flow of 1 m^3/s. The initial seepage was muddy water and became clear later. After the lake water level gradually rose, on May 30th, at around elevation 700 m of the upstream dam slope, 8 more seepage points were found, with overall seepage flow of 5 m^3/s. The seepage was clear water. At the moment, there were over 650 technical experts and CAPF Hydropower Force military personnel. The impact of seepage on the stability of barrier dam was crucial to the safety of the personnel on the dam.

7.3.2 Countermeasures for Engineering Risks

Due to the numerous uncertainties during emergency handling, the engineering decision-making and construction was accompanied with enormous risk. In order to guarantee the safety of downstream citizens and construction personnel on dam, the headquarter established a sound fast-respond system of decision-making, implemented on-site safety protection measures, prepared alternative schemes, prepared the scheme for transfer and emergency evacuation at downstream, and successfully achieved the target of "zero death".

7.3.2.1 Fast-Response, In-Time Decision-Making and Efficient Implementation

Due to the urgency on construction period, in order to avoid too many signal transmission links, a fast-response system covering all participating units was established on site. The fast-response system includes the command decision-making system and the dam area emergency communication guarantee system. The command decision-making system was mainly composed of staff from Ministry of Water Resources, CAPF Hydropower Force, Government of Sichuan Province, Government of Mianyang City, etc. This system focuses on the in-time adjustment and efficient implementation of scheme according to site conditions. In order to meet the requirement of on-site communication, China Mobile Sichuan Branch had set up a temporary base station that can accommodate 30 communications at the same time on the site. The CAPF Hydropower Force had set up 16 short wave radios and multiple satellite phones on and off the site.

7.3.2.2 Safe Construction, Safety Monitoring and Warning System

A temporary hydrological station was set up in the upstream of the barrier dam to monitor the water level in front of the dam once every 1 h, and the monitoring frequency would be intensified in case of large flow. The remote real-time video monitoring system was set on the hillside on the left and right banks of the barrier dam, and data of the barrier dam were collected and transmitted to the rear headquarter by satellite in time.

A safe construction system was established on construction site on the dam. During construction, protective clothing was required. Operators at the waterside should wear life jackets and work in accordance with the safety operation protection regulations to ensure personal safety.

An evacuation command system was established, and emergency evacuation routes and temporary refuge sites were planned. The emergency evacuation route was arranged along the path on the left side of the living camp, and the temporary refuge was located on the highland at the downstream side of the left dam abutment. In case of emergency, the on-site headquarter shall promptly notify all units to evacuate to the temporary refuge site at the downstream of the left bank of the barrier dam in an orderly manner.

7.3.2.3 Preparation of Alternative Schemes

Alternatives schemes include scheme of blasting method for channel excavation, scheme of channel bottom blasting, scheme of manually carrying petrol to site, scheme of excavating a branch channel, etc.

7.3.2.4 Scheme for Transfer and Emergency Evacuation at Downstream and Drill

According to the flood process and risk assessment results calculated by different barrier dam-break modes, the population within the inundation area of 1/3 barrier dam-break scheme shall be transferred in advance, and the personnel transfer plan for 1/2 barrier dam-break scheme and the complete barrier dam-break scheme were be formulated at the same time.

Chapter 8
Emergency Monitoring and Guarantee System

Tangjiashan barrier lake, formed due to “5.12” earthquake, was an extreme high-risk grade barrier lake, featured by high water head and enormous water volume, posing a serious threat to the safety of 1.3 million people downstream. The barrier dam was under adverse condition of continuous aftershocks, continuous rolling stones on the slopes both banks, complex weather and unstable water regime thus made the emergency handling increasingly urgent and challenging. During the implementation of emergency handling construction on the dam and evacuation of people at downstream, the establishment of an emergency monitoring and guarantee system for forecast, early warning and providing basis for decision-making was a necessary guarantee in facing large-scale disaster events. For this reason, necessary construction safety guarantee system shall be established at the construction site of Tangjiashan barrier dam for the construction implementation on the dam. Meanwhile, it was of great necessity to establish hydrologic emergency monitoring system, information transmission and communication system, hydrological information analysis and processing system, flow and rain forecast and outburst flood warning system, and broadband wireless video monitoring system.

8.1 Hydrologic Emergency Monitoring System

8.1.1 Purpose and Principles

Tangjiashan barrier lake was an extreme high-risk grade barrier lake. In addition, there were four barrier lakes at the downstream of Tangjiashan barrier lake, including Kuzhu dam, Xinjiecun, Baiguoshu and Yanyangtan barrier lake. They were posing serious threat to the safety of 1.3 million people in downstream villages including Beichuan, Dengjia, Tongkou, Xiangshui, Qinglian and the second largest city in Sichuan Province, Mianyang City.

Q. Yang and W. Yang, *Emergency Handling of the “5.12” Tangjiashan Barrier Lake*,
https://doi.org/10.1007/978-981-99-1706-8_8

According to the requirements of emergency handling of Tangjiashan barrier lake and the evacuation of downstream citizens, it was necessary to monitor the water regime changes in the upstream and downstream of the barrier lake and the flood process in and out of the lake during the whole process. When Tangjiashan barrier lake discharged, it was necessary to monitor the flood changes in the main towns and water-related engineering facilities along the Tongkou River basin during the whole process, so as to find the hidden danger in time and carry out necessary countermeasures. Therefore, the establishment and effective implementation of the hydrologic emergency monitoring system were very necessary and urgent.

In order to ensure the scientific and accurate decision-making of risk elimination and avoidance, the key point of hydrologic emergency monitoring was the geometry survey of Tangjiashan barrier lake and four downstream barrier lakes, as well as the emergency real-time water regime monitoring and reporting of all concerned points along the river during the discharge process of Tangjiashan barrier lake.

As the upstream and downstream areas of Tangjiashan barrier lake were seriously damaged by the earthquake, the monitoring conditions were very poor with personal safety assurance, and the emergency monitoring system must respond quickly in case of sudden outburst or massive artificial flood discharge, the following principles shall be followed in the operation of the hydrologic emergency monitoring system:

(1) Convenient, fast, safe
(2) Making full use of the existing hydrological station network and historical hydrological data in the basin
(3) Making full use of advanced observation equipment and technology
(4) Stable and reliable information transmission
(5) Coordinated application of multiple disciplines and schemes.

8.1.2 Items of Monitoring and Monitoring Scheme

8.1.2.1 Station Network Layout Plan

Hydrological (precipitation) stations including Baishi, Jiangjunshi and others were successively arranged in Tongkou River basin. Basic information of the stations were shown in Table 8.1.

In order to strengthen the monitoring of hydrological features along the Tangjiashan barrier lake when the barrier dam-break occurred, temporary observation points were added on the basis of the existing hydrological stations in the Tongkou River basin (Table 8.2).

8.1.2.2 Monitoring and Reporting Scheme

According to the layout of stations in Tongkou River basin, the monitoring and reporting schemes of each monitoring station were summarized in Table 8.3.

Table 8.1 Basic information of the hydrological (precipitation) stations

Station name	Type	Catchment area (km^2)	Data period	Measured maximum discharge (m^3/s)	Investigated maximum discharge (m^3/s)
Baishi, Piankou	Precipitation		1962 till present		
Zhicheng	Water level	3134	1962 till present		
Beichuan	Hydrological	3575	1998 till present	1940	
Maoba	Hydrological	3577	1961–1964	3710	6720
Tongkou HPP	Hydrological	4102	1988–1997	5390	
Jiangjunshi hydrological station	Hydrological	4162	1957–1987	4820	8060

Table 8.2 Added temporary observation points

Number	Location	Station name	Purpose
1	Tail of barrier lake	Fushun, Zhicheng and Youfang	Grasp the precipitation condition of the basin
2	Zhicheng	Zhicheng	Control the inflow flood condition
3	Discharge outlet of Tangjiashan barrier lake	Tangjiashan	Control the discharge process condition
4	Around 4.6 km downstream of dam site	Beichuan	Flood routing process monitoring
5	Tongkou town at upstream of Beichuan hydrological station	Tongkou HPP	Flood routing process monitoring
6	Downstream of Tongkou temporary observation station	Xiangshui	Flood routing process monitoring
7	Fujiangqiao	Fujiangqiao	Flood routing process monitoring

8.1.2.3 Instruments and Equipment of Hydrologic Emergency Monitoring

Hydrologic emergency monitoring was the hydrologic observation under special environmental conditions. Because its instruments and equipment will be subject to various adverse factors, it puts forward higher requirements for instruments and

Table 8.3 Summary of monitoring and reporting schemes of each monitoring station

Station name	Data collection method	Transmission method and frequency of monitoring and reporting	Discharge and other items
Baishi, Piankou, Fushun, Youfang	Tipping bucket rain gauge	Beidou satellite, incremental automatic transmission	Establish the stage-discharge rating curve
Zhicheng	Pressure water level sensing	Beidou satellite, every 30 min	Estimate the inflow of the barrier lake through the water level in Zhicheng
Tangjiashan	Pressure water level sensing	Beidou satellite, every 15 min	The non-prism total station and infrared distance measuring instrument measure the water crossing section, the electric wave flow meter measures the water surface velocity, and the satellite telephone reports. Remote video monitoring device
Beichuan	Automatic monitoring and reporting of precipitation and water level	Beidou satellite, every 15 min	The discharge was directly checked from the water level using the comprehensive water level-discharge rating curve of the station over the years
Tongkou HPP	Automatic monitoring and reporting of precipitation and water level	Beidou satellite, every 15 min	Comprehensive stage-discharge rating curve was directly checked and calculated by water level. Slope-area method
Xiangshui	Manually observe water level through non-prism total station	Satellite telephone, intensify the frequency during discharge	The measured hydrological data of Jiangjunshi hydrological station from 1957 to 1987 were used to establish the stage-discharge rating curve for calculation

(continued)

Table 8.3 (continued)

Station name	Data collection method	Transmission method and frequency of monitoring and reporting	Discharge and other items
Fujiangqiao	Float Gauge	GSM and PSTN backup each other, every 20 min	The discharge was checked by the comprehensive stage-discharge rating curve of the station since 1954

equipment. To ensure the smooth implementation of the above scheme, the equipment configuration of each monitoring station in the emergency real-time hydrological forecasting system was shown in Table 8.4.

Table 8.4 Equipment configuration of each monitoring station

Station name	Monitoring feature	Instruments and equipment
Baishi, Piankou, Fushun, Youfang	Precipitation	Tipping bucket rain gauge
Zhicheng	Water level, precipitation	Ultrasonic/pressure water level gauge
Tangjiashan	Water level, precipitation, discharge	Non-prism total station, electric wave flow meter, infrared distance meter, pressure water level gauge, bubble water level gauge, tipping bucket rain gauge, video acquisition instrument, GPS
Beichuan	Water level, precipitation	Tipping bucket rain gauge, bubble water level gauge, video acquisition instrument
Tongkou HPP	Water level, precipitation, discharge	Tipping bucket rain gauge, pressure water level gauge, non-prism total station, electric wave flow meter, infrared distance meter, video acquisition instrument, GPS
Xiangshui	Water level	Non-prism total station, infrared distance meter, video acquisition instrument, GPS
Fujiangqiao	Water level, precipitation, discharge	Tipping bucket rain gauge, float gauge, non-prism total station, electric wave flow meter, infrared distance meter, video acquisition instrument

The information collected from each station in Table 8.4 was transmitted to Mianyang water regime subcenter through the selected channel, and then transmitted to the national center, the water regime center of the Changjiang Water Resources Commission and the earthquake relief command and decision-making authorities at all levels through the wide area network of Sichuan hydrologic center.

8.1.3 Emergency Survey on the Geometric Parameters and Typical Sections of the Barrier Lake

8.1.3.1 Emergency Survey Items

The basic situation of the barrier lake, the shape and features of the barrier lake, the flow in and out of the barrier lake, the storage capacity, the deposits and their volume were surveyed in the river section of the survey area. The typical sections of Tangjiashan barrier lake and its downstream reach were surveyed, and the hydrologic monitoring was carried out during the implementation of emergency handling of the barrier lake. Main items of the survey were as follows:

(1) Basic information, including the name of the barrier lake, the name of the river of the barrier lake, the geographical location and coordinates, the river bank of landslide mass entering the river, the catchment area, etc.
(2) Features, including backwater length, average width of water surface, average water depth, height from water surface in front of reservoir to weir crest, etc.
(3) Flow and storage capacity in and out of the barrier lake, including the flow in and out the barrier lake, total storage capacity, measured water level and storage volume.
(4) Deposits and the volume, including the material composition of the barrier dam, weir height, weir width, weir thickness and weir volume.
(5) Survey of typical section of Tangjiashan barrier lake and its downstream reach.

8.1.3.2 Instruments and Equipment of the Emergency Survey

After study and analysis of a large number of technical indicators of instruments and equipment, the following key instruments and equipment were determined to be used in the emergency survey.

(1) Electric Wave Flow Meter

Due to the great variation of flow velocity, rapid development of the breach, and the danger of collapse at any time, the use of conventional instruments was greatly limited. The electric wave flow meter was to measure the water surface velocity. Through the conversion of the water surface velocity coefficient, the purpose of measuring the river flow can be achieved. The instrument adopts non-contact flow measurement, which would not be affected by sediment concentration and floating objects. It had the advantages of safe operation, short measurement time, fast speed, etc., and could meet the requirements for monitoring the flow velocity of the breach.

(2) Non-prism Total Station

With the continuous strengthening of the retrogressively scouring of the discharge channel, the breach size was increasing and the flow velocity was accelerating. In order to ensure the safety of the monitoring personnel, the personnel must be as far away from the breach as possible. Therefore, traditional methods were difficult to achieve the goal of safety and efficiency. The mature unmanned ruler measurement technology, coupled with high-precision non-prism total station, could safely monitor the width of the breach and the water level at the breach.

(3) High-precision GPS/Beidou

After the earthquake, the elevation and coordinate systems had changed greatly. It was necessary to establish a relatively unified elevation system and leveling system in the survey area and must be equipped with high-precision GPS/Beidou.

8.2 Information Transmission and Communication System

8.2.1 Emergency Communication Channel Selection

Due to the severe earthquake, the satellite channel became the only communication channel not affected by the earthquake in this area. The satellite channel was the key communication networking channel in this survey area. At then, the available satellite communication modes in China mainly included Beidou satellite communication, synchronous meteorological satellite (OmniTRACKS) communication and maritime satellite C(Inmarsat_C) communication.

The Beidou system had a single satellite communication capacity of more than 200 earth stations per second, more than 720,000 service times per hour and a system reliability indicator of 99.993%. The system adopts CDMA code division multiple access and PN code spread spectrum technology to communicate in L-band, and had both the advantages of station Inmarsat_C and OmniTRACKS all line satellite communication terminals, high communication speed, strong concurrency, high confidentiality and security, which were suitable for the application of hydrological data acquisition system with short data transmission.

At the same time, according to the requirements of emergency remote video monitoring image information transmission at the main observation points, remote video images had high requirements for network bandwidth, while the observation stations currently did not had the conditions to lay high-speed dedicated lines, and conventional PSTN, GPRS, CDMA, etc. could not meet the installation and transmission requirements. Therefore, it was considered to use satellite channels to transmit remote video image information to the satellite ground center station, and then access Mianyang water regime monitoring and reporting center through broadband Internet to meet the requirements of real-time image monitoring. Through technical research, using China Satcom's Zhongxing series image satellites and two-way broadband resources (bandwidth greater than 20 MB), remote image downlink and remote video PTZ command uplink control could be achieved. In order to ensure the reliable information communication of important observation points during emergency discharge, it was considered to establish a small capacity GSM temporary base station at the dam site of Tangjiashan barrier lake to realize the dual channel communication mode of mutual backup. Channel selection of each observation point was shown in Table 8.5.

Table 8.5 Communication channels of Tangjiashan barrier lake emergency water regime remote monitoring station

Station name	Reporting item	Main channel	Standby channel	Reporting type
Baishi	Precipitation	Beidou satellite	GSM message	Automatic
Piankou	Precipitation	Beidou satellite	GSM message	Automatic
Youfang	Precipitation	Cable phone communication	Mobile phone communication	Manual
Tumen	Precipitation	Beidou satellite	GSM message	Automatic
Zhicheng	Precipitation, water level	Beidou satellite	GSM message	Automatic
In front of Tangjiashan dam and discharge outlet	Precipitation, water level, discharge	Beidou satellite	Satellite telecommunication/GSM	Automatic/manual
Beichuan	Precipitation, water level	Beidou satellite	GSM message	Automatic
Tongkou HPP	Precipitation, water level, discharge	Beidou satellite	Satellite telecommunication/GSM	Automatic/manual
Xiangshui	Water level	Mobile phone communication	Satellite Telecommunication	Manual
Fujiangqiao	Precipitation, water level, discharge	Cable phone communication	Satellite phone communication	Manual

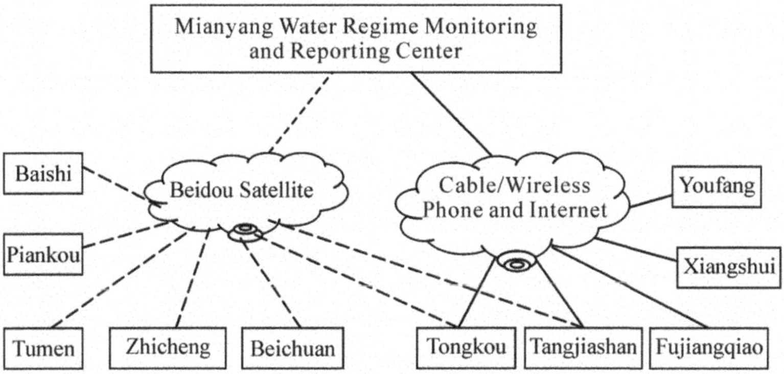

Fig. 8.1 Network structure from observation points to Mianyang water regime monitoring and reporting center

8.2.2 *Network Structure of Emergency Monitoring and Reporting Data Transmission*

8.2.2.1 Network Structure from Observation Points to Water Regime Center

According to the channel selection scheme, except Baishi, Piankou and Tumen precipitation stations, other observation points had dual channel transmission guarantee. The communication network structure from every observation points of Tangjiashan barrier lake emergency water regime forecast system to Mianyang water regime monitoring and reporting center was shown in Fig. 8.1.

8.2.2.2 Network Structure from Mianyang Water Regime Monitoring and Reporting Center to Earthquake Disaster Relief Command Authorities at All Levels

After receiving the water regime data from the reporting stations, the Mianyang Water Regime Monitoring and Reporting Center would process the data manually and by automatic computer processing, generate standard water regime reports according to the *Water Regime Information Coding Standard* and transmit them to the Tangjiashan barrier lake emergency handling command office located in Fuleshan, Mianyang and to Sichuan province water regime center through internet (with PSTN as the standby channel), and forward them to relevant disaster relief departments of Sichuan province, national center, Changjiang Water Resources Commission and hydrology bureaus of neighboring provinces. At the same time, it would be forwarded to the earthquake disaster relief command authorities at all levels, the disaster relief departments of Guangyuan, Mianyang, Suining, etc. and the disaster relief offices in

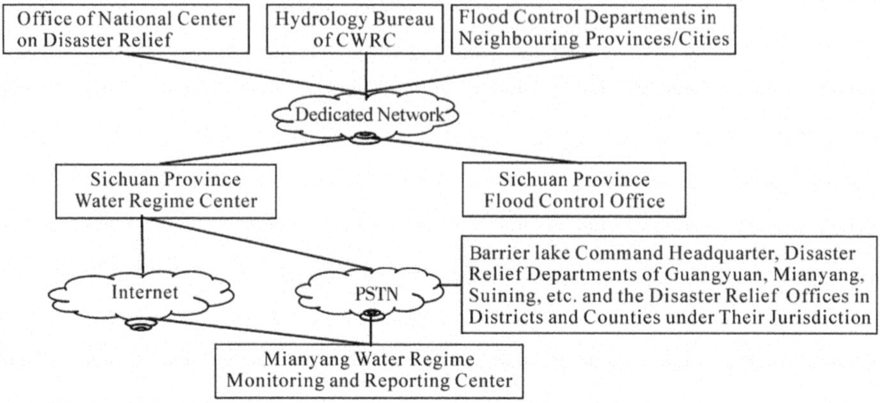

Fig. 8.2 Network structure

districts and counties under their jurisdiction in the form of PSTN (or through phone calls). The network structure was shown in Fig. 8.2.

8.3 Hydrological Information Analysis and Processing System

8.3.1 Difficulties in Monitoring Calculation and Analysis of Measured Hydrological Data Under Special Conditions

The hydrologic monitoring stations at the upstream and downstream of Tangjiashan barrier lake were seriously damaged, and the hydrologic monitoring and reporting facilities at most of the hydrologic monitoring stations were destroyed, which could not meet the requirement normal hydrologic monitoring and reporting work. In order to meet the needs of earthquake relief, emergency measures must be taken to quickly restore and rebuild some stations to ensure the collection of basic hydrological information.

The emergency hydrologic monitoring of Tangjiashan barrier lake was a hydrologic monitoring work under special working conditions at an extraordinary time. During the whole emergency handling period, the hydrologic river course had a wide monitoring range, large number of stations, intense monitoring frequency, large amount of data and information and high timeliness requirements. It was greatly affected by aftershocks, geographical environment and construction. There were many potential safety hazards and severe risks. Hydrologic measurement work was extremely difficult.

8.3.2 Difficulties in the Analysis and Processing of Measured Data

(1) The change of hydrologic emergency monitoring methods brought uncertainty of errors. At this specific time after the earthquake, the error caused by the change of monitoring methods could not be determined due to the failure of conventional synchronous monitoring. This was a major difficulty in the analysis and processing of hydrologic monitoring data in emergency situations.
(2) It was impossible to accurately analyze the changes in hydrological features of the station. The catastrophic earthquake had changed the hydrological features of the river channel, the station control and section control of most hydrologic monitoring stations had changed, and there were problems in the interconnection between the old and new data series, which made it difficult to analyze and check the rationality of data of a single station using historical data.
(3) There were difficulties in upstream and downstream comparison and rationality check. The underlying surface conditions of the basin had changed greatly. The river was no longer a natural river. Different river sections were affected to different degrees, which affected the upstream and downstream comparison and rationality check.
(4) There was a contradiction between the high timeliness requirement of emergency monitoring and the rapidity of monitoring objects. The hydrologic emergency monitoring work of Tangjiashan barrier lake was featured with large number of stations, intense monitoring frequency, large amount of data and information, and high timeliness requirements, thus made the workload for calculation and analysis on hydrological data huge. With the change of time, the water flow conditions and features of each monitoring station changed rapidly, which increased the difficulty of hydrological data calculation and analysis. At the same time, it was difficult to verify the calculation and analysis results.

8.3.2.1 Measures Taken for Calculation and Analysis of Hydrologic Monitoring Data

(1) The Zhicheng water level station in the lake collected a water level data every 15 min and calculated the inflow according to its telemetry water level change process. The discharge of Zhicheng water level station from 14:00 on May 12th to 8:00 on May 18th was calculated by using the discharge process of Beichuan hydrological station from May 1st to 12th.
(2) For the discharge of Tangjiashan barrier lake, the electric wave flow meter was used to measure the water surface velocity from June 7th to 9th, the ultrasonic depth meter was used to measure the channel water depth, and the handheld infrared distance meter measuring the starting point distance was used to measure the section flow, and empirical coefficient was taken. After the channel discharge increased significantly on June 10th, due to the large flow, fast velocity, and deep water, personnel could not get close to it, and there was no

section measurement facilities and equipment, so only the width of its discharge section could be measured, and the change process of water depth could not be measured. The water level difference and velocity measurement results of the dam water level and the channel flow measurement section were used to estimate the non-cohesive riverbed formula during the flow measurement period, and the maximum scouring depth was estimated as the water depth of the channel flow section. Rectangular section was used to calculate the section area.

(3) The water level of Tongkou hydrological station was monitored by the water gauge of the old Tongkou station. The flow velocity was measured by natural buoy for 10 times, and the empirical coefficient was adopted.

(4) In the early stage of discharge, the flow velocity of Fujiangqiao was measured with a flow meter. The natural buoy method was used to measure the flow due to the impact of a large number of floating objects in the late discharge period. Flow measurement was conducted for the whole discharge process. The large section was measured on June 14th after the discharge.

8.3.3 Compilation and Calculation Analysis of Measured Hydrological Data

8.3.3.1 Tangjiashan Barrier Lake Inflow

(1) Precipitation

The precipitation of Zhicheng water level station was observed manually and by self-recording. After the backwater of Tangjiashan barrier lake flooded Zhicheng town, it was impossible to observe manually, so the "Beidou 1" satellite telemetry rain gauge was used to observe the precipitation.

(2) Water Level

The water level was observed manually at Zhicheng water level station until 16:00 on May 26th. On May 25th, "Beidou 1" satellite telemetry water level gauge was installed in Xuanping town, 11.5 km downstream of Zhicheng, and was compared with the water level of Zhicheng water level station. Comparative observation results were shown in Table 8.6. The water level relationship between Zhicheng water level station and Xuanping water level station was shown in Fig. 8.3.

According to the comparison of observation data, the self-recorded water level of Xuanping water level station had a positive relationship with Zhicheng water level station, which could fully represent the water level of Zhicheng water level station.

After 08:00 May 18th, according to the storage capacity curve of Tangjiashan barrier lake, the flow process was back calculated by using the change process of water inflow during the period.

Table 8.6 Observation results of lake water level comparison

Time	Xuanping water gauge number	Xuanping zero point elevation (m)	Xuanping (yellow sea height system) water level (m)	Zhicheng water level (water resource group) (m)	Zhicheng water level (estimated) (m)
19:00, May 25th	0.73	723.27	724.00	63.51	63.54
21:00, May 25th	0.85		724.12	63.65	63.66
23:00, May 25th	0.97		724.24	63.79	63.78
00:00, May 26th	1.05		724.32	63.85	63.86
04:00, May 26th	1.32		724.59	64.13	64.13
06:00, May 26th	1.46		724.73	64.28	64.27
08:00, May 26th	1.62		724.89	64.43	64.43
08:30, May 26th	1.66		724.93	64.47	64.47
09:00, May 26th	1.71		724.98	64.51	64.52
09:30, May 26th	1.75		725.02	64.56	64.56
10:00, May 26th	1.78		725.05	64.59	64.59
11:00, May 26th	1.87		725.14	64.67	64.68
14:00, May 26th	2.09		725.36	64.91	64.90
15:00, May 26th	2.18		725.45	64.98	64.99

8.3.3.2 Tangjiashan Barrier Lake Discharge

The flow of Tangjiashan barrier lake was measured continuously throughout the discharge process, with high frequency. Therefore, the flow of the entire discharge process was calculated using the continuous measured flow process.

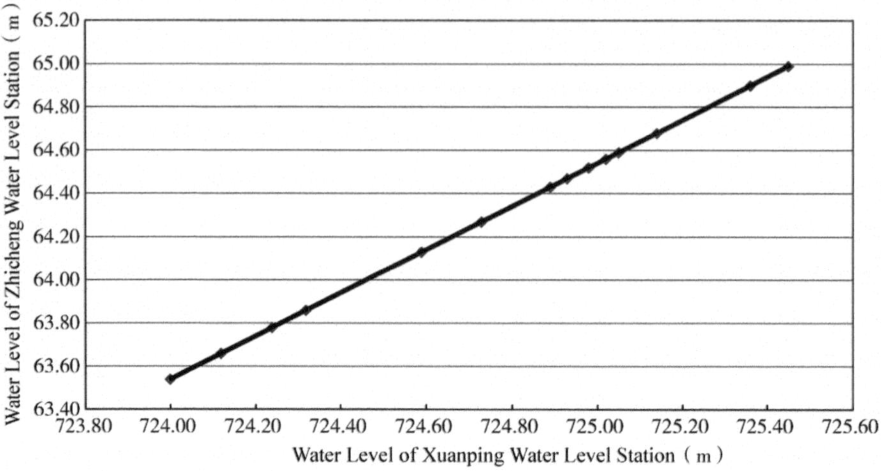

Fig. 8.3 Water level relationship between Zhicheng water level station and Xuanping water level station

8.3.3.3 Beichuan Hydrological Station

According to the water level-discharge rating curve of Beichuan hydrological station in year 2007, the cutoff water level of Beichuan hydrological station was calculated according to the cutoff water level calculation formula $Z_0 = \frac{Z_a Z_c - Z_b{}^2}{Z_a + Z_c - 2Z_b}$. A total of five groups of data were selected for calculation. Finally, according to the arithmetic mean value, the cutoff water level of Beichuan hydrological station before the earthquake was determined to be 86.80 m.

The measured maximum discharge of Beichuan hydrological station in 2005 was 1940 m^3/s, and the corresponding water level was 90.99 m. The measured discharge of Maoba hydrological station, about 1000 m downstream of Beichuan hydrological station, from 1962 to 1964, and the data of three major floods investigated in 1902, 1934 and 1964 were used to determine the stage-discharge relationship by integrating the original measured discharge results of Beichuan hydrological station.

According to the analysis and comparison of the cutoff water level of Beichuan hydrological station before and after the earthquake, the extended stage-discharge rating curve of the original Beichuan hydrological station was moved to the post-earthquake according to the water level translation, and then the data of Beichuan hydrological station were compiled and calculated according to the stage-discharge rating curve of Beichuan hydrological station before and after the earthquake.

8.3.3.4 Tongkou Hydrological Station

The water level of Tongkou hydrological station was observed manually, the precipitation was observed manually in dry season, the diary-type self-recording rain gauge was used in flood season, flow meter was used to measure flow at medium and low water level, and buoy was used to measure flow at high water level. The stage-discharge relationship over the years was stable. The measured maximum flow over the years was 5390 m^3/s on August 11, 1995. Historical floods mainly occurred in 1902 and 1934, but there were no flow results.

(1) Water Level

The water level of Tongkou hydrological station was monitored by the water gauge of the original Tongkou hydrological station. Before the discharge of Tangjiashan barrier lake, the four-segment system was used for observation. On June 10th, during the discharge of Tangjiashan barrier lake, when the water level changed gently, the eight-segment system was used for observation, and when the water level changed rapidly, the water level was observed every 6 min for intensive observation. The measured highest water level was 549.73 m and was observed at 14:24 on June 10th.

(2) Discharge

The section was measured on May 24th before discharge. After the Tangjiashan barrier lake discharged, the flow velocity was measured by natural buoys for 10 times. The empirical coefficient was taken for buoy coefficient.

(3) The Determination of Stage-Discharge Relationship

After the earthquake, all the water in the reservoir of downstream XiangshuiHPP was discharged. The water retaining structure of XiangshuiHPP was 2000 m away from the observation section of the Tongkou, and all its gates were opened. Its river reach was consistent with the natural river channel of the original Tongkou hydrological station. Therefore, the stage-discharge relationship of Tongkou hydrological station was determined by analysis based on the combination of the comprehensive stage-discharge rating curve of the original Tongkou hydrological station and the measured flow results on June 10, 2008.

8.3.3.5 Qinglian Hydrological Station

Qinglian water level station was located at the upstream of the water intake of Qinglian HPP on Tongkou River. The monitoring section overlapped with the basic section of the original Jiangjunshi hydrological station.

For the discharge measurement of Jiangjunshi hydrological station, flow meter was used for the flow at medium and low water level, and buoy was used for the flow at high water level. Flow data of 25 years were collected. The measured maximum peak discharge of Jiangjunshi hydrological station over the years was 4890 m^3/s (July 21, 1964), and the corresponding water level was 18.05 m.

(1) Water Level

From May 31st, non-prism total station was used for water level observation at this section. The instrument observation point was set on a solid rock at the left bank commanding point, with a wide field of vision, which could monitor the river conditions hundreds of meters upstream and downstream. According to the *Flood Investigation Manual*, the base level of Jiangjunshi hydrological station and the Yellow Sea Height System adopted a conversion method of: H (Yellow Sea) = H (Jiangjunshi) + 512.253. The elevation of instrument points was measured by non-prism total station from the benchmark of the original Jiangjunshi hydrological station. When the Tangjiashan barrier lake was not discharging, the four-segment system was used for observation. During the discharging period, the observation times were increased according to control the water level change process. The water level data from 10:00 May 31st to 08:00 June 13th were completely collected.

(2) The Determination of Stage-Discharge Relationship

The stage-discharge rating curve of Jiangjunshi hydrological station was drawn by using the measured discharge data and the surveyed flood data. After the formation of Tangjiashan barrier lake, according to the unified requirements of the water resources working group, all the flood gates of the water intake of Qinglian HPP were opened. According to the actual measurement, the original Jiangjunshi hydrological station measured that the river reach had an average siltation of about 2 m thickness, so the comprehensive stage-discharge rating curve of the original Jiangjunshi hydrological station was adjusted by about 2 m as the result of the stage-discharge relationship of Qinglian water level station, and the data of Qinglian water level station were compiled according to the rating curve.

8.3.3.6 Fujiangqiao Hydrological Station

The water level of Fujiangqiao hydrological station was observed manually. The precipitation was observed manually in dry season, and the diary-type self-recording rain gauge was used in flood season.

The observation section of Fujiangqiao hydrological station was about 1000 m long along the river, the main channel was about 100 m long, the medium and high flow swung to the right, the low flow was controlled by the overflow dam and beach mouth of Kaiyuan HPP 1000 m downstream, and the medium and high flow was controlled by Baoji-Chengdu railway and double-track piers. The flow measurement section overlapped with the basic water gauge section, and the section was in "*W*" shape. The change of erosion and deposition was within ± 3%. The stability was good. The middle section of the buoy was 30.26 m below the base. The basic water gauge section was used for the calculation of buoy discharge.

(1) Water Level

From May to June 2008, after the formation of Tangjiashan barrier lake, Fujiangqiao hydrological station observed the water level before the discharge of Tangjiashan barrier lake using the manual four-segment system and densified it during the discharge period. A temporary water level observation point was set on the top of the mountain, and the non-prism total station was used to observe the water level. At 17:18 on June 10th, the highest water level of 97.76 m was observed at Fujiangqiao hydrological station.

(2) Discharge

On May 27th, Fujiangqiao hydrological station measured the data of large section. Flow meter was used to measure the flow at medium and low water level, and buoy was used to measure flow at high water level. Since the station was built, the measured maximum flow of Fujiangqiao hydrological station was 10,300 m^3/s, and the corresponding water level was 98.98 m. Affected by a large number of floating objects during the discharge of Tangjiashan barrier lake, flow meter was used to measure the flow in the early stage of discharge, and natural buoy method was used to measure the flow in the later stage. The whole discharge process was measured, and the large section data were measured on June 14th.

(3) The Determination of Stage-Discharge Relationship

The stage-discharge rating curve was determined according to the measured discharge data of Fujiangqiao hydrological station, and the discharge of the station was calculated according to the water level data of Fujiangqiao hydrological station.

The maximum flow and its occurrence time of each monitoring station during the discharge process was shown in Table 8.7.

Table 8.7 Maximum flow and its occurrence time of each monitoring station during the discharge process

Station name	Maximum flow (m^3/s)	Occurrence time (Year-Month-Day h:min)
Tangjiashan	6500	2008-06-10 12:30
Beichuan	6540	2008-06-10 13:00
Tongkou HPP	6210	2008-06-10 14:24
Qinglian Town	6160	2008-06-10 14:48
Fujiangqiao	6100	2008-06-10 17:18

8.4 Flow and Rain Forecast and Outburst Flood Warning System

8.4.1 Difficulties of Forecast

8.4.1.1 Above Tangjiashan Barrier Lake Was a Small Watershed Mountain River, Which Was Difficult for Accurate Rainfall Forecast

Because the forecast object was located in the mountain area with large topographic changes, the rainfall in this area was particularly affected by the terrain, and the spatial distribution of rainfall was extremely uneven. Tangjiashan barrier lake had a catchment area of 3550 km^2. Before June 5th, there were only two rainfall stations, namely Piankou station and Beichuan station, with low density and poor representativeness. The rainfall at Tangjiashan, Youfang, Tumen and Baishi stations began to increase after June 6th. Some observation stations were affected by earthquakes, and the local meteorological data was sometimes unavailable. Therefore, the local meteorological information was obviously insufficient, and it was difficult to accurately monitor the local actual rainfall. The error of rainfall forecast would directly lead to the error of lake inflow calculation.

8.4.1.2 Uncertainty in Reservoir Storage Curve

The reservoir storage curve was calculated according to the topographic data before the landslide. Relevant departments had successively made five reservoir storage curves according to various methods, with a large difference between them. The difference between each curve at 740 m elevation could reach 80 million m^3. The inaccuracy of the storage curve would directly lead to the error of forecasting the inflow, and the error of the storage capacity curve might lead to deviation to inflow calculation system.

8.4.1.3 Lack of Hydrological Data to Verify the Forecast Results

In the water regime forecast of Tangjiashan barrier lake, the only "true" and reliable data were the water level of Zhicheng (Tangjiashan), and the rest data of inflow and storage curves were calculated values. Due to the lack of verification of hydrological observation data in the prediction of inflow, it was difficult to make real-time improvement of the forecast scheme.

8.4.2 *Numerical Precipitation Forecast Technology*

The precipitation forecast in Tangjiashan barrier lake area mainly adopted two forecast methods, which were mainly based on the experience of forecasters and numerical weather forecast.

8.4.2.1 Forecast Method Mainly Based on Experience of Forecasters

Based on the weather data from conventional ground and upper air detection, combined with remote sensing monitoring information such as satellite cloud map, on the basis of comprehensive analysis and revision of the forecast data products from various authoritative numerical weather forecast models at China and abroad, and mainly based on the experience of the forecasters, the quantitative precipitation forecast range and tendency value (quasi quantitative) of the region were determined through comprehensive consultation. This method was focused to realize the precipitation forecast on the forecast period of 1–2 days.

8.4.2.2 Forecast Method Mainly Based on Numerical Weather Forecast

The precipitation forecast method based on the comprehensive application of numerical weather forecast technology, using the MM5 meteorological model facing the Yangtze River basin at 08:00 and 20:00 (Beijing time) every day, could directly provide the quantitative forecast information of Tangjiashan area for 1–3 days. In July 2004, the Hydrology Bureau of CWRC introduced and built a mesoscale numerical weather forecast system for the Yangtze River basin. In addition, the 3–7d precipitation forecast of the region was realized based on the interpretation and application of the forecast data products of various authoritative numerical weather forecast models at China and abroad, such as the numerical forecast data products (such as precipitation, atmospheric height field, wind field, etc.) released by the European Center for Medium-Range Weather Forecasts (ECMWF), Japan, Germany and China.

The above two methods were used to realize daily precipitation forecast in Tangjiashan barrier lake area. Among them, the forecast information of areal precipitation every 12 h was provided within the forecast period of 2d (48 h), and the range and tendency value of 24 h areal precipitation was mainly provided for forecast period of 3–7d.

8.4.3 *Water Regime Forecast Scheme*

According to the forecast scheme of Beichuan hydrological station at the downstream of Tangjiashan before the formation of Tangjiashan barrier lake and the reservoir storage curve provided by the hydrological expert group in front, the water inflow forecast was carried out. According to the error between the predicted Zhicheng (Tangjiashan) water level and the actual observation value, the forecast scheme was revised and the forecast was issued after comprehensive consideration.

8.4.3.1 Data and Water Regime Station Network

Two precipitation stations were selected in the preparation of the precipitation runoff forecast scheme for the river basin above Beichuan, using the data of seven years from 1998 to 2004. The areal average precipitation of the basin was calculated by the arithmetic mean method, and the average runoff depth of each period was the arithmetic mean of the runoff depth of each precipitation station. See Fig. 8.4. The distribution of water regime stations in the river basin above Beichuan was shown in Fig. 8.4.

Since the evening of June 5th, Tumen, Youfang and Baishi precipitation stations had been successively added. The weight adopted of precipitation stations was shown in Table 8.8.

Fig. 8.4 Distribution of water regime stations in the basin above Beichuan

Table 8.8 Weight adopted of precipitation stations in the basin above Tangjiashan

Station name	Tumen	Youfang	Piankou	Baishi	Zhicheng (Tangjiashan)
Weight	0.164	0.302	0.181	0.249	0.104

8.4.3.2 Description of Forecast Scheme

The traditional precipitation runoff forecast method was used for flood forecast in the basin above Beichuan. Two precipitation stations were arranged in this section. The average areal precipitation of the basin was calculated by the arithmetic mean method. The average runoff depth in each period was the arithmetic mean of the runoff depth of each precipitation station. The flood forecast scheme of the basin above Beichuan was a 6 h forecast scheme.

(1) Runoff Forecast

P-Pa-R correlation diagram was used for runoff forecast. The P-Pa-R correlation diagram was determined after partial adjustment based on the original runoff forecast scheme of Jiangyou, Ganxi, Beichuan and Fujiangqiao. The main parameters were Im = 80 and $K = 0.9$. The parameter distribution was shown in Fig. 8.5.

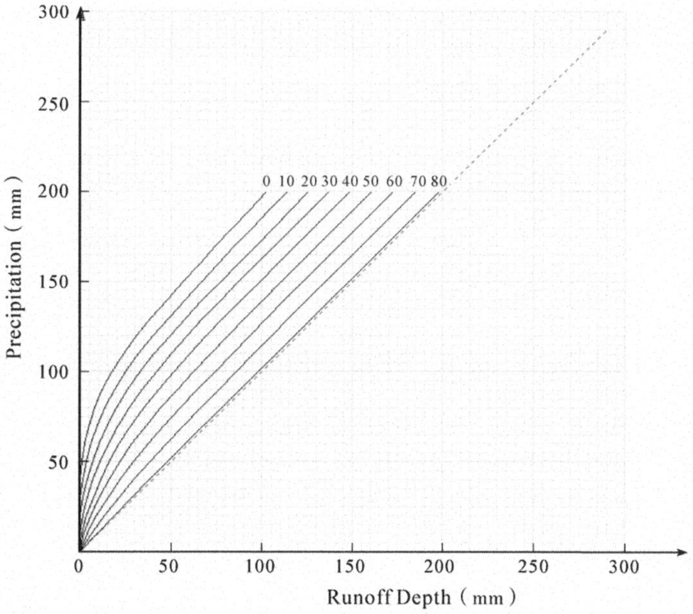

Fig. 8.5 P-Pa-R correlation diagram of river basin above Beichuan

Table 8.9 Unit line nodes and discharge for confluence above Beichuan

Node	0	1	2	3	4	5	6	7
Discharge (m^3/s)	0	264	642	380	184	77	24	0

(2) Confluence Forecast

Sherman unit hydrograph was adopted for confluence forecast. The unit line was partially adjusted on the basis of the confluence unit line between Jiangyou, Ganxi, Beichuan and Fujiangqiao. The unit line was determined as $R = 10$ mm/6 h and $\Delta t = 6$ h. The unit line nodes were shown in Table 8.9.

(3) Evaluation and Verification of Forecast Scheme

According to the provisions of *Code for Hydrological Information Forecast (SL250-2000)*, 20 flood samples in 7 years from 1998 to 2004 were verified. The results showed that the qualified rate of runoff depth was 70%, the qualified rate of peak discharge was 85%, the qualified rate of peak time was 80%, and the qualified rate of water volume was 90%. The qualified rates of all evaluation elements in this scheme were above Grade B.

Since May 20th, relevant departments had successively provided five storage curves for reference of forecast. The actual forecast work mainly adopted the storage curve agreed by experts of Changjiang Water Resources Commission and POWERCHINA Chengdu in the early morning of May 22nd. On May 23rd, the hydrologic expert group in front provided the storage curve to all units.

8.4.4 Main Forecast Method

The water level forecast of Tangjiashan barrier lake mainly adopted the static storage flood regulation calculation method that was the predicted lake water level was calculated according to the predicted water inflow based on the known initial lake water level and the storage curve. When water started to flow through discharge channel, because the parameters such as the depth of the discharge channel's flow section and the width of the channel could not be obtained, in order to estimate the discharge of the discharge channel, the open channel uniform flow calculation method was used to estimate the discharge of the drainage trough, and the trial-and-error method was used to reverse calculate the lake water level. The principles of the two methods and their specific applications in the forecast work were as follows:

8.4.4.1 Static Storage Flood Regulation Calculation Method

Static storage flood regulation calculation method was adopted for flood regulation calculation, which was to obtain the reservoir water level and the discharge at the end of the time period through trial-and-error method, based on the storage curve, the discharge capacity curve of the discharge channel, the inflow discharge process, the initial reservoir water level and the outflow at the beginning of the time period.

In order to rectify the forecast results, this forecast scheme had established a correlation diagram of water level in front of the dam, daily rise of water level in front of the dam, and inflow for the reference of real-time forecast.

8.4.4.2 Hydraulic Calculation Method of Flow Discharge in Discharge Channel

After water started to flow through the discharge channel, the section changed constantly, and the flow discharge also changed accordingly. It was very difficult to analyze and forecast the flow, so it was necessary to use the hydraulic model of unsteady flow in moving bed for calculation. However, the real-time flow section or underwater terrain of the discharge channel was difficult to be obtained in time, and the boundary conditions of the flow, especially the lower boundary, were also difficult to be observed in time. Therefore, when the calculation conditions of hydraulic model of unsteady flow in moving bed were difficult to be satisfied, in order to carry out real-time forecast, it was necessary to simplify the calculation and use the calculation formula of open channel uniform flow to calculate the flow.

The basic formula for the hydraulic calculation of uniform flow in open channels was the Chezy formula, and the hydraulic gradient was replaced by the water surface gradient.

$$\left.\begin{aligned} v &= C\sqrt{Ri} \\ Q &= CA\sqrt{Ri} = K\sqrt{i} \end{aligned}\right\}$$

When using Manning formula $C = \frac{1}{n}R^{1/6}$ to calculate Chezy coefficient C, the above formula could be written as:

$$\left.\begin{aligned} v &= \frac{1}{n}R^{1/2}i^{1/2} = \frac{A^{2/3}i^{1/2}}{nX^{2/3}} \\ Q &= \frac{A}{n}R^{2/3}i^{1/2} = \frac{A^{5/3}i^{1/2}}{nX^{2/3}} \end{aligned}\right\}$$

In the formula: X—wetted perimeter.

Both the above two formulas were used in the forecast to mutually verify the reliability of the parameters, which were mainly used to estimate the roughness and water surface gradient.

In the process of calculating the flow discharge of the discharge channel, due to the continuous changes in the section and water depth of the discharge channel, it was necessary to review and adjust in time according to the limited flow, section, water depth and other measured data. The extension and estimation of the future section and water depth changes were carried out, and then the analysis and forecast of the discharge of the discharge channel were carried out.

8.4.5 Hydrometeorological Forecast

On May 20th–31st, the water level process forecast of Zhicheng station and the daily increase of the water level of Tangjiashan barrier lake were released, with inflow estimated mainly based on precipitation. From June 1st, the specific water level forecast of Tangjiashan (Zhicheng) was released. The precipitation runoff model was used to forecast the inflow, and the static storage flood regulation calculation method was used to forecast the reservoir water level.

8.4.5.1 Forecast of Main Meteorologic Process

There had been three precipitation processes since the May 21st when the forecast began. Among them, the analysis and forecast for May 24th–26th and June 5th–6th on the occurrence of precipitation, the time distribution of the precipitation process and the total precipitation magnitude of the process were relatively successful. The above precipitation information was generally forecasted 3–4 days in advance. The precipitation process from May 24th to 26th was a critical forecast to the emergency response of the barrier lake discharge.

The Hydrology Bureau of CWRC started the precipitation forecast in the Tangjiashan barrier lake area at 01:00 on May 21st and made the first precipitation forecast. At that time, according to the analysis of the European central height field at 20:00 on May 20th, it was found that on May 25th, a low trough on the 500 hPa westerlies belt moved eastward, affecting the upstream of the Tangjiashan barrier lake area, accompanied by cold air intrusion: the Japanese precipitation forecast forecasted that there would be heavy rainstorm in the barrier lake area, with the process areal precipitation exceeding 100 mm; the German model product forecasted that the intensity was not strong, mainly small to moderate rainfall. The forecasters referred to the products of the above numerical weather model and combine the local topography to release the precipitation forecast of 24th–26th was moderate rain, and the accumulated areal precipitation was about 30 mm. Eventually, there were local showery rainfall on the 24th, moderate rainfall on the 25th, local heavy rainfall and sporadic light rainfall on the 26th in the upstream of Tangjiashan.

8.4.5.2 Forecast of Precipitation Process and Water Storage on May 25th

At 08:00 on May 25th, it was forecasted that there will be a moderate rainfall process in Beichuan from May 25th to 26th. There would be no rainfall on 27th–28th and light rainfall on 29th–31st. The accumulated precipitation was 38 mm, and the total water inflow into the lake was about 70 million m^3. At that time, the water level in Zhicheng was 723.58 m. If calculated according to the predicted inflow, the water level of the barrier lake would rise sharply. According to the post event statistics, the total actual water inflow into the lake from 08:00 on May 25th to 08:00 on May 31st was about 53.44 million m^3. The forecast inflow was 31% larger than the actual.

8.4.5.3 Forecast of the Discharge Time of Tangjiashan Barrier Dam Discharge Channel

Due to the influence of the elevation change of the discharge channel and the damage of the water level observation equipment in Tangjiashan and Zhicheng in certain periods, the forecast objects varied from time to time.

At 08:00 on May 30th, according to the water regime at that time and the precipitation in the forecast period, it was for the first time to forecast that the water level of the barrier lake Zhicheng station would reach 740 m on June 4th.

At 20:00 on June 1st, the forecast was clearly released that "it was impossible for the water level of Tangjiashan barrier lake to reach or exceed 740 m before 08:00 on June 5th, and when it will reach 740 m after 08:00 on June 5th will need be further clarified according to the precipitation condition at 20:00 on June 4th".

At 20:00 on June 3rd, according to the development of water and precipitation conditions, the 17th forecast further clarified that the Tangjiashan barrier lake would started to discharge at 02:00 on June 7th, with a forecast period of 78 h. This forecast conclusion was maintained in the following five consecutive forecasts: 08:00 on the 4th, 20:00 on the 4th, 08:00 on the 5th, 20:00 on the 5th and 08:00 on the 6th. The eventual result was that the water level in Tangjiashan reached 740 m at 00:00 on June 7th.

8.4.5.4 Forecast of in and Out Balance of the Lake

At 20:00 on June 9th, it was forecasted that from 20:00 on June 10th to 2:00 on June 11th, the inflow and outflow of Tangjiashan barrier lake would basically reach balance. On the evening of the 9th, due to relatively significant changes, the external conditions for the balance of in and out of the lake changed fundamentally.

8.4.5.5 Information Sharing and Consultation Mechanism for Three Places Multi-units

In order to safeguard the hydrological information forecast service of Tangjiashan barrier lake, the hydrological professional group of the earthquake relief headquarter of the Ministry of Water Resources had established and improved the mechanism for information sharing and smooth support in Beijing, Wuhan and Mianyang, with participating units including the hydrological forecast office of the Ministry of Water Resources, Tsinghua University, the forecast office of the Hydrology Bureau of CWRC, the hydrological professional group, and the Hydrology and Water Resources Survey Bureau of Sichuan Province.

8.5 Broadband Wireless Emergency Communication System

As the most direct and powerful monitoring method, video monitoring had been widely used in many occasions. Due to the large amount of information data that video monitoring needs to transmit, fixed lines were usually used to realize communication transmission. However, in some occasions where fixed communication lines had not been set up or were difficult to set up, wireless communication technology must be used to transmit video monitoring data. However, the traditional wireless communication technology belongs to narrowband communication, which was mainly used in voice services. It was difficult to realize services such as transmission of videos, pictures and other services that need to transmit large amounts of data. To realize broadband wireless video monitoring, the key technology was the "broadband wireless communication technology" in the system.

In the emergency handling project of Tangjiashan barrier lake, before the broadband wireless video monitoring system was set up, the Tangjiashan barrier lake dam and some important downstream areas had no broadband communication capability at all. Officers and experts could only go to the scene or watch the delayed video of aircraft aerial photography to understand the specific situation of the barrier lake. Mianyang command center was also unable to obtain urgently needed video information in real time. The establishment of video monitoring system had become an urgent matter. After the earthquake, the local cable broadband communication network had been completely paralyzed. With the restrictions of the Tangjiashan barrier lake dam and the risky terrain condition of the downstream basin, it was impractical to temporarily set up a cable video monitoring system. In contrast, the broadband wireless video monitoring system had many advantages, such as easy to set up, fast networking speed and free from terrain restrictions. In the urgent need of flood discharge, the use of broadband wireless video monitoring system to deploy Tangjiashan barrier lake area had become the only option.

The Chinese Academy of Sciences had urgently deployed MiWAVE broadband wireless emergency communication system in Tangjiashan barrier lake dam and its downstream important areas, realizing high-definition real-time broadband wireless multi-point video monitoring of Tangjiashan barrier lake and its downstream basin. Mianyang command center could continuously monitor the situation of Tangjiashan barrier lake basin for 24 h, thus successfully solved the problem.

The key technology of MiWAVE broadband wireless emergency communication system was a new generation of broadband wireless communication technology with independent intellectual property rights—MiWAVE broadband wireless emergency communication technology. MiWAVE broadband wireless emergency communication technology was a new generation of broadband wireless mobile communication technology developed with the strong support of the Chinese Academy of Sciences, "863" project of Ministry of Science and Technology, major projects in Shanghai, the

Ministry of Industry and Information Technology, etc. In combination with the application needs of modern emergency communication, it could quickly establish broadband wireless mobile networks in various scenarios and different scales of natural disasters, military operations, man-made accidents, and group activities, and support voice, data, images, video and other applications. And could make quick access to the internet through cable, wireless, satellite and other ways. MiWAVE broadband wireless emergency communication system combined the technical features of the new generation broadband wireless communication technology and emergency communication, which not only had the technical features of modern broadband wireless communication technology to support wide coverage, high capacity, high mobility, multimedia, all IP, enhanced broadcast functions, but also had the features of emergency communication, such as supporting portable, reliable, flexible, practical, self-contained power supply individual/backpack communication system that does not rely on external communication equipment, etc. In addition to the normal main electricity supply mode, it also supported a variety of power supply modes. The base station could use a portable generator or an automobile engine for power supply, and the terminal could use a backpack battery or an automobile inverter for power supply. It was featured by low power consumption and power-saving.

The key technologies of MiWAVE broadband wireless emergency communication system included the following aspects:

(1) Realizing interconnection between broadband emergency communication system and existing multiple access networks

MiWAVE system could be interconnected with various existing access technologies. Wired mode includes ADSL, cable, Ethernet, optical fiber, etc.; wireless mode includes microwave relay, satellite relay, etc. In the case of blocked roads, individual communication systems could also be used to build a local broadband wireless network in places deep in the disaster area, and connected to the backbone network through one of the above modes according to local conditions.

(2) Low power consumption, small size, adaptive to terminal and base station equipment with fast, mobile and flexible network deployment

MiWAVE broadband wireless emergency communication equipment was featured by modularity, easy installation, easy disassembly, flexible networking mode and low requirements for network erection environment. A broadband wireless network covering a wide range could be quickly built in a few hours. The individual/backpack communication system was portable, reliable, flexible, practical, with its own power supply, independent of external communication equipment (such as optical cables) and could be used in the worst disaster areas and roads. Although the device had small capacity, it had good mobility, long communication distance and wide coverage and could convey the disaster situation in a timely manner. The base station system was compact and power-saving, and the portable power generator could ensure its normal operation.

(3) Support the configurable "communication in motion" of any terminal

The traditional emergency communication system used mobile satellite communication to achieve "communication in motion", which was generally in the form of vehicle, with high cost, and could not be used in areas with inconvenient traffic. The MiWAVE system used TDD technology, which could improve the uplink throughput through the configuration of the uplink and downlink transmission ratio to support two-way broadband multimedia services. In this way, any mobile terminal could be used as the "on-site command center", and the mobile terminal could be carried by a single soldier, convenient for rapid and flexible network layout.

(4) Multiple power supply forms

The power supply infrastructure in Tangjiashan barrier lake area had been seriously damaged, and at the same time, the requirements of continuous 30-day unattended monitoring proposed by hydrology and water conservancy departments needed to be satisfied. The MiWAVE broadband wireless emergency communication system overcame many difficulties through repeated experiments and multi-party authentication. During the whole monitoring process, it used a variety of power supply methods, such as battery packs, various generators, solar energy, main electricity supply, etc., to solve the problem.

(1) At observation points such as observation point at left side of behind the dam and observation point in front of the dam, the generator and oil barrel were refitted, and automatically supply oil to the generator through siphon effect, ensuring the continuous operation for more than 20 days.
(2) In the relatively steady and open environment of the base station, such as the observation point of Beichuan hydrological station, solar power supply was selected, with a certain number of batteries, which could support several days of work in cloudy weather when enough batteries were equipped and fully charged.
(3) For observation point at right side of behind the dam, numerous sets of high-capacity lithium battery packs were used to ensure continuous unmanned monitoring for 30 days.
(4) In places where conditions permitted, such as the observation point of Fujiangqiao, mains electricity supply were used.
(5) The terminal could be powered by backpack battery or automobile power inverter.

(5) The emergency communication system could cover the disaster area from point to surface, so that the information transmission could be extended from the regional network to the entire IP backbone network

The remote broadband wireless video monitoring system consisted of three parts: front-end wireless video unit, base station and rear monitoring command center (Fig. 8.6).

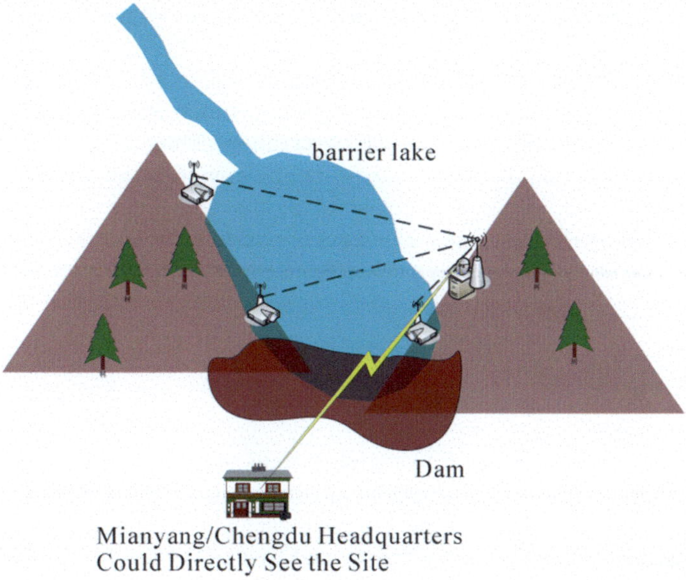

Fig. 8.6 Remote wireless video monitoring network deployment of barrier lake

The front-end wireless video unit included front-end video acquisition components (including remote zoom camera, PAN/TILT, video encoder), broadband wireless multimedia terminal, emergency power supply equipment (including solar or diesel generator, battery pack, power inverter). The data/video service flow of multiple front-end wireless video units was accessed to broadband wireless multimedia base station equipment through wireless terminals. The base station forwarded the video signal to the rear monitoring command center and backbone network through satellite, broadband wireless metropolitan area network, ground fixed network and other access methods. The command center could directly control the PAN/TILT direction and image focus of the front-end video acquisition equipment, and could conduct voice/data interaction with the front site through the internet. The rear command center was composed of internet access gateway, video display equipment, computer, command and dispatching software system. By centralizing the data and video of the site in all directions, the command center could understand the site situation in time, coordinate the site resources and make dispatching command according to the actual situation.

The broadband wireless multimedia system transmitted real-time video monitoring images to Mianyang command center through broadband wireless transmission technology, and relevant departments and officers could also learn about the on-site situation in real time through the Internet and private network.

On June 3rd, premier Wen Jiabao, in Zhongnanhai, through the wireless broadband special line set up by the rescue team, watched the 12 route video monitoring images of Tangjiashan barrier lake, and found the location of the discharge channel in the monitoring images. He was very satisfied with the quality of the video monitoring images.

After the system was set up, it could automatically record the hydrological data during the discharge process. No manual on-site operation was required. The monitoring screen could be zoomed, and the observation points at the downstream of the Tangjiashan dam and the barrier lake could be monitored all-weather to ensure that dangerous situations could be found at the first time. The system run stably and had excellent performance in the process of real-time monitoring the disaster of Tangjiashan barrier lake, provided a new effective method for disaster monitoring.

8.6 Summary

The emergency monitoring and guarantee system was an unconventional hydrological monitoring system under special environmental conditions. It was greatly affected by aftershocks, geographical environment and construction, with tight time requirements, many potential safety hazards and great risks. According to the features of emergency monitoring, it mainly used mature and advanced instruments, equipment and technical means, and used new instruments and technologies, which enriched the contents of emergency monitoring, provided support for emergency decision-making and played an important role in the process of emergency handling and risk avoidance of Tangjiashan barrier lake.

8.6.1 Hydrologic Emergency Monitoring System

Under the condition that the elevation system was uncertain, the elevation system of Tangjiashan barrier dam was quickly measured by GPS, and the geometric shape and features of the barrier dam were measured by non-prism total station. The basic data of Tangjiashan barrier dam was obtained, which provided a basis for the formulation of the project scheme.

On the basis of the original two precipitation stations in the basin above Tangjiashan, three precipitation monitoring points were added to basically control the temporal and spatial distribution of precipitation. The water level monitoring of Zhicheng and Beichuan was resumed, the water level monitoring points of Tangjiashan, Tongkou and Xiangshui was added, and the stage-discharge relationship was proposed to calculate the flood discharge process. The precipitation and water level were automatically monitored, and Beidou satellite communication was used to transmit data to basically monitor the whole process of precipitation, water level change and discharge. The flood routing process monitored by each station had been checked and compared with the upstream and downstream peak discharge and total flood balance, and the results were reasonable.

8.6.2 Information Transmission and Communication System

Making full use of the existing communication resources to quickly build an efficient and practical communication guarantee network to ensure the collection and real-time transmission of hydrological information. The main channel of the hydrological telemetry station was Beidou satellite, and the standby channel was GSM. The manual flood reporting station used wireless mobile phones as the main channel and satellite phones as the standby channel. Broadband wireless video used MiWAVE technology to realize the interconnection between emergency broadband communication system and existing multiple access networks.

8.6.3 Information Analysis and Processing System

For the difficulties encountered in emergency monitoring, the existing resources, including station network and data, were fully utilized, and the combination of advanced technology and traditional technology was adopted to ensure the efficient implementation of monitoring and the quality of results.

8.6.4 Meteorological and Hydrological Forecast System

(1) Meteorological Forecast

During the period of emergency handling of the barrier dam, from May 21 to June 10, 2008, Hydrology Bureau of CWRC had issued a number of 34 precipitation forecasts for the Tangjiashan barrier lake area. The average accuracy of 24 h and 48 h areal precipitation forecast was 79% and 80% respectively, which was relatively high accuracy.

(2) Hydrological Forecast

The water and precipitation forecast of Tangjiashan barrier lake and its downstream observation points had been released for more than 40 times, with 98% of them having good accuracy. Among them, the water level forecast issued at 09:00 on June 6th that the water level would reach 740 m elevation at 00:00 on 7th, which was completely consistent with the actual situation. The balance point of inflow and outflow water released on June 8th was 02:00 on June 10th, with the highest water level of 743.28 m, which was close to the actual situation and played an important role in the decision-making of emergency handling.

(3) Remote Consultation Mechanism

Establishing a multi-party remote consultation mechanism to analyze, discuss and forecast the water regime. From May 20th to June 9th, a number of 28 water and precipitation analysis of Tangjiashan barrier lake were released. According to statistics, the prediction error of average water level in 12 h, 24 h, 48 h and 72 h prediction periods were 0.08 m, 0.09 m, 0.14 m and 0.23 m respectively. It was predicted 78 h in advance that Tangjiashan barrier lake would reach the discharge water level of 740 m at 02:00 on the 7th. The hydrological forecast had high accuracy and a long forecast lead period, provided timely and effective emergency response warning information for the emergency handling of the barrier lake.

8.6.5 Extensive Application of Broadband Wireless Video

In the process of emergency handling and risk relief of Tangjiashan barrier lake, the traditional voice emergency communication system was difficult to meet the needs of various information collection and scientific processing in the emergency handling. Remote broadband wireless emergency communication systems such as MiWAVE could transmit video, image, data and voice at the same time, provided more, comprehensive and three-dimensional multimedia information in emergency handling, and could play a greater role in emergency handling practice in water resources and other fields in the future.

Chapter 9
Water Discharge Process and Disaster Relief Effect Evaluation

9.1 Scouring Process in Discharge Channel

The discharge of Tangjiashan barrier lake began from 7:08 on June 7, 2008. Till 08:45 on June 11th, inflow and outflow of the barrier lake were basically balanced. Retrogressively scouring and lateral erosion are obviously observed in the channel during the 4-day discharge process. The side slopes of the channel underwent a process of stability-creep-stability.

9.1.1 Retrogressively Scouring Process

The channel was heavily flushed during discharge. The flow downcut and deepened the channel, and lengthened it toward upstream as it eroded away the rock and soil in the channel. The intensity of retrogressively scouring was relevant to the material composition, flow volume and velocity.

At 320–380 m from the inlet along the channel, cataclasite rocks fell to form a NEE 20 m-long 80 m-wide rock ridge. The rock ridge, with a crest elevation of 750–760 m, is connected to the mountain on the right bank eastward and to the barrier westward. The rock strike was in compliance with the rock ridge and vertical to the discharge channel, with strong resistance to scouring (see Fig. 9.1). A scarp with waterfall was shaped from scouring at the bottom of the channel (see Fig. 9.2). The scarp gradually moved upward at a slow pace as the water eroded the rocks away along with time. The ridge on lower section of the discharge channel is critical for controlling the discharge process and duration. It helped reduce the velocity of the discharge to a stable speed.

The channel section on upstream of the ridge is about 320 m long, being loose geologically with yellow gravel soil and dark gray fine sand with soil particles. The channel section is vulnerable to scouring and will present a strong retrogressively scouring with obvious fast retreat speed.

Q. Yang and W. Yang, *Emergency Handling of the "5.12" Tangjiashan Barrier Lake*,
https://doi.org/10.1007/978-981-99-1706-8_9

Fig. 9.1 Rock ridge is shaped from thin cataclasite resistant to water scouring

On June 7th, slope at bottom of discharge channel was small with small velocity, small flow and weak retrogressively scouring;

On June 8th and June 9th, retrogressively scouring was strengthened and channel was deepened along with the increase of velocity and discharge;

On June 10th, bottom elevation of the channel dropped to 714 m and retrogressively scouring stopped.

Due to retrogressively, the bottom elevation of the channel dropped from 740–742 to 714 m. Meanwhile, and collapse on both side widened the channel.

9.1.2 Lateral Erosion Process

In lateral erosion, the side slopes underwent backward movement due to collapse of the channel side slopes. Compared with the design axis of the channel, the actual axis deviated toward the right bank after the drainage. Deviation at various section is demonstrated in Table 9.1. The curve of the discharge channel moved downstream by 115 m (Fig. 9.3).

The cross section of the channel after discharge is totally not symmetric. The distance from right bank to the axis is much larger than that from the left bank. The

Fig. 9.2 Waterfall-shaped proved strong resistance against scouring

Table 9.1 Difference between the design axis of the channel and the actual axis after discharge

The straight channel section on upstream of the curve (m)	Channel section near the curve (m)	The straight channel section on downstream of the curve (m)	Tail section
35–70	60–130	60–110	Same

distance between right bank and design axis at the curve and on the upstream section of the curve is more than twice of the distance between left bank and design axis.

Table 9.2 shows that lateral erosion on right bank and downstream section is stronger during discharge process. As the discharge increased, the channel was expanded toward both sides. The ridge on downstream of the discharge channel help mitigated the lateral erosion. The expansion of the upstream channel section with gravels became more severe as the movement of the ridge. Once the ridge was eroded away, the lateral erosion was borne by the mountain on right bank downstream to the ridge.

Fig. 9.3 Collapse of soil slope on right bank of the discharge channel (upstream section)

Table 9.2 Distance between each bank to design axis after discharge (m)

Bank	The straight channel section on upstream of the curve	Channel section near the curve	The straight channel section on downstream of the curve
Left bank	70–75	20–45	50–90
Right bank	120–165	150–240	60–110

9.1.3 Deformation of Side Slope During Discharge

9.1.3.1 Deformation and Stability of Side Slopes During Discharge

The stress on both side slopes kept changing as the channel was deepened and expanded during the discharging. The side slopes underwent a process of stability-creep (collapse)-stability. Different geological structure on both banks resulted in different consequences of souring and deformation.

A 100–230 m wide flat area with gentle slope was developed on the channel section near and on upstream of the ridge, which is located at the tail of the landslide. Landslide bed, with accumulation from the collapse, was just behind this flat area. The landslide bed rock, which is basically stable, has a dip slope along the river with same angle as that of the stratum. Elevation of the flat area is 755–785 m with a slope of 8″–10″. On upstream of the ridge is 10–15 m thick yellow gravel soil partly covered by dark gray fine sand on the surface. In the area near and on downstream of the ridge is cataclasite. The discharge channel was scoured downward during drainage, thus heightening the side slope which was gradually destroyed and moved backward due to small collapses. After the discharge, width of the flat area was reduced to 20–100 m. No obvious impacts of discharge were observed on the stability of mountain behind the flat area and the collapse accumulation thereon.

An insidious landslide located at downstream of the ridge and right bank of the discharge channel. On the 6th and 7th of June, there was small discharge through the channel with little impact on the landslide. Since June 9th, the discharge had increased to flush the foot of the landslide, leading collapse of the sliding mass (Fig. 9.4). On the morning of June 10th, the discharge further increased, knocking down. The landslide slope and gradually destroyed the stability of the slope. After 12:00 p.m. of the same day, the lateral erosion on right bank was intensified and the line of maximum velocity in the channel continued shifting rightward. Scouring of the discharge exposed the rock bed of the landslide slope from its upstream side (Fig. 9.5); thus, the stability of the landslide was not further reduced.

The left bank slope of the discharge channel is the east slope of barrier dam. Deformation of this side slope during discharge was mainly through collapse and creep. At early stage of discharge, the water flow was small and the scouring of side

Fig. 9.4 Since 9th June, the discharge had increased to flush the foot of the landslide

Fig. 9.5 Scouring of the discharge exposed the rock bed of the landslide slope from its upstream side (avoiding further scouring of this part)

slope was weak, which led to creep of the yellow gravel soil of the side slope at small scale. The slope remains stable in general. As the flow increased since June 9th, the earth body collapsed with the side slope crest retreating, leading to a larger scale of creep within a reach of more than 50 m.

On the morning of June 10th, the discharge through the channel rose to its peak. As a result, creep occurred on the whole side slope in the east. Penetrating tension cracks of more than 20-m at longest appeared on the side slope, creeping from the foot to the top of the slope (Figs. 9.6 and 9.7). The width of the crack kept extending to a maximum of 3.2 cm. Image from monitoring on #L4 crack at tail skirt of the landing field and #L3 crack on eastern slope of the landing field is shown in Figs. 9.8 and 9.9.

The creep on the slope affected the landing field during a certain period until 13:30 p.m. on June 10th, when width of none crack on the slope continue increasing and the creep of the slope came to a stop on the whole.

9.1.3.2 Geological Analysis and Advice During Discharging

The retrogressively scouring and lateral erosion became distinct from the evening of June 9th to June 10th due to the increasing flow through discharge channel. Stress

Fig. 9.6 #L4 crack at tail skirt of the landing field

Fig. 9.7 #L3 crack on eastern slope of the landing field

on both side slopes surged, causing a rapid expansion of side slope deformation and making psychological distress of staff working on the barrier dam. Professional advice based on deformation data monitored and timely analysis on side slope stability was offered by geologist on site to ease local staff's mental stress and to provide guidance for hedging.

(1) Analysis on slope stability and timely adjustment of security line

The retrogressively scouring and lateral erosion were not obvious during June 9th to June 10th due to small flow through discharge channel. The side slopes were reshaped only gently with macroscopical deformation at very small scale.

From small hours of June 10th, discharge became significantly larger with its maximum flow reaching to 6500 m^3/s at noon. Meanwhile, the side slopes were reshaped distinctly with obvious macroscopical deformation. Penetrating cracks appeared at middle, upper and top areas of the side slope on left bank, endangering

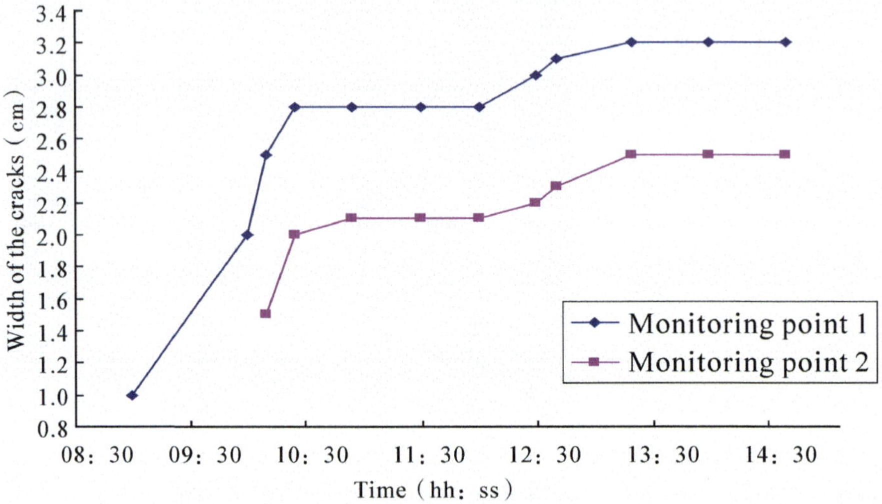

Fig. 9.8 Changing curve of the width of #L4 crack (monitored on June 10, 2008)

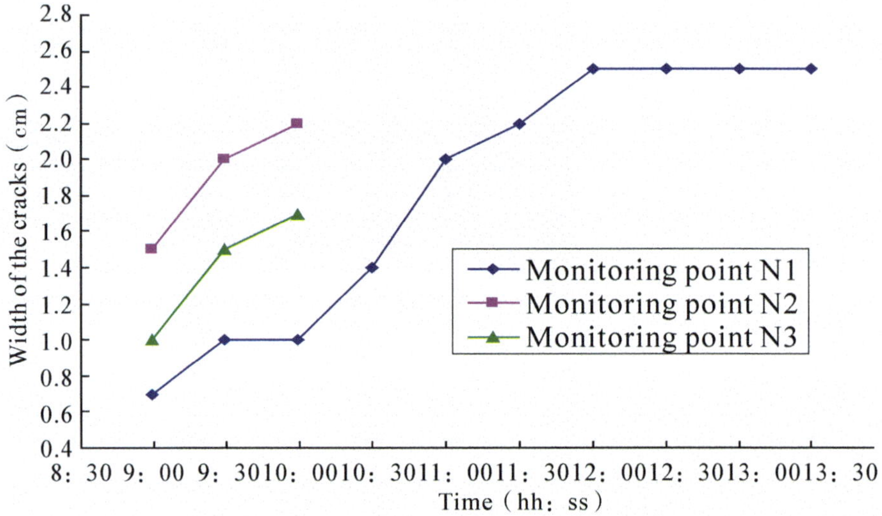

Fig. 9.9 Changing curve of the width of #L3 crack (monitored on June 10, 2008)

the emergency staff on barrier dam. Following the advice of geologist on site, location of security line on the left of the discharge channel was adjusted timely. At around 09:00 a.m., the security line was shifted from the flat area at middle and upper part of the slope to the front edge of the landing field. This line was further moved backward to the tail edge of the landing field and withdrew from the field when cracks were observed on it. No one was allowed to step over the security line in case of emergency, except monitoring staff, ground staff on duty upon landing of the helicopters and people to be evacuated by the helicopters.

(2) Analysis of bearing capacity of the landing field for Mi-26 heavy-lift helicopters

In the rescue process of Tangjiashan barrier lake, S-70C helicopters were selected for transporting the emergency staff and light equipment from Mianyang Airport to Tangjiashan barrier dam. The taking off and landing of such helicopters were all arranged on the landing field on site. Mi-26 with larger transport capacity was adopted for lifting of large equipment to Tangjiashan barrier dam. Such helicopters would return once the equipment got unloaded, without landing on site.

Evacuation of personnel was started at around 9:00 a.m. on June 10th, when the cracks were found. The time for evacuation was postponed since 3 S-70C helicopters were firstly mobilized to cope with 10 t of TNT explosive. During this period, monitoring data demonstrated that creep on the side slope were becoming intensively serious.

Under such urgent occasion, the rescue headquarters on site decided to evacuate personnel with Mi-60 helicopters. However, it was too risky and insecure to use the simple and crude landing field on top of the barrier dam without any measures taken on the foundation for landing of Mi-26.

After careful assessment, geologists reached to the conclusion that the landing field on site would remain stable in short time with enough bearing capacity for Mi-26. Based on such assessment, Mi-26 helicopters immediately landed on top of the barrier dam and evacuated most of the rescue staff from Tangjiashan barrier dam by about 13:00 p.m.

9.1.4 Effects of Engineering Measures

The discharge channel became wide after drainage (Fig. 9.10). An arc-shape bending toward the right bank can be seen on the surface of the channel, with a center line being 890 m long. The cross section of the channel is in the shape of a trapezoid, 145–235 m wide on top and 80–10 m wide on bottom. The slope inclined at 35″–50″ on right bank and 45″–60″ on left bank. The height of the slope is 10–60 m. The water surface width of the new channel is 18–3 m at normal water levels.

Site geological investigation showed that the height of the side slopes is higher on middle sections of the channel and lower on upstream and downstream section. The scour depth reached 60 m at maximum. The geological structures on both banks are symmetrical having 20–40 m thick yellow gravel soil on upper part and 10–20 m thick cataclasite in middle of the slope. Dark gray silicious cataclasite was exposed at foot of both slope and at the bottom of the channel, where original rock strata were maintained partly. The right bank of the channel after discharge was shaped finally under obvious influence of scouring and slope collapse, with most rock and soil flushed away. The left bank with gravels also underwent collapse not as strong as

Fig. 9.10 Discharge channel after drainage

that on right bank. The foot of slope on downstream of the barrier dam was severely flushed as well, yet no collapse happened in large scale. The barrier dam was mostly located on left bank in arc-shape on plane, 600 m along the channel and 300 m long horizontal to the channel. The height of the barrier dam is 80–120 m, with the residual earth rock quantity about 15 million m^3. Height of the rugged barrier dam ranges from 30 to 130 m. The barrier dam horizontal to the channel is high on both banks and low in middle part. The max. elevation of the barrier on left bank is 739.9 m. The barrier dam section in parallel to the channel is high in middle and lower on upstream and downstream side. About 1/4, i.e., 5 million m^3, of the barrier dam was rushed away along with the discharge through the channel.

Upon completion of the disaster relief works, the discharge channel was shaped into a 700 m long slightly curved waterway, with an elevation of 736 m at the inlet, 740 m through the discharge channel and 5–7 m wide on bottom. Longitudinal profile of the channel is in trapezoid with higher elevation in inlet and outlet, lower in middle section of the channel. The discharge began at 7:08 a.m. on June 7th, with continuous scouring and erosion against the channel. Consequently, the channel was gradually moved toward the right bank. The profiles of the channel before and after the discharge are compared and demonstrated in Fig. 9.11.

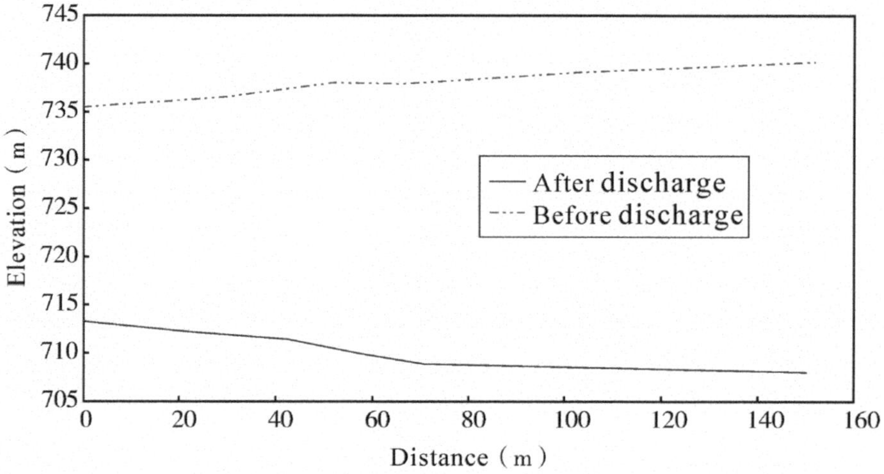

Fig. 9.11 Profiles of the channel before and after the discharge

9.2 Barrier Lake Discharge Process and Back Analysis

9.2.1 Discharge Process of Barrier Lake

Based on the hydrological data acquired during barrier lake discharge process, the stage-discharge process at major hydrological stations and gauging stations is listed as below.

9.2.1.1 Zhicheng Gaging Station

After May 12, 2008, water level at Zhicheng gaging station rose due to heading-up of water. At 2:00 a.m. on May 18th, the water level at the gaging station rose to El. 707.02 m, and the tail of water dammed by the barrier dam reached Zhicheng. With the water from upstream, the water level at Zhicheng gaging station rose by 1–2 m daily. At 7:08 June 7th, discharge through the channel began. On June 10th, the discharge process was sped up and water level at Zhicheng gaging station reduced drastically. The stage-discharge curve is shown in Fig. 9.12.

9.2.1.2 Tangjiashan Gaging Station

During 12th–21th May, average daily inflow into Tangjiashan barrier lake was 58–175 m^3/s. During May 22th to June 1st, average daily inflow into Tangjiashan barrier lake was 90–120 m^3/s and the water level was lifted at a speed of 1.5–2.5 m/d. On

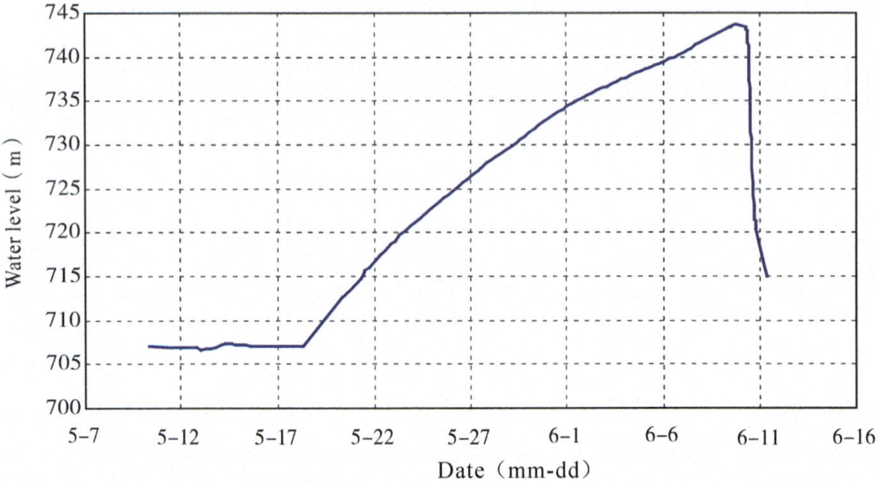

Fig. 9.12 Stage-discharge curve at Zhicheng gaging station

June 10th, average daily inflow into the barrier lake was 70–170 m^3/s, and the water level was lifted at 0.9–1.3 m/d. The increase of water level gradually slowed down.

At 7:08 a.m. on June 7th, the discharge through the channel began, when the water level on upstream of the barrier dam was El. 740.39 m and the lake volume was 226 million m^3. The water had been increasing unceasingly with inflow exceeding outflow. At 1:30 on June 10th, the water level in the barrier lake reached its peak, i.e., El. 743.23 m, with corresponding lake volume of 245.6 million m^3. The water level was reduced to El. 714.13 m and lake volume reduced to 86.1 m^3.

The flow through the discharge channel at 18:00 on June 9th was 113 m^3/s, which continued to increase later on. The flow was 209 m^3/s at 20:30 p.m. and 401 m^3/s at 1:30 a.m. on Oct. 1st. This discharge reached its peak, 6500 m^3/s at 12:30 a.m. on June 10th and then began to reduce. At 8:45 on June 11th, the flow through discharge channel was 79 m^3/s, with the inflow and outflow basically in balance. The discharge process at the barrier dam site and water level process on upstream of the barrier was measured and recorded (see Fig. 9.13).

Figure 9.12 shows the scouring of the channel during discharge recorded on site. Data on width of water surface, depth, water level and discharge process in Fig. 9.12, the flow in the channel grew slowly and maintained no more than 600 m^3/s before 6:00 a.m.

The flow grew from 574 to 811 m^3/s during 6:00–8:00 a.m., with an increase of about 120 $m^3/(s.h)$. The water level dropped and the flow grew during 8:00–12:00 a.m. when the increase of flow was about 1310 $m^3/(s.h)$ (Table 9.3).

Based on the data observed regarding the width of water surface in cross section of the discharge, the surface was widened by about 30 m during 18 min from 11:12 to 11:30 a.m., showing an expansion of scouring in the channel during this time period. The bottom was deepened at a speed of 2.2 m/h through scouring from 12:00 to 14

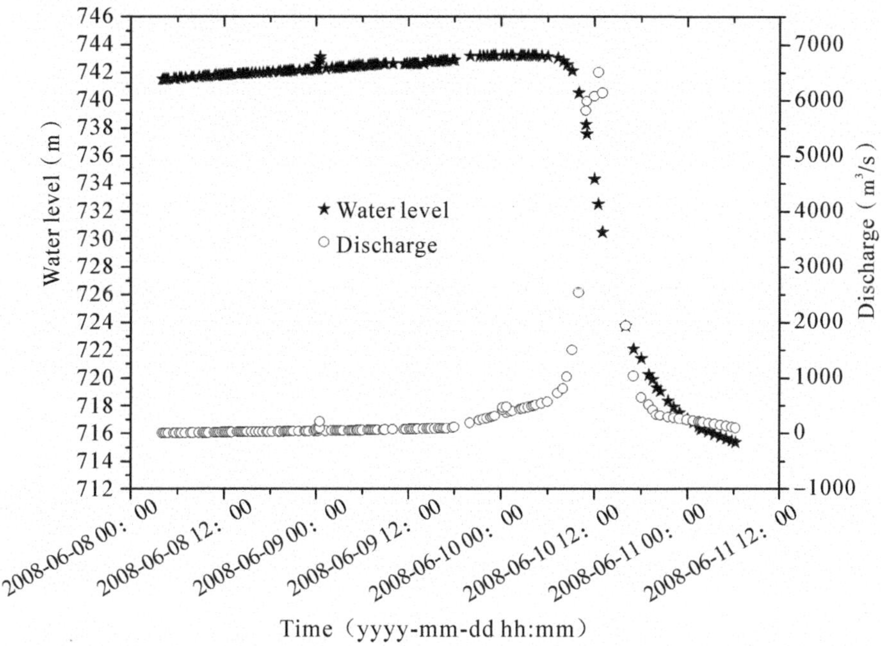

Fig. 9.13 Discharge process at the barrier site and water level process on upstream of the barrier was measured and recorded

a.m. The width of water surface remained unchanged basically. The bottom of the channel was deepened at 0.36 m/h during 14:00–18:00 p.m., which is much slower than previous periods.

Based on analysis above and barrier dam break mathematical modeling, the discharge peak at the barrier dam site was usually the same as or earlier than the dam break time set by the model. The longer it takes for barrier dam break to happen, the earlier discharge peak occurs. A preliminary conclusion was then drawn by the expert team that the dam break would happen during 8:00 a.m.–14:00 p.m., taking around 6 h in total.

9.2.1.3 Beichuan Hydrological Station

No flow had been recorded at Beichuan hydrological station since May 12th. At early stage of discharge from the Tangjiashan barrier lake, the water was discharged into Kuzhuba barrier lake, which was filled and overflowed at 14:00 on June 9th. The discharge from Kuzhuba barrier lake continued to grow after 23:00 p.m. and the water level at Beichuan hydrological station reached El. 614.22 m at 1:10 a.m. on June 10th with a flow of 915 m^3/s being the first flood peak. The discharge flow fell to the minimum at 4:20 and then rose to El. 620.3 m at 13:00 with a peak discharge

Table 9.3 Development process of the cross section of the discharge from Tangjiashan barrier lake on June 10, 2008

Time	Width of water surface	Water depth	Water level (El. m)	Elevation of the channel bottom	Flow
00:45			743.22		372
01:30			743.23		401
02:30			743.22		439
04:00			743.21		497
06:00			743.13		574
08:00			742.80		811
10:00			740.51		2530
11:12	100	11			
11:30	129	11	737.10	726.10	
12:00	130	8.5	734.30	725.80	6070
12:30			732.51		6500
13:00	132	8.5	730.48	721.98	6130
14:00	145	6.5	727.94	721.44	
15:00	145	4.6	725.80	721.20	
16:00	145		723.75		1940
17:00	145	1.8	722.08	720.28	1040
18:00	145	1.4	721.40	720.00	653
19:00			720.25		524
20:00	80	1.15	719.30	718.15	430

flow of 6540 m. The discharge and water level process measured and recorded at Beichuan hydrological station are shown in Fig. 9.14.

9.2.1.4 Discharge Process Along the Channel

At 12:30 June 10th, the maximum flow in the discharge channel of Tangjiashan barrier lake was 6500 m^3/s, which is 79 times the long-term average of Tangjiashan section in early June and exceeds the design peak flood with return period of 100 years by 6040 m^3/s. The discharge from Tangjiashan barrier lake ran through Beichuan station, Tongkou HPP, Qinglian Township and Fujiangqiao hydrological stations with the peak flow of 6540 m^3/s arriving at Beichuan at 13:00 under the maximum water level of El. 620.3 m. The discharge flow reached Tongkou temporary gaging station with the maximum water level of 549.73 m and peak flow of 6210 m^3/s; The discharge flow reached Fujiangqiao at 17:18 with the maximum water level of 465.28 m and peak flow of 6100 m^3/s, which is slighter larger than 5680 m^3/s, the long-term average measured at the bridge. The transition of peak flow from Tangjiashan to Fujiangqiao located on mainstream of Fujiang River took 4 h 42 min. Data on water level and

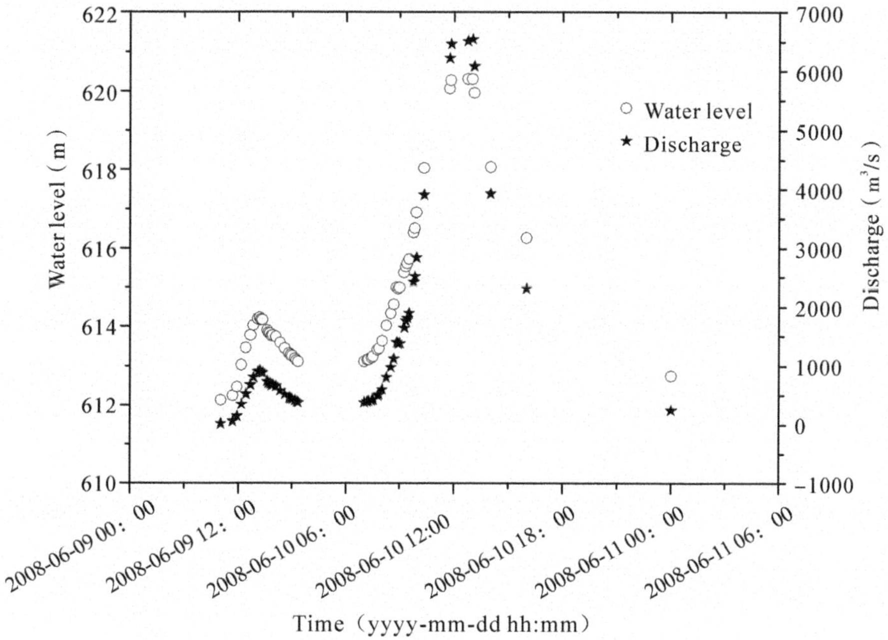

Fig. 9.14 Discharge and water level process at Beichuan hydrological station

flow measured at major stations are shown on Table 9.4. The discharge and water level process at Tongkou, Qinglian and Fujiangqiao hydrological station are shown in sequence in Figs. 9.15, 9.16 and 9.17.

Table 9.4 Discharge and water level process at major stations

Location	Peak flow (m^3/s)	Max. water level (El. m)	Time of peak flow (HH:MM dd/mm/yyyy)
Upstream side of Tangjiashan barrier		743.23	02:15 06/10/2008
Tangjiashan barrier site	6500		12:30 06/10/2008
Beichuan station	6540	620.30	13:00 06/10/2008
Tongkou HPP	6210	549.73	14:24 06/10/2008
Qinglian township	6160	532.90	14:48 06/10/2008
Fujiangqiao	6100	465.28	17:18 06/10/2008

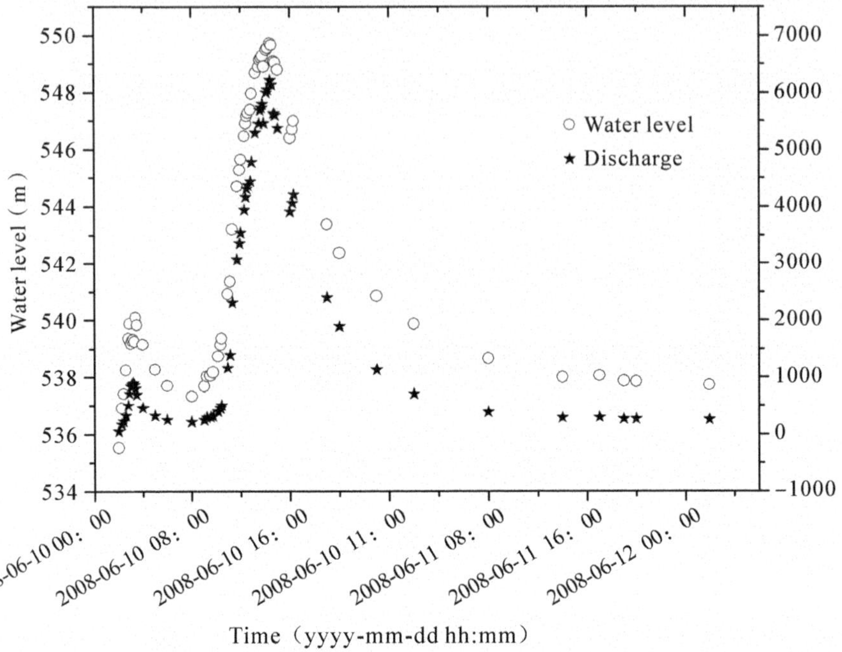

Fig. 9.15 Discharge and water level process measured at Tongkou station

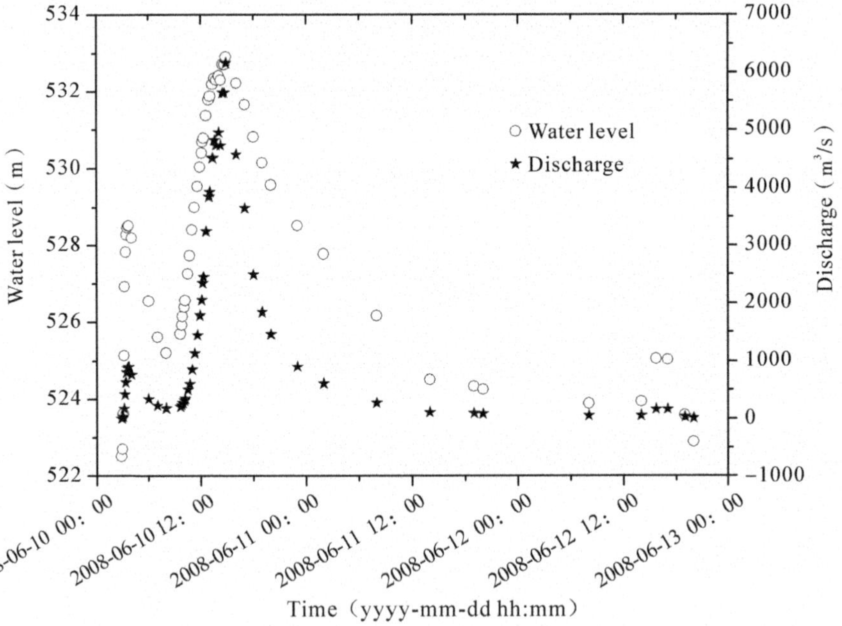

Fig. 9.16 Discharge and water level process measured at Qinglian station

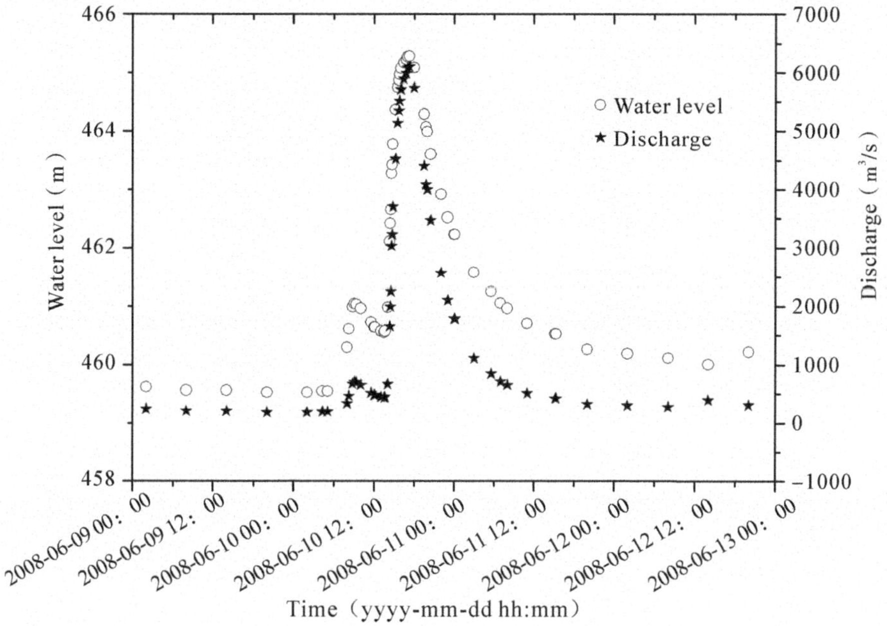

Fig. 9.17 Discharge and water level process measured at Tongkou station

9.2.2 *Barrier Lake Discharge Process and Back Analysis*

BRESZHU modeling, MIKE1 model and DAMBRK model are adopted to simulate the lake discharge process on the basis of actually measured data during the discharge of barrier lake.

9.2.2.1 Back Analysis Results of BRESZHU Model

1. Modeling of discharge channel section change and flood process at dam site

BRESZHU modeling was selected to simulate the change of discharge channel cross section and the flood process at barrier site. The discharge channel was continuously deepened and widened through scouring during discharge of lake water. This process can be understood as the breach extension of the gradual break process of earth-rock-dam. In the model, at early stage of the scouring, the cross section of the channel is 7–10 m wide at the bottom with an elevation of 740 m and the gradient of the side slopes is 1:1.5. The parameters of the barrier dam were calculated with given data on water level process on upstream side of the dam. These parameters are further applied in the calculation of the cross-sectional change of the channel as well as discharge flow process. The cross-sectional change of the channel during the discharge are subject to interaction between the discharge flow and the materials shaping the barrier dam. The width and bottom elevation of the ultimate discharge channel are calculated automatically once the program ran.

(1) Water level process on upstream side of the barrier dam

Figure 9.18 demonstrates a comparison between calculated and measured water level data on upstream side of the barrier dam. The figure shows that the calculated value matched well with the measured value.

(2) Discharge flow process at barrier site

Figure 9.19 shows a comparison on discharge flow process in the channel between data calculated with the model and that measured on site. The figure shows both type of data matched well, especially on peak flow, time for occurrence of peak and the discharge flow process after the peak. The measured peak flow discharged was 6500 m^3/s while the calculated number is 6616 m^3/s., with a discrepancy of 116 m^3/s, the calculated being 1.8% larger than the measured. The recorded peak flow discharged was measured at 12:30 on June 10th while the calculated time for the peak was 12:31 on June 10th, with discrepancy of only 1 min. The measured discharge flow grew slowly when the discharge just started scouring the channel; however, the calculated speed was fast. The reason for this discrepancy is the use of supporting

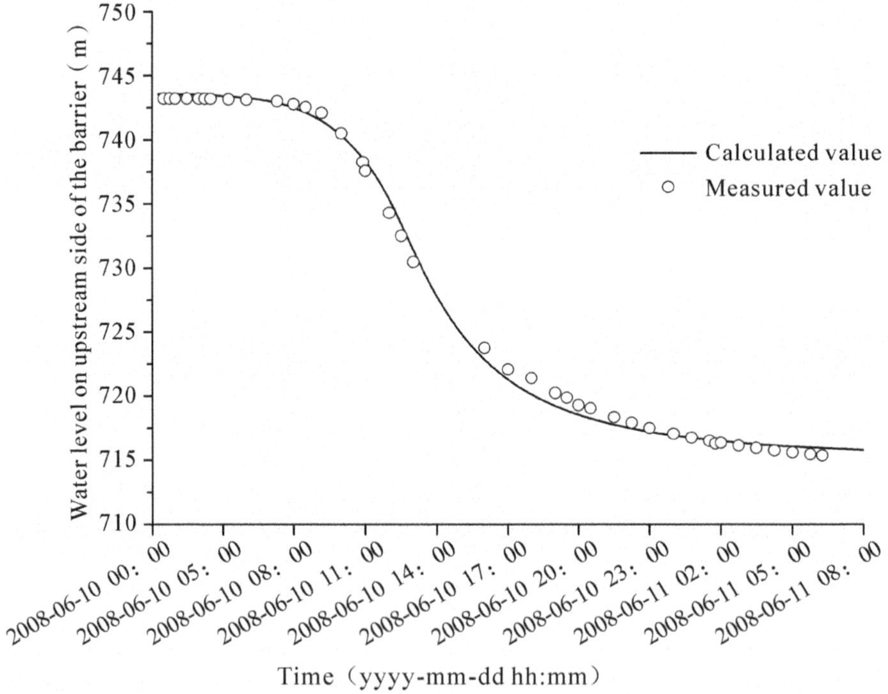

Fig. 9.18 Comparison between calculated and measured water level data on upstream side of the barrier

wire boxes on certain sections of the channel's side slope, which slowed down the extension of channel width and decreased the discharge flow. As the scouring became stronger, the supporting wire boxes were rushed away, after which the widening of the discharge channel was distinctly faster and the discharge flow bigger.

(3) Change process of the channel's cross-section

According to calculation by the modeling of the channel's cross-sectional change process, the water surface in the channel was 166.8 m wide and the channel bottom 100.1 m wide, which is in compliance to the on-site measured data. The calculated average water depth during discharge through the channel also matched with the actual numbers (Fig. 9.20). The maximum average water depth of 11 m was measured at 6:10 a.m. and 11:30 a.m. on June 10th, while the maximum depth was calculated to occur at 11:25 a.m. on June 10th, which is in line with the actual data.

BRESZHU modeling was selected to simulate and calculate the barrier lake discharge process and the development of channel's cross section. Based on analyses and verifications above, the calculation by the model is in compliance with measured data (see Table 9.5).

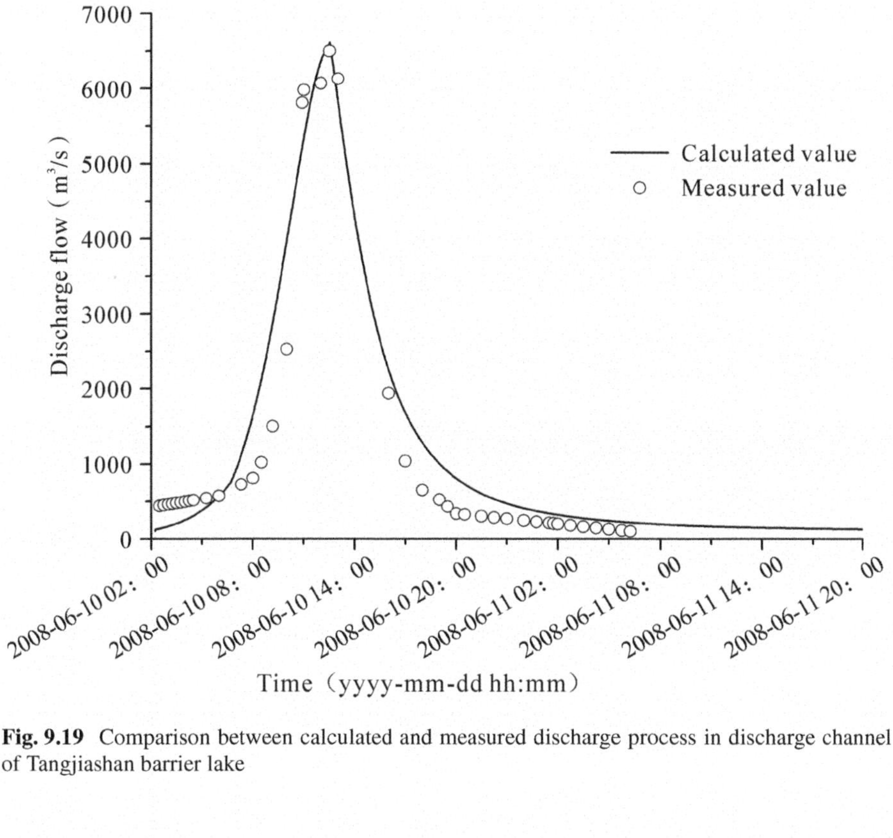

Fig. 9.19 Comparison between calculated and measured discharge process in discharge channel of Tangjiashan barrier lake

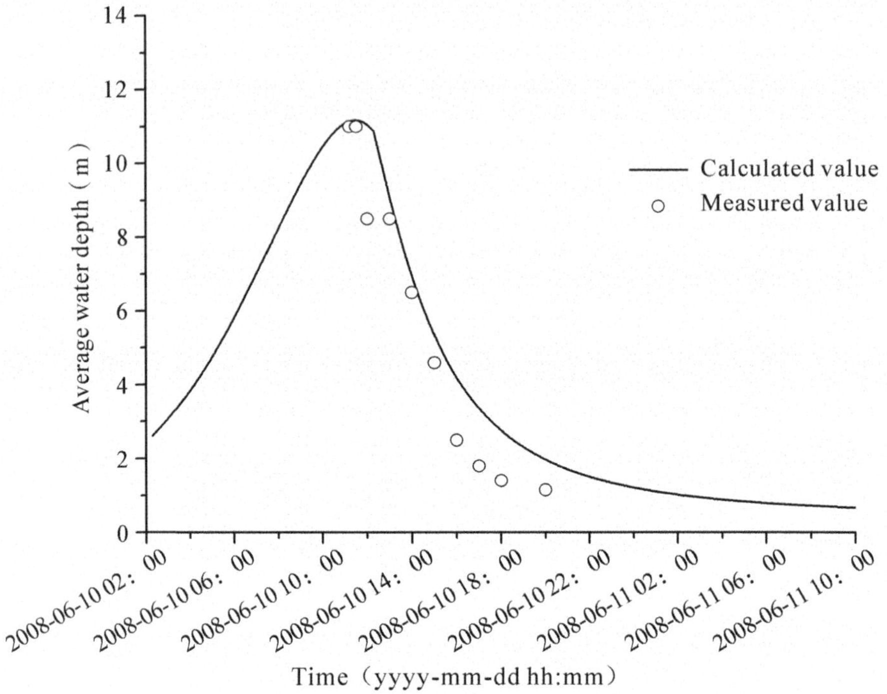

Fig. 9.20 Comparison between calculated and measured average water depth in discharge channel of Tangjiashan barrier lake

Table 9.5 Comparison between calculated and measured data

Type of data	Peak flow		Final size of discharge channel (m)		Max. water depth in the channel	
	Flow (m^3/s)	Time of occurrence (h:min)	Top width	Bottom width	Depth (m)	Time of occurrence (h:min)
Measured	6500	12:30	145–235	80–100	11	11:00, 11:30
Calculated	6616	12:31	166.8	100.1	11.18	11:25

2. Back analysis of the discharge process on downstream of the barrier dam

The discharge process along downstream of the barrier was simulated with MIKE11HD module. The scope of calculation and modeling covered two river sections on Tongkou River and Fujiang River. The 59 km section on Tongkou River started from Kaiping Township 23 km upstream to the barrier lake to Huikou, while the 30 km on Fujiang River started from the location 11 km upstream to Huikou reaching out to Fujiangqiao hydrological station in Mianyang City. Inflows in Tongkou River and Fujiang River were adopted as upper boundary; while stage-discharge relationship at Fujiangqiao hydrological station was considered as lower boundary. The reservoir operation of Tongkou HPP 24.5 km downstream to Tangjiangshan barrier lake was taken into account in the model. Flushing gate and overflow orifices of Tongkou hydropower station were all assumed to be open at the beginning while modeling.

Time period selected for modeling was 00:00 on June 10th–00:00 June 12th (48 h in total). The initial conditions for modeling of Fujiang river section were obtained through warm start, i.e., adopting an unchanging flow calculated based on boundary conditions upstream and downstream. Initial conditions on upstream of Tongkou River section are given based on measure water levels upstream of the dam, with dry wash downstream to the dam. The process of barrier lake water discharge can be understood as gradual dam break. The barrier breach size was given based on cross section of the discharge channel, the roughness factor was set 0.035 with reference to previous studies and time interval was 5 s.

(1) Water level process on upstream side of the barrier dam

The recorded data showed the water level on upstream side of the barrier was El. 740.51–743.23 m at 0:00–10:00 a.m. on June 10th, which was 0.5–3.23 m above the channel bottom. Later on, the water level on upstream of the barrier dropped obviously as the discharge channel was scoured and widened. Figure 9.21 showed a comparison between calculated and measured data of water level process upstream to the barrier. The figure suggested the model accurately simulated the changing process of the water level.

(2) Discharge process at the barrier site

Calculated discharge process was verified in comparison with actual measured discharge process (see Fig. 9.22). Calculated discharge flow at barrier break reached its peak 6800 m^3/s at 12:30 on June 10th, which is slightly larger than measured flow. The calculated and actual flow process are in good conformity.

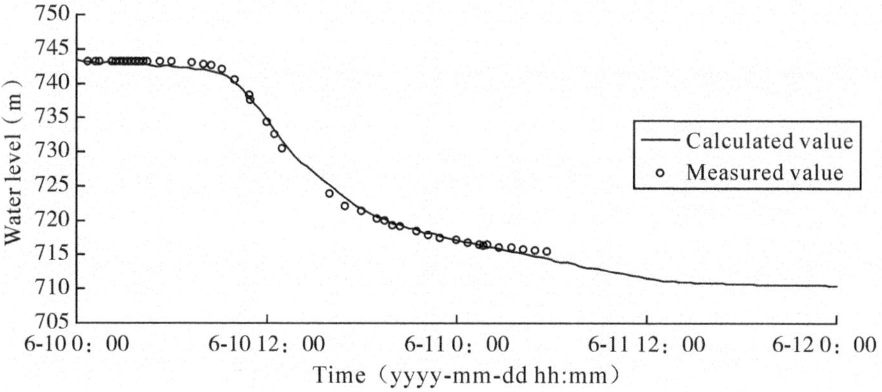

Fig. 9.21 Comparison between calculated and measured data of water level process upstream to the barrier

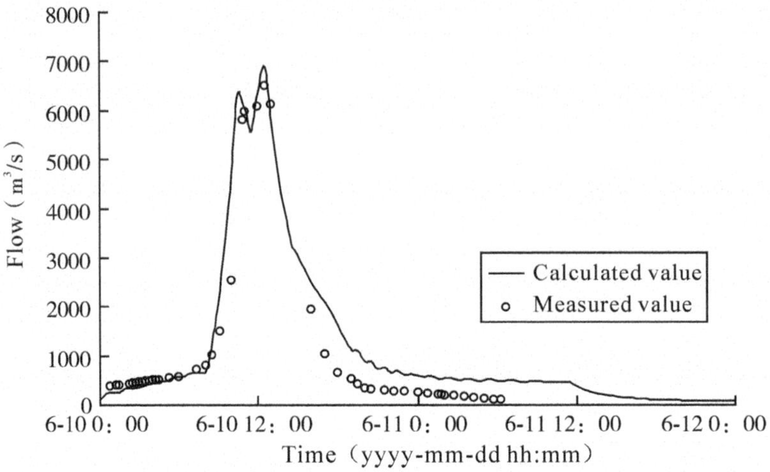

Fig. 9.22 Calculated discharge process was verified in comparison with actual measured flow process

(3) Discharge process on downstream of the barrier dam

Calculated discharge process was verified in comparison with actual measured discharge process at four hydrological stations, i.e., Beichuan, Tongkou, Qinglian and Fujiang (see Figs. 9.23, 9.24, 9.25, and 9.26). Little discrepancy was observed regarding the peak flow and time of occurrence. The calculated and actual flow processes are in good conformity.

Table 9.6 presents the comparison between calculated and recorded data on peak flow and time of its occurrence. The calculated peak flows at Tangjiashan, Beichuan and Tongkou are larger than measured ones, while the calculated peak flows at Qinglian and Fujiangqiao are smaller than the measured ones. Nevertheless, the error in each case is within 5%. Calculated times of peak flow occurrence are also in compliance with the actual ones measured, with errors no longer than 0.5 s.

(4) Water level process on downstream of the barrier dam

Calculated water level process was verified in comparison with measured water level process at Fujiang hydrological station (see Fig. 9.27). The calculated and actual water level process are basically in conformity, with the calculated value being slightly smaller than the measured one.

BRESZHU modeling was selected to simulate and calculate the flood routing on downstream of the barrier. Based on analyses and verifications above, the calculation by the model is in compliance with measured data.

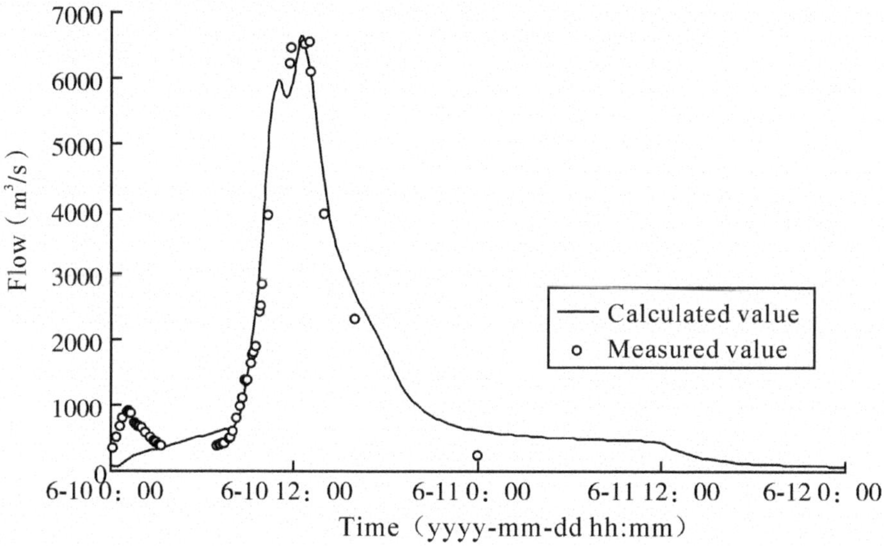

Fig. 9.23 Calculated discharge process verified with actual measured on at Beichuan hydrological station

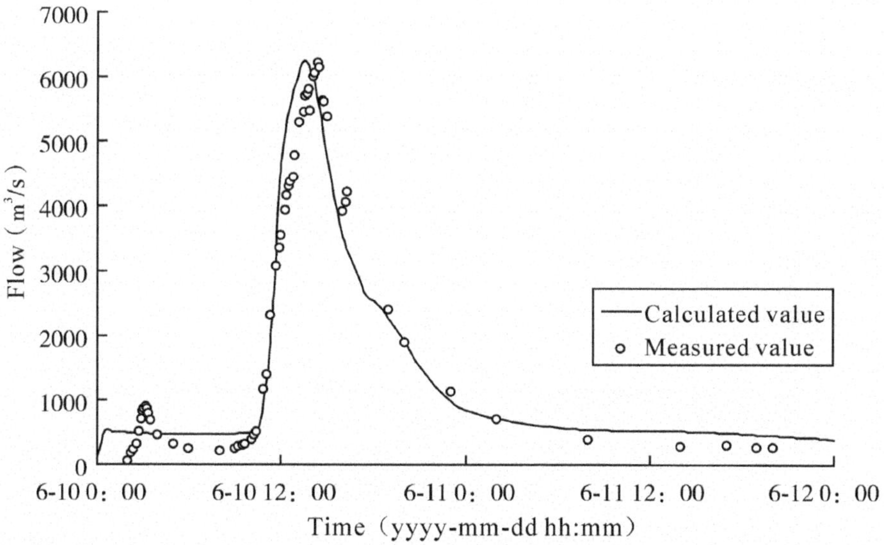

Fig. 9.24 Calculated discharge process verified with actual measured on at Tongkou hydrological station

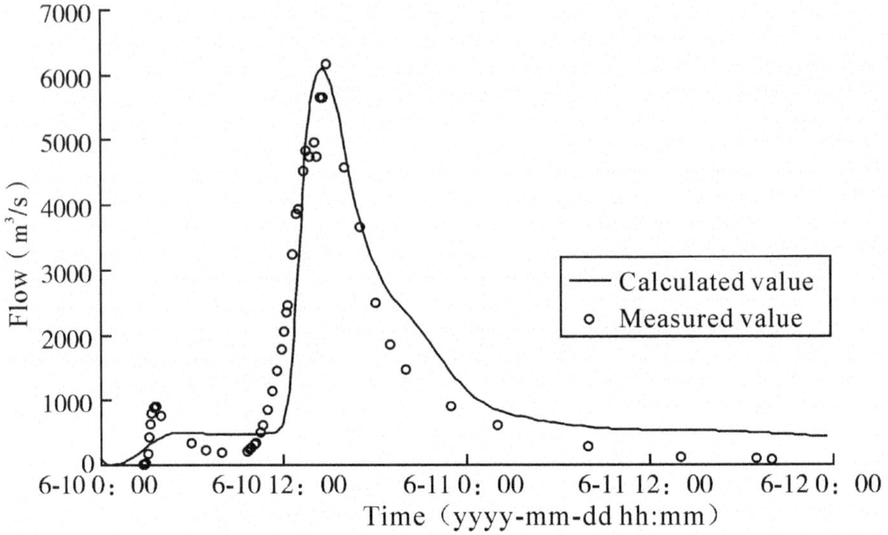

Fig. 9.25 Calculated discharge process verified with actual measured on at Qinglian hydrological station

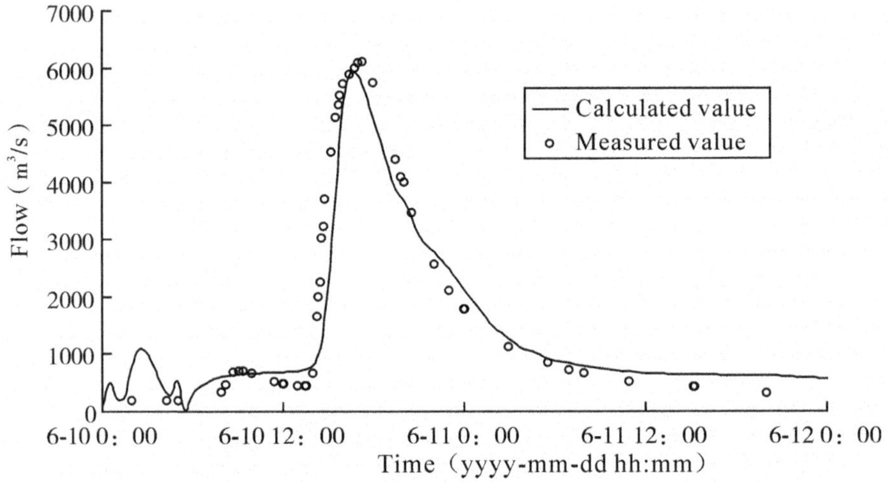

Fig. 9.26 Calculated discharge process verified with actual measured on at Fujiang hydrological station

9.2.2.2 Back Analysis Results of MIKE11 Model

MIKE11 dam break and flood model was used for simulation of discharge flow on downstream of the barrier and flood routing. Known conditions for the simulation include the stage capacity cure, measured breach width and depth. Starting time point

Table 9.6 Comparison between calculated and measured peak flow and time of occurrence

Location	Distance to the barrier (km)	Peak flow (m^3/s)			Time of occurrence (h:min)	
		Measured	Calculated	Error (%)	Measured	Calculated
Tangjiashan station	0	6500	6800	4.6	12:30	12:30
Beichuan station	4.6	6540	6630	1.4	13:00	12:50
Tongkou station	26.0	6210	6230	0.3	14:24	13:50
Qinglian station	46.5	6160	6110	−0.8	14:48	14:20
Mianyang city	68.9	6100	6000	−1.6	17:18	16:50

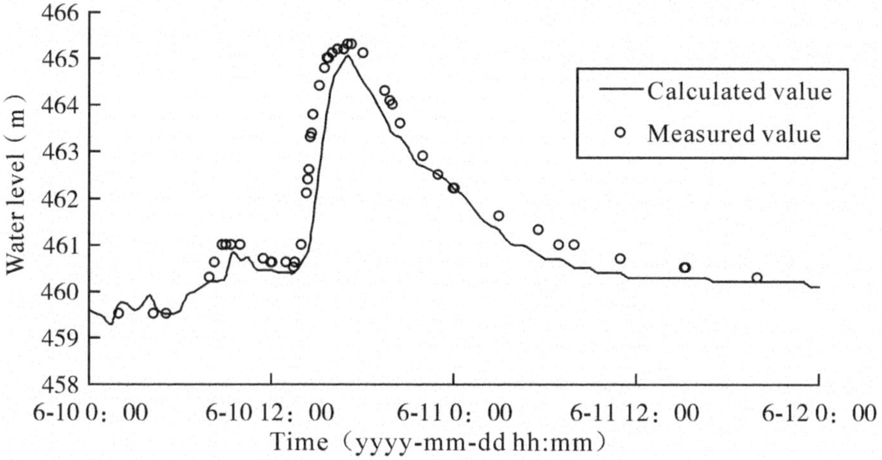

Fig. 9.27 Calculated water level process verified with the measured ones

for the simulation was set as 6:00 on June 10, 2008, and the water level at this time point was set as El. 743.13 m.

Boundary conditions for simulation of discharge flood routing include measured flow at barrier site, measured inflow from upstream of Fujiang River and measured water level at Fujiangqiao hydrological station. Inflow into reservoir of the barrier was set at 100 m^3/s, and inflow from upstream of Fujiang River was set at 196 m^3/s.

(1) Back analysis of discharge process at the barrier dam site

Simulation of the discharge process at the barrier site and water level of the barrier lake was verified (see Figs. 9.28 and 9.29). As shown in the figures, the flood process and time for flow peak calculated with the model basically match the measured flood process based on the actual width and depth of barrier dam breach. The calculated flood peak of 6400 m^3/s with relative error of only 1.5% compared to the measured peak of 6500 m^3/s. Certain discrepancy was observed between measured and calculated water level of the lake. Given that the calculated water level of the lake was based on cross-sectional downstream to Tangjiashan cross section, the error of calculated value should be smaller than the actual value as shown in Fig. 9.29. According to calculation elaborated above, the adopted model provided good simulation of the flood process of barrier dam break.

(2) Back analysis of downstream flow process of the barrier dam

The calculated flow processes at Beichuang, Tongkou, Qinglian and Fujiangqiao hydrological stations were compared with measured ones (see Figs. 9.30, 9.31, 9.32

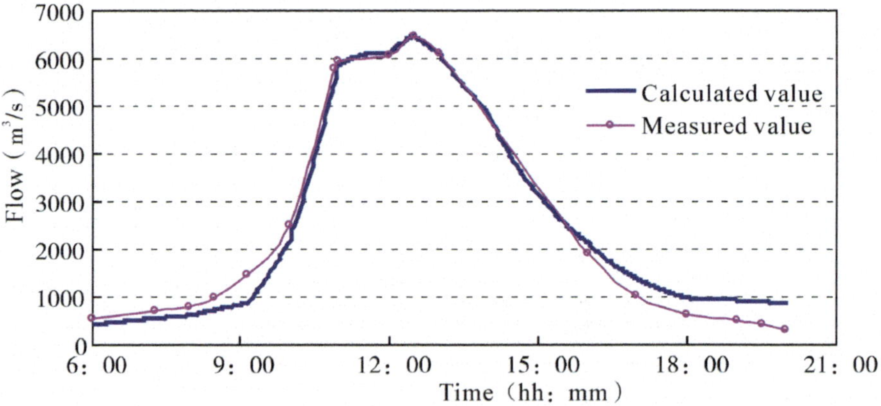

Fig. 9.28 Verification of the flow process of discharge at Tangjiashan barrier dam site

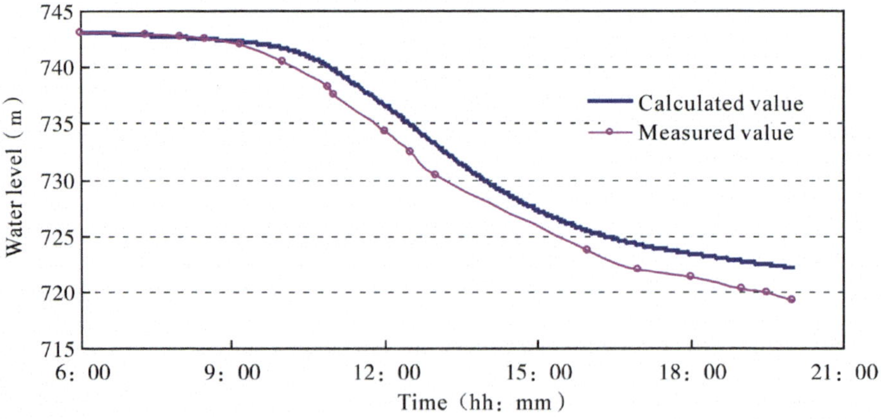

Fig. 9.29 Verification of the water level in reservoir of Tangjiashan barrier

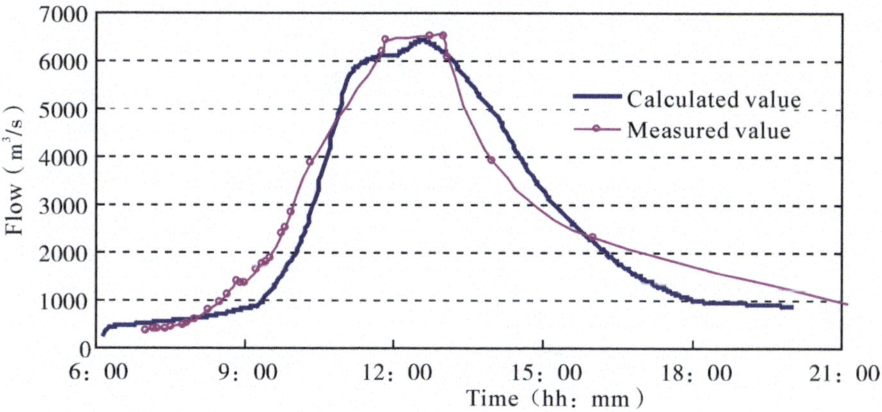

Fig. 9.30 Calculated and measured flow at Beichuan station

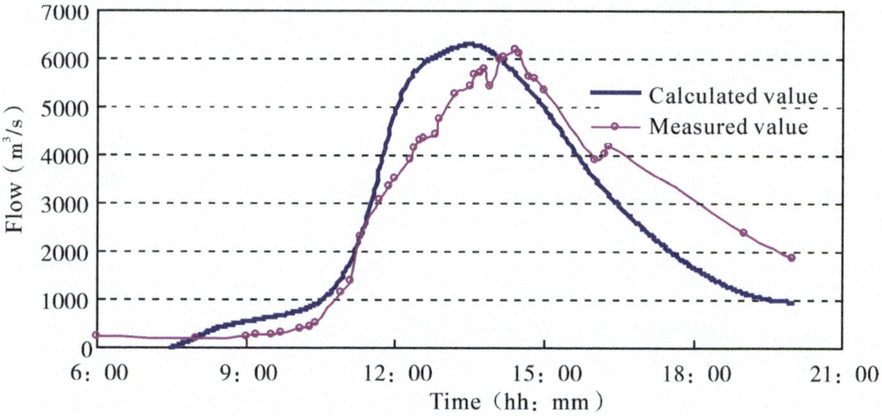

Fig. 9.31 Calculated and measured flow at Beichuan station

and 9.33). Table 9.7 shows calculated flood peak and time of peak occurrence at abovementioned stations in comparison with measured actual values.

The calculated flow processes are verified with and proved to match the actual values. Few errors were observed in the calculated peak flow. However, the calculated time of peak flow occurrence is earlier than actual since the flood detention effects of the smaller barrier lake on downstream were not considered in the mode.

9.2.2.3 Back Analysis Results of DAMBRK Model

DAMBRK programmed by National Weather Service (NWS) of the US was applied with adaptation and improvement for simulation of barrier lake discharge.

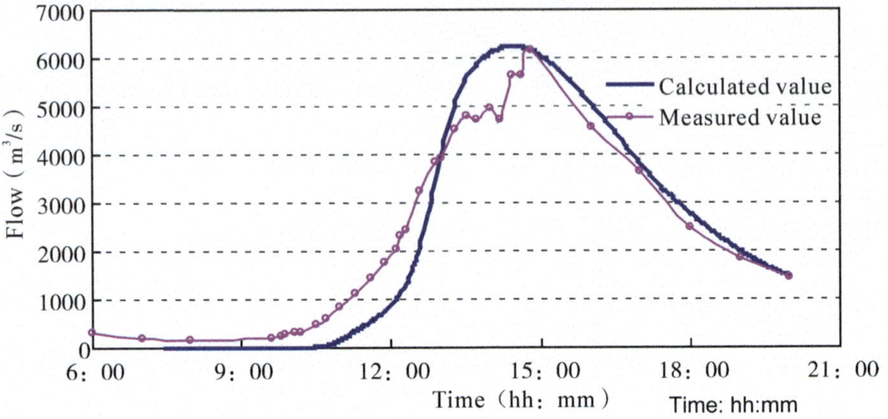

Fig. 9.32 Calculated and measured flow at Qinglian station

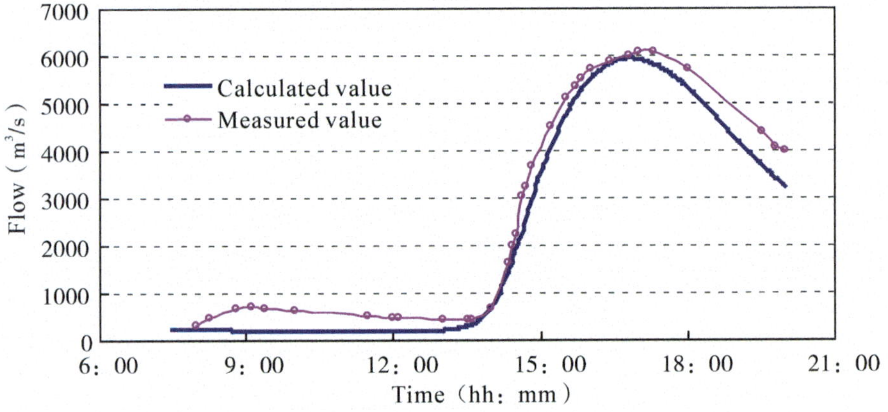

Fig. 9.33 Calculated and measured flow at Fujiang station

Table 9.7 Comparison between calculated and measured peak flow and time of occurrence

Location	Peak flow			Time of occurrence		
	Measured value (m^3/s)	Calculated value (m^3/s)	Error (%)	Measured value (hh:mm)	Calculated value (hh:mm)	Error (%)
Barrier site	6500	6400	−1.5	12:30	12:29	0.00
Beichuan	6540	6350	−2.9	13:00	12:35	−0.40
Tongkou HPP	6210	6306	1.6	14:24	13:29	−0.90
Qinglian Township	6160	6248	1.4	14:48	14:27	−0.33
Fujiangqiao	6100	5942	−2.6	17:18	16:47	−0.50

Note "−" was used to indicate that the calculate peak occurrence time is earlier than the actual time

(1) Back analysis of discharge flow at the barrier dam site

The break of Tangjiashan barrier dam started at 4:00–5:00 on June 10th as measured. A huge rectangle breach of 145 m wide with bottom elevation of 721.4 m had been shaped by 14:00 on the same day. On June 10th, inflow of 50–100 m^3/s was measured from the natural river channel, based on which this inflow coming from Tongkou River was set as 100 m^3/s in the model. The flow measured at Fujiangqiao at 8:00 a.m. on June 10th was measured as 192 m^3/s, which was adopted as the natural inflow from Fujiang River in the model.

As calculated by the model, the maximum peak flow was 6637m^3/s, quite close to the measured value of 6500 m^3/s with difference of only 2.1%. Figure 9.34 shows the calculated flow process at barrier dam breach in comparison with the measured value. It can be seen in the figure that the time for measured increase and drop of flow at barrier dam breach, though slightly lagging behind the calculated situation, could well match the process in the model. Such lagging behind was due to the use of linear evenly break in modeling while the barrier was not homogeneous dam. The barrier dam material consisted of gravel soils with large pebbles which better resisted scouring and impeded the fast discharge at early stage of barrier break.

(2) Back analysis of downstream flood routing of barrier dam

The scope for back analysis of flood routing was from the dam break site to Fujiang hydrological station on mainstream of Fujiang River, covering river section of 68.88 km. River channel roughness rate was set as 0.036–0.08, based on measured water levels and time composite rate of flood routing along the channel.

(1) Peak flow

On June 10th, the flood peak flow measured was 6210 m^3/s at Tongkou hydrological station and 6040 m^3/s at Fujiangqiao station. The value calculated in modeling was

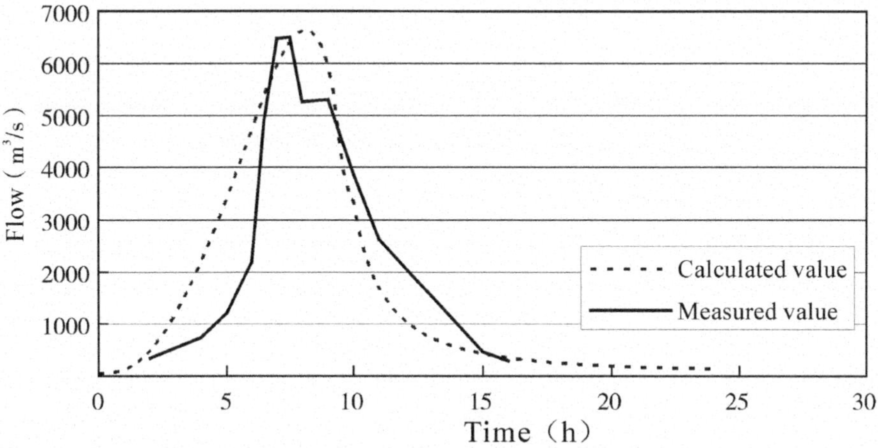

Fig. 9.34 Calculated flow process at barrier break in comparison with the actual measured value

Table 9.8 Flood water level discharged from Tangjiashan barrier

Distance to the barrier (km)	Location of the cross section	Max. water level measured (El. m)	Max. water level calculated (El. m)	Difference in water level (m)
0.00	Barrier break			
29.50	Tongkou HPP	549.73	549.90	0.17
36.10	Jiangjunshi hydrological station	534.34	534.13	−0.21
38.50		522.94	523.11	0.17
41.73		516.07	516.25	0.18
43.88		512.40	512.39	−0.01
46.51	Qinglian township	502.17	502.32	0.15
68.88	Fujiangqiao	465.28	465.35	0.07

6561 m^3/s at Tongkou while 6251 m^3/s at Fujiang, correspondingly 5.7 and 3.5% higher than measured value.

(2) Maximum water level

Upon barrier dam break on June 10th, the measured max. water level at Tongkou station and Fujiang station were El. 549.73 m and El. 465.28 respectively, with calculated value being 549.9 and 465.35 m accordingly. The water level difference between measured and calculated value are 0.17 and 0.07 m accordingly. During June 28th and July 10th, POWERCHINA Chengdu carried out survey and measurement on discharge water level within a 20 km river section from Tongkou to Hekou hydrological station. Table 9.8 shows comparison between the site investigated water level and the calculated water level, with a max. difference of 0.21 m and min. difference of 0.01 m.

9.2.3 *Social and Economic Benefit of the Emergency Handling Measures*

Analysis on measured rainfall data shows that a total of 47 million m^3 of water inflow into Tangjiashan barrier lake from the river basins on its upstream during the period of 8:00 a.m. June 7th–8:00 a.m. June 11th. The discharge volume from the barrier lake reached 186 million m^3 during the same period. The upstream water level of the barrier dam reached its maximum of El. 743.23 m at 1:30 a.m. on June 10th, which reduced to El. 714.62 m at 8:00 a.m. on June 11th. Due to the emergency engineering measures taken on site, water level of the barrier reservoir was reduced from El. 752.2 m (reservoir volume of 318 million m^3) to El. 740 m (reservoir volume of 224 million m^3). By 1:30 a.m. June 10th, the maximum water level in the barrier lake was El. 743.23 m, under which the reservoir capacity was 2.48 m^3. The water head was reduced by 9.0 m and reservoir volume reduced by 70 million m^3.

Without the emergency handling measures, in case of long-term average inflow during corresponding time period, the water level on upstream of the barrier could have reached El. 746.62 m (reservoir volume 274 million m^3) at 8:00 a.m. June 11th and El. 752.20 m at about 10:00 a.m. June 17th leading to overtopping of the barrier lake by the water. With the emergency handling measures taken efficiently, the upstream water level of the barrier dam had dropped by 32 m to El. 714.62 m by 8:00 a.m. June 11. It can be concluded that the discharge channel as an emergency measure had helped reduced water head and barrier reservoir volume, thus mitigating the threat of barrier lake flood to the life and property of residents on the downstream.

Had the discharge channel not been excavated, the water level in the barrier lake could have been lifted to approximately El. 752 m in accordance with the flood calculation. In case of barrier dam break under such a water level, the peak flow and water level on downstream of the barrier could have been much bigger than the measured value.

To better analyze the efficiency of the discharge channel, MIKE11 dam break and flood routing model simulated scenarios to calculated the peak flows on downstream of the barrier with and without the discharge channel. The time period for calculation was 6:00 a.m.–20:00 p.m. on June 10th 2008. In with-discharge-channel scenario, the barrier break occurs at water level of El. 743.1 m. The width and depth of the breach are enlarged following the calculation of the model. The water level in the reservoir dropped to El. 719.3 m by 20:00 p.m. In without-discharge-channel scenario, the barrier break starts at the water level of 752 m and expands following the rule set in the model. The water level in the reservoir dropped to El. 719.3 m by 20.00 p.m.

Table 9.9 shows peak flood at cross section of various locations under two different scenarios. It can be concluded that discharge channel helped lower the reservoir water level for barrier break and reduce discharge water volume. The peak flood flow at barrier site was reduced by about 3400 m^3 m^3/s and that at Fujiangqiao reduced by 3000 m^3/s. In summary, the discharge channel helped mitigate the peak flood flow of various hydrological stations along the channel by 33.6–34.6%, mitigating the flood threat to downstream.

9.3 Initial Stability Assessment of the New River Channel

Regarding the geological composition of both bank, both side slopes are symmetric after scouring, with 20–40 m thick yellow gravel soils partly on the surface and 10–20 m thick cataclasite in middle and lower part of the slope. Metamorphic bedding structures were maintained partly (see Fig. 9.35).

Table 9.9 Peak flood at cross section of various locations under two different scenarios

Location of the cross section	Peak flood flow		
	Scenario 1: with-discharge-channel (m^3/s)	Scenario 2: without-discharge-channel (m^3/s)	Peak flow scenario 1/peak flow scenario 2 (%)
Barrier site	6400	9785	65.4
Beichuan	6350	9650	65.8
Tongkou HPP	6306	9578	65.8
Jiangjunshi hydrological station	6296	9562	65.8
Qinglian town	6248	9487	65.9
Fujiangqiao station	5942	8945	66.4

Fig. 9.35 New river channel and side slops

9.3.1 Remaining Barrier Dam Body on Left Bank

After the discharge of barrier lake with large water volume, most of the gravel soil in trough on the right of the barrier dam were washed away. The cataclastic rock beneath that was also scoured. The major part of the barrier dam on the left bank was cataclasite. Metamorphic bedding structures were maintained partly at front edge of the slope, extending for long distance along the channel being able to resist the

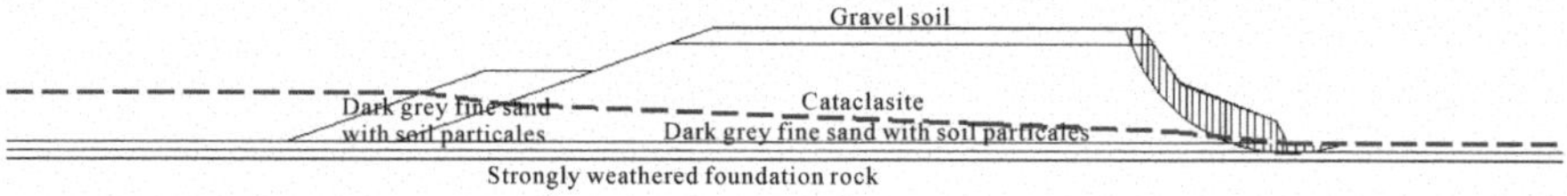

Fig. 9.36 Stability analysis result of the remaining barrier body (upstream water level is 714.13 m. the min. safety coefficient (Fs) is 1.29)

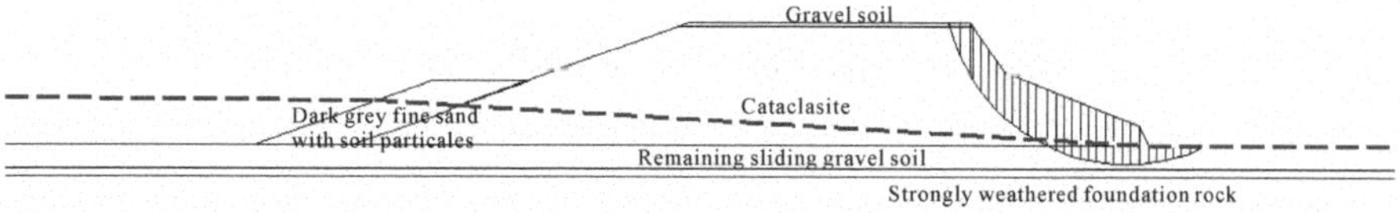

Fig. 9.37 Stability analysis result of the remaining barrier body (upstream water level: 714.13 m; min. safety coefficient: Fs = 1.34)

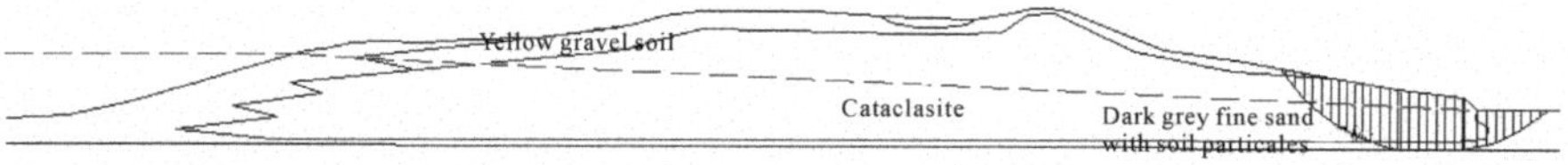

Fig. 9.38 Stability analysis result of the remaining barrier body (upstream water level upstream: 714.13 m; min. safety coefficient: Fs = 1.96)

destruction of seepage and scouring for better stability of the slope. The water level drastically dropped after discharge. As a result, the remaining barrier dam body is in a better stable state.

At 14:00 p.m. on June 11th, the upstream water level of the barrier dam reduced to El. 714.13 m. Under such conditions, calculation was carried out to study the stability of remaining barrier body. Results are demonstrated in Figs. 9.36, 9.37 and 9.38.

9.3.2 Slope Stability of the New River Channel

(1) Discharge channel

The river bed after discharge was mainly composed of cataclastic rock. The water is currently contained within the cataclastic channel. The bed rock under the channel can well resist scouring and is stable under natural conditions.

(2) Side slope on right bank

On upper reaches of the new river channel, the mountain has a dip slope with a slope angle (60″) basically in line with the rock strata being stable. On surface of the slope are colluvium and deluvium, which are loose in structure. Debris flows in small scale are likely to happen in case of rainstorm with little influence on the new

Table 9.10 Safety coefficient under various scenarios

Conditions	Water level at the cross section: El. 714.13 m	
	Operation under normal conditions	In case of earthquake of W intensity
I–I′ cross section on left bank	1.37	1.15
II–II′ cross section	1.00	0.86
III–III′ cross section	1.35	1.02

river channel. On lower reaches of the new channel, the mountain on the right bank has a dip slope with large amount of debris loosely deposit on the top. The debris could slide in case of rainstorm or earthquake.

(3) Side slope on left bank

Due to scouring, lots of cracks can be observed on the eastern side slope of the barrier body near the new channel. The strike of the cracks are basically in conformity with the side slope. There is a possibility for occurrence of slope collapse. The side slope on northeast of the landing field is especially liable to suffer from rainwater seepage or earthquake with its existing 1–3 cm cracks.

Cross sections of the side slope are selected for stability calculation with a given water level of 714.13 m. Cross sections are selected from the side slope on upstream, middle stream and downstream. They are numbered as II–II′, I–I′ and III–III′. Results of the calculation are presented in Table 9.10 and Figs. 9.39, 9.40, 9.41 and 9.42.

It can be learn from the calculation that the side slopes of the new channel are unstable with safety coefficient of 1.0–1.3. In case of aftershocks, slides of small scales could happen in small slip circles instead of large ones. The slopes are steep with earth and gravel stratum on the surface, hard to resist flushing of water. Many tensile fissures can be seen in 20–40 m at the edge of the slope, leading to instability. Once large flood and continuous rainfall occur, slope collapse could be further

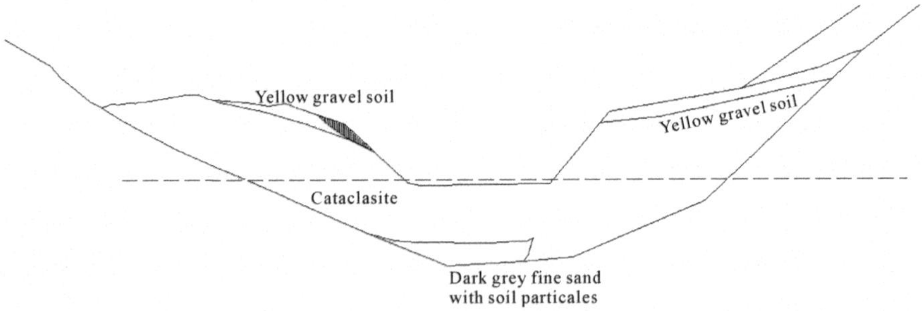

Fig. 9.39 Stability analysis result of cross Sect. 9.1 (left bank, Min. safety coefficient (Fs) is 1.37)

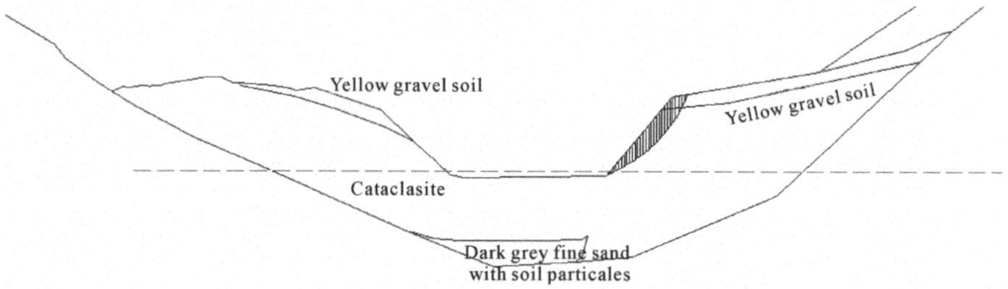

Fig. 9.40 Stability analysis result of cross Sect. 9.1 (right bank, Min. safety coefficient (Fs) is 1.32)

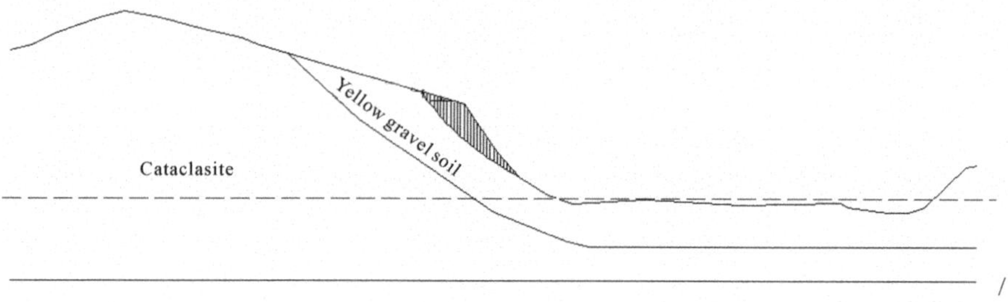

Fig. 9.41 Stability analysis result of cross Sect. 9.2 (left bank, Min. safety coefficient (Fs) is 1.0)

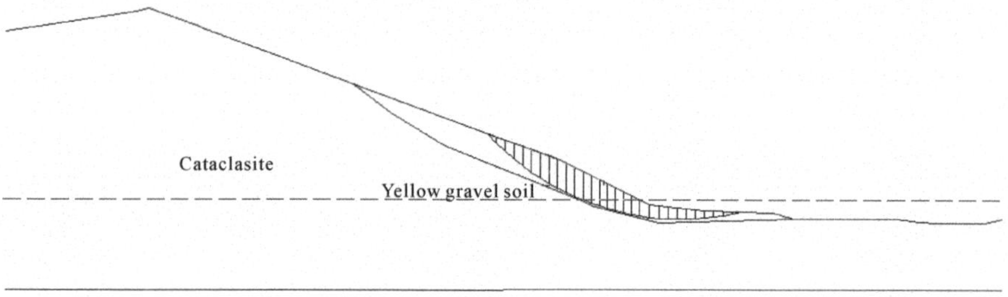

Fig. 9.42 Stability analysis result of cross Sect. 9.3 (left bank, Min. safety coefficient (Fs) is 1.35)

induced but in small scales. Even if the new channel gets blocked with deposit, which cannot well resist the scouring. In case of flow increase, the deposit will be rushed away without obvious reduction of the flood discharge capacity.

On back side of the landslide on Tangjiashan right bank, the materials are broken rock structure and loose overburden deposits. In case of the rainfall and aftershock, further collapses in small scales could occur making the slope unstable. The natural side slopes along the small bays downstream are steep with the slope foot being flushed and cut. This part of the slopes is deposits from sliding of overburden, with collapse in some parts. Once continuous rainfall, aftershocks or further scouring occur, slides will continue happening. However, deposits loose in structure would not block the river channel even in case of sliding. Neither will it impact the flood discharging capacity.

In conclusion, the remaining part of the barrier dam body is large in size with gentle slopes and had been shaped from high-speed landslide of bed rocks. There is no possibility of damage by soil flow or piping. With good stability, no collapses in large scale will happen. In case of flood, barrier could collapse partially yet in a gradual manner instead of an instant outburst. Under the influence of scouring on the slope foot on downstream of the barrier, slow collapse and slides could also happen, leading to a partial deformation and loss of stability on upper part of the slope. Nevertheless, the barrier dam body will not be unstable entirely.

All the analyses above proved to confirm with reality, which can be justified based on the flushing during discharge of flood on June 11th. After discharge of flood, most parts of the overburden on left of the former barrier body and the gravel stratum on its right were flushed away. Small quantity of the cataclasite on lower stratum was also scoured. The remaining barrier body is comprised of cataclasite rock which can better resist seepage, scouring and slides. After discharge, the water level in the reservoir declined sharply. The remaining barrier body thus kept good stability.

9.4 Flood Discharge Capacity of the New River Channel

9.4.1 Cross Section of the New River Channel and Hydraulic Calculation

(1) Cross section of the new river channel

After the barrier lake was discharged from the channel, the discharge channel was reshaped into a new one due to the scouring of water. The plane graph of the new channel is shown in Fig. 9.43.

(2) Flood discharge capacity of the new channel

Channel water surface curve method was adopted by CISPDR to calculate the flood discharge capacity of the new channel. The results are presented in Table 9.11.

POWERCHINA Chengdu had also calculated and analyzed the flood discharge capacity with wide crest weir discharge formula. Their calculation was based on the water regimen notice realized on June 10 and June 11, 2008, which included the measured flow, water levels, the topographic condition of the barrier body near the new channel and the geographic structures. Currently, the bottom of the new river channel is higher on the left and lower on the right. There is a deep groove beneath the stream on the right side of the channel. It was preliminary presumed that the bottom elevation of the deep groove was about 711 m. The cross section of the new channel for hydraulic calculation was in trapezoid shape with bottom width of 145 m. The hydraulic calculation was carried out based on the presumed section of the flow

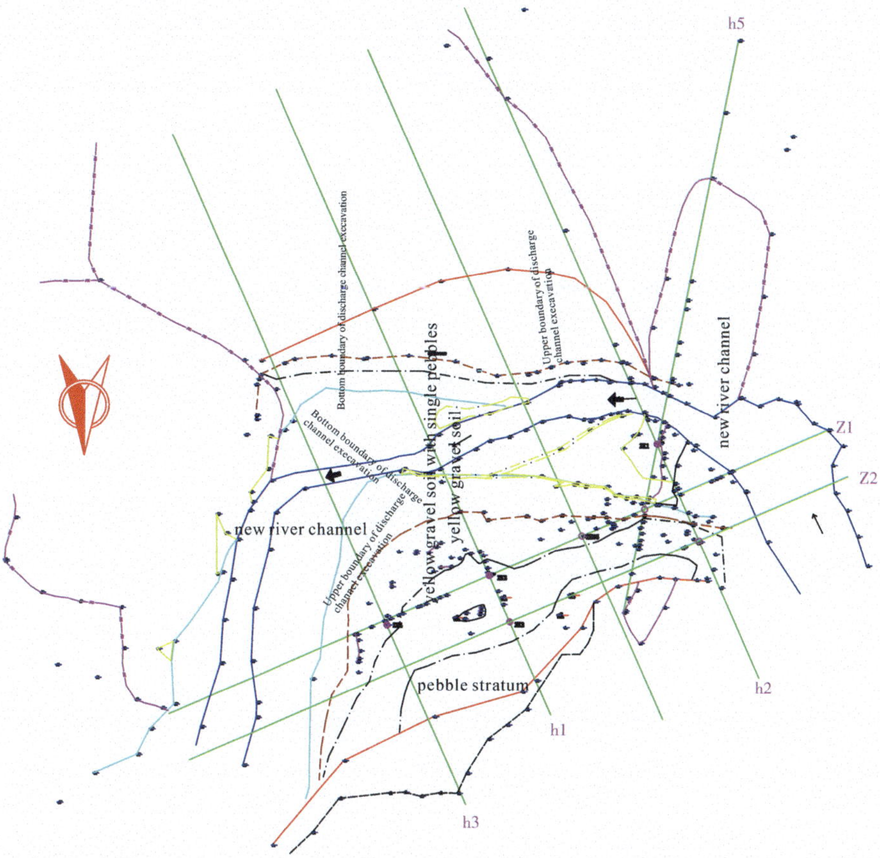

Fig. 9.43 Plane graph of the new channel after discharge of the barrier lake

and elevation of the sill. The stage-discharge relationship and hydraulic elements are given in Tables 9.12 and 9.13.

During flood season when discharge flow is 3040 m^3/s, equal to the peak flow under flood with return period of 10 years, the velocity in the new channel is 6.37 m/s. When the discharge flow is 3920 m^3/s, i.e., the peak flow under flood with return period of 20 years, the velocity in the new channel is 6.97 m/s. When the discharge flow is 5120 m^3/s, i.e., the peak flow under flood with return period of 50 years, the velocity in the new channel is 7.64 m/s. When the discharge flow is 6040 m^3/s, i.e., the peak flow under flood with return period of 100 years, the velocity in the new channel is 8.08 m/s. When the discharge flow is 6970 m^3/s, i.e., the peak flow under flood with return period of 200 years, the velocity in the new channel is8.50 m/s. In case of the velocity above, the cross section of the channel will be enlarged and the channel will be cut more deeply. The side slope composed of gravel soil on right bank will be rushed and collapse.

Table 9.11 Calculation on the flood discharge capacity of the new river channel

Water level (m)	Flow (m^3/s)	Velocity at the inlet cross section (m/s)	Water level (El.)	Flow (m^3/s)	Velocity at the inlet cross section (m/s)
712.50	245.60	2.69	720.00	1934.70	5.90
713.00	308.90	2.95	720.50	2154.50	6.10
713.50	372.20	3.19	721.00	2386.00	6.30
714.00	445.50	3.44	721.50	2641.00	6.43
714.50	526.80	3.68	722.00	2911.80	6.49
715.00	609.70	3.92	722.50	3206.50	6.44
715.50	706.80	4.17	723.00	3529.00	6.41
716.00	804.30	4.41	723.50	3871.40	6.37
716.50	913.00	4.64	724.00	4238.80	6.33
717.00	1024.10	4.87	724.50	4627.10	6.19
717.50	1144.60	5.08	725.00	5053.60	5.88
718.00	1271.10	5.29	725.50	5500.00	5.77
718.50	1403.20	5.46	726.00	5946.40	5.72
719.00	1561.90	5.61	726.50	6415.10	5.72
719.50	1738.20	5.75	727.00	6902.00	5.74

Table 9.12 Stage-discharge relationship of the new river channel

Water level (El. m)	713	716	718	720	721.5	722.9	724	725.3	726.2	727
Flow (m^3/s)	77	412	826	1467	2190	3040	3920	5120	6040	6970

Table 9.13 Calculation of hydraulic parameters of the new river channel

Flood frequency (%)	Flow (m^3/s)	Normal water depth (m)	Max. velocity (m/s)
20	2190	3.33	5.66
10	3040	4.09	6.37
5	3920	4.81	6.97
2	5120	5.71	7.64
1	6040	6.36	8.08
0.5	6970	7.0	8.50

Note The flood frequency values are calculated without taking into account the operation of the reservoir

9.4.2 Flood Regulation Calculation

1. Basic data

   (1) Storage-capacity curve developed by POWERCHINA Chengdu was adopted in the calculation
   Design flood data: the design flood process curve at Tangjiashan barrier site provided by Hydrology Bureau, CWRC (see Table 9.14) were adopted.

(2) Calculation on flood regulation by CISPDR

The mitigation of flood discharged from Tangjiashan barrier lake was carried out through open discharge. It was assumed that the cross sections keep unchanged during the discharge process. Flood mitigation calculation was conducted under scenarios of $P = 0.5\%$, $P = 1\%$and $P = 2\%$. Based on data mentioned above and rules for flood mitigation, the initial water level is set as 708.4 m, same as the cross section's min. elevation during discharge through the barrier. The flood parameters under above three scenarios are given in Table 9.15.

Results of the calculation suggest that in case of flood with return period of 200 years, due to the flood detention function of the barrier body, the max. water level at the barrier site is El. 725.44 m and max. discharge flow in the new channel is

Table 9.14 Design flood process at Tangjiashan barrier dam site (m^3/s)

S. No.	Flow ($P = 0.5\%$)	Flow ($P = 1\%$)	S. No.	Flow ($P = 0.5\%$)	Flow ($P = 1\%$)
1	2050	1790	19	3480	3030
2	2530	2210	20	3350	2920
3	3010	2630	21	3210	2800
4	3490	3050	22	3080	2690
5	4000	3490	23	2950	2570
6	4630	4030	24	2840	2480
7	5470	4770	25	2750	2390
8	6970	6040	26	2670	2320
9	5650	4930	27	2590	2260
10	5010	4370	28	2510	2190
11	4650	4060	29	2430	2120
12	4530	3950	30	2330	2030
13	4370	3810	31	2250	1960
14	4210	3670	32	2170	1890
15	4050	3530	33	2090	1830
16	3880	3380	34	2010	1760
17	3750	3270	35	1920	1670
18	3610	3150			

Table 9.15 Flood parameters under three scenarios during discharge through the new channel

Flood frequency (%)	0.5	1	2
Peak flood flow (m^3/s)	6970	6040	5120
Max. water level (El. m)	725.44	724.4	723.28
Max. discharge flow (m^3/s)	5440	4540	3720

5440 m^3/s. The flood could be mitigated to the level of those with a return period of 100 years. In case of flood with return period of 100 years, the max. water level at the barrier site is El. 724.4 m and max. discharge flow in the new channel is 4540 m^3/s. The flood could be mitigated to the level of those with a return period of 50 years. In case of flood with return period of 50 years, the max. water level at the barrier site is El. 723.28 m and max. discharge flow in the new channel is 3720 m^3/s. These numbers well prove that the newly shaped channel can endure floods with return period of 200 years.

(3) Flood regulation calculation by POWERCHINA Chengdu

According to flood survey, the historical peak flood flow was 6720 m^3/s at Tangjiashan barrier dam site, which is quite close to the peak flow under flood frequency of 0.5%. The calculation was conducted under four scenarios with various flood frequency. As had been measured, the water level in barrier lake was El. 714.51 m and corresponding discharge flow was 68.4 m^3/s at 10:00 a.m. on June 11th. Preliminary analysis concluded that when the lake water level drops to El. 714 m, the discharge almost equals inflow from upstream thus the water level maintained unchanged. Therefore, the initial water level was set as 714 m for the calculation. Results are shown in Table 9.16.

Calculation on typical flood in 1995 and 1964 at various frequency shows a maximum water level of El. 721.46–722.17 m at the barrier dam body and a maximum

Table 9.16 Flood mitigation scenarios at Tangjiashan barrier lake

Schemes	Flood frequency (%)	Max. water level (El. m)	Peak flow during discharge (m^3/s)	Peak inflow into reservoir (m^3/s)	Flood detention duration (h)	Flow mitigation (m^3/s)
Typical flood 1995 (double peak)	0.5	724.88	6132	6970	3	838
	1	724.11	5276	6040	3	764
	2	723.43	4523	5120	3	597
	5	722.17	3471	3920	3	449
Typical flood 1964 (single peak)	0.5	724.48	5691	6970	3	1279
	1	723.70	4823	6040	3	1217
	2	722.83	3928	5120	3	1192
	5	721.46	2985	3920	3	935

discharge flow of 2985–3471 m^3/s, which is 449–935 m^3/s less than the peak inflow into the reservoir with a time lag of 3 h.

Upon occurrence of flood with return period of 50 years, the maximum water level at the barrier dam is El. 722.83–723.43 m with a maximum discharge flow of 3928–4523 m^3/s, which is 597–1192 m^3/s less than the peak inflow into the reservoir, with a time lag of 3 h.

Upon occurrence of flood with return period of 100 years, the maximum water level at the barrier dam is El. 723.70–724.11 m with a maximum discharge flow of 4823–5276 m^3/s, which is 764–1217 m^3/s less than the peak inflow into the reservoir, with a time lag of 3 h.

Upon occurrence of flood with return period of 200 years, the maximum water level at the barrier body is El. 724.48–724.88 m with a maximum discharge flow of 5691–6132 m^3/s, which is 838–1279 m^3/s less than the peak inflow into the reservoir, with a time lag of 3 h.

Even in the event of flood with return period of 200 years, the max. water level at barrier body maintains smaller than El. 743.10 m, the measured max. water level before discharge from the barrier lake.

9.5 A Summary of Experience

The Tangjiashan barrier lake was formed by the earthquake on May 12, 2008. The discharge of water in the barrier lake had been finished by June 11, 2008. The relief of the disaster took 1 month. To cope with the emergency in a safe, scientific and fast manner so as to safeguard the life of people as a primary goal, as required by the State Council, the expert team under the direct leadership of MWR of PRC worked out a scientific disaster relief plan to relieve disaster with full support from the army, police, residents as well as the local governments. The Earthquake Relief Headquarters under State Council issued a congratulatory letter praising the relief action as "a world wonder in coping with large barrier lake".

The relief action was a success against the disaster through unconventional engineering measures taken scientifically at special time under extreme conditions. Chinese people's courage in front of earthquake was shown in implementation of the action, which also showed the wisdom and competence of those working in the field of water resource technology. It was with the support of such technology that we witnessed a good practice of barrier lake disaster relief. The experience gained, the advanced engineering philosophy and scientific technological work plan will serve as a model with great significance in dealing with future similar emergency events.

1. An efficient mechanism for decision making and collaboration

    (1) The foremost guarantee for the success of disaster relief action was the strong leadership of the central committee of Communist Party of China and the State Council as well as the discipline of coping with the emergency in a safe, scientific and fast manner established by them. The on-site instruction by

Tangjiashan Barrier Lake Disaster Relief Headquarters. The then Premier Wen Jiabao made three site visit to Tangjiashan barrier lake; the disaster relief was well managed with leadership and efficient decision making by MWR, PLA Headquarters Central Staff and Armed Police Forces.

After field investigation on May 21st, a work plan was made through consultation comprising two aspects, i.e., engineering measures for disaster relief and measures helping residents on upstream and downstream for emergency rescue.

During making the disaster relief work plan, experts advised two MI-26 helicopters were needed for transportation of construction equipment. Immediately support was offered domestically with 1 MI-26. Another was transported from overseas to Mianyang city within 3 days through coordination and arrangement by International Science, Technology and Cooperation Division, MWR, with full support from MOFA, General Administration of Customs, the Headquarters of the General Staff. With the leadership and command on site as well as timely decision making, more time and supply were guaranteed for implementation of disaster relief.

(2) Close collaboration among parties with different responsibilities ensured the success implementation of the emergency work plan. A well-organized work mechanism was established incorporating collaboration through different levels of governments, between the army and residents, and among different authorities as well as regions. The immediate actions taken by all relevant parties ensured implementation of the work plan. CAPF Hydropower Force worked continuously of high intensity and over fulfilled the tasks of discharge channel construction ahead of schedule. Chengdu Military Area Command ensured the air transport channel as scheduled. Land air corps from both Chengdu Military Area and Jinan Military Area working in high intensity ensured air transportation. Fast channel was offered by MOFA, Ministry of Public Security, the Headquarters of the General Staff, Civil Aviation Administration of China and relevant parties to make sure the MI-26 helicopters going into action as quick as possible. Experts from Chinese Academy of Science installed remote wide band video monitoring system over barrier dam body. Various levels of governments in Sichuan Province guaranteed the safe rescue and transfer of residents upstream and downstream, through arduous and careful work.

(3) All-round technical supports in water resources industry made sure the disaster relief work conducted in scientific way. Experienced experts were dispatched from MWR to join the Tangjiashan Barrier Lake Disaster Relief Headquarters and on-site expert team. Meanwhile, expert consultation mechanism was established both in Mianyang and on barrier site to offer professional advice for decision making. Full support was also offered to the site team from organizations in water resources industry through technical seminars and optimization of the work scheme proposed on site. Various organizations carried out flood discharge and dam break calculation in parallel with the team on site, thus providing all-round technical support

to the disaster relief measures and emergency rescue. About 170 scenarios were simulated in dam break flood calculation and alternative studies by CWRC, POWERCHINA Chengdu, CAS, etc. In lack of data, the experts led studies on the rock composition on site and innovatively applied Retrogressive scour theory in gravel channel. These two technical tasks were critical to the success in disaster relief of Tangjiashan Barrier Lake.

2. People-oriented philosophy and scientific decision making

   (1) Government at different levels work with advanced emergency decision-making mechanism and administrative philosophy. No secondary disasters were allowed to be induced as a result of mistreatment under emergency situation so that residents' life and property could be safeguarded. All risks were handled technically and politically.
   (2) Sound mechanism of risk sharing was established. The risks confronted by disaster-hit area should be handled by the government. The risks needed to be shared in such a sound manner that residents should take the responsibility to avoid risks while the government should try to mitigate the risks.
   (3) The rule of scientific decision making in large-scale disasters was strictly followed. Engineering measures and non-engineering measures were carried out simultaneously. The rescue work plan of local residents were made based on assumption of 1/3 dam break. Meanwhile, emergency response plans were prepared for 1/2 or entire dam break. These arrangements reflect how the rule was followed.

3. Excavation of the discharge channel to take advantage of local circumstances enables a scientific engineering scheme to relieve disasters with nature's power.

   (1) Basis data were collected and analyzed fast to make the engineering scheme. The acquisition of existing data was an important method for fast and accurate judgment of the disaster circumstances. In making of the emergency treatment work plan, the site team had collected the existing hydrological, meteorological, topographic and geological data by May 22nd, which well supported the identification of the disaster circumstances, preparation of relief work plan and further technical work.
   (2) The excavation scheme of discharge channel was to excavate along the low trough on right side of Tangjiashan barrier dam. It involved smaller excavation quantity and easier construction with gravel soil on upper section of the trough. Without access for large equipment to land on barrier, this scheme can lower the channel's bottom elevation as quickly as possible to accelerate the discharge process.
   (3) One preferred scheme was selected with backup alternative in dynamic adjustment. A basic strategy to handle the disaster and uncertainty in construction duration was to prepare for the worst construction conditions, to compare and select the best scheme and strive for the highest effectiveness. This is also a shared philosophy in relief of similar disasters. The

preferred construction alternative was to excavate the channel with equipment, and blasting was a backup alternative. The channel was designed to have constant unchanged excavation boundary and different channel bottom. Construction scheme could be dynamically adjusted based on the actual construction period.

(4) The Tangjiashan barrier body was huge in size with gentle slope on upstream side of the barrier with soil overburden and being compact underneath, without any possibility of sudden dam break. Considering the characteristics of barrier body's shape and particle size distribution, the break of it would occur partially, with lateral erosion and retrospectively scouring in parallel. The hydro-energy was utilized for expansion and deepening of the channel. The strongly weathered rock stratum, large rock blocks and rock ridges mitigated the dam break process. It was speculated based on acquired geological data that the bottom elevation of the barrier break was most likely to be 720 m. The width of the break on the barrier was estimated to be 340 m based on empirical formula and geological composition of the barrier.

4. Dynamic adjustment of the design and construction scheme

(1) The selection of a preferred scheme with a backup alternative helped handle the uncertainty in emergency rescue. The construction of discharge channel was conducted mainly through excavation with machinery, supplemented with blasting in small scales. The main equipment adopted were bulldozer and backhoe excavators which operated continuously with high intensity to ensure progress of the construction. This arrangement helped speed up the disaster relief. The above mentioned two construction design strategies were of significance as a reference for similar disaster relief practice. The emergency rescue of Tangjiashan barrier lake started on May 26th and ended in early morning of June 1st, when the discharge channel was completed 3 days ahead of schedule. During the construction, CAPF Hydropower Force worked restlessly for 24 h each day, overcoming difficulties brought by rainfall, aftershocks and landslides. The earth rock quantity in excavation was 135,555 m^3, with a daily excavation quantity of 23,000 m^3, setting a new record for operation intensity of small equipment on small working planes.

(2) With navigation and land transportation both cut off, air transportation was the major external access for conveying goods. This is a critical guarantee in treating the barrier body, as had been proved. Air transportation is fast with large carrying capacity. It's recommended to be a conventional transportation choice in emergency rescue practice.

(3) Dynamic adaptation and optimization were applied during construction in disaster relief of Tangjiashan barrier body. Five times of optimization to the design of the discharge channel were made to speed up construction and save more time for further actions based on actual geological conditions and construction capacity. On-site dynamic optimization of the design by experienced engineers was a golden rule in construction of rescue works.

(4) There were safety risks in construction in earthquake-hit areas. To ensure safety of site staff, a series of guarantee measures were carried out on construction site including a complete decision making system, safety monitoring system, emergency rescue response plan, evacuation plan, etc. All these measures were taken on a people-oriented basis to ensure the disaster relief to be safe, scientific and fast. Such experience can be shared and adopted in similar projects.

5. Adaptation of new high-tech in data acquisition and establishment of alert system to support disaster relief actions

(1) Extensive use of new high-tech better ensured the fast relief of disasters. In searching and treating Tangjiashan barrier lake, technologies such as DEM data, satellite and aerial image, aerial survey to acquire topographic condition in the channel, verify the volume curve and to acquire cross sections of flood routing. The data were acquired in a more efficient way of higher accuracy.

(2) To ensure the efficiency of engineering and non-engineering measures, several systems were established including water regimen and rainfall forecasting system, remote real-time video monitoring system on the barrier body, safety monitoring system at barrier site, communication guarantee system at barrier site and dam break consultation and decision making mechanism. All these actions were necessary to make sure the disaster relief work conducted in scientific way.

(3) Wide-band wireless communication technology was adopted for monitoring of risks on site so as to offer direct visual data of the site to support scientific decision making. The technology can be further promoted for adaptation in similar projects.

(4) Meteorology and water regimen forecasting system were adopted based on both numerical weather forecasting and experience of forecasters. Consultation among various parties remotely and water regime forecasting supported the emergency treatment at Tangjiashan barrier lake through provision of timely and accurate forecast data. This is also an efficient work mechanism to handle the risks brought about by unpredictable weather conditions.

Appendix
Timeline of Key Events

At 14:28 on May 12, 2008, a magnitude-8 earthquake hit Wenchuan, Sichuan Province.

On May 14th, Tangjiashan barrier lake was found on the aerial image taken after the earthquake by the technical staff from the Institute of Remote Sensing and Digital Earth, CAS. The river section near Tangjiashan mountain was severely damaged by the earthquake, with land and water transportation both cut off.

On May 15th, a working group led by senior official from Office of State Flood Control and Drought Relief Headquarters, MWR was set up in Mianyang, Sichuan Province. Head of CWRC arrived in Mianyang in the evening to guide on earthquake relief.

During May 15th–May 18th, four groups of water resource experts were organized by MWR earthquake relief command on site to travel to Tangjiashan area overland yet failed due to road cutoff.

On May 16th, Air Force helicopter tried to send water resource experts to Tangjiashan area but failed to arrive at the barrier dam due to the weather conditions.

On May 17th, a Water Resource Team was set up under the Earthquake Relief Headquarters under the State Council.

On May 18th, water resource experts made close observation of the barrier dam for the first time with support from the Air Force. A meeting was convened late that night in Chengdu on treatment of barrier lakes, especially Tangjiashan barrier lake, formed by the earthquake. Senior officials and experts from MWR, Office of State Flood Control and Drought Relief Headquarters and CWRC attended the meeting.

On May 19th, senior officials and experts from MWR and CWRC arrived in Mianyang guiding the emergency treatment of Tangjiashan barrier lake. Soon a team of water resource experts gathered in Mianyang specifically studying and supporting the disaster relief at Tangjiashan barrier lake.

During May 20th–May 21st, the water resource experts landed on barrier dam crest by helicopter. Field investigation and water level measurement had been carried out.

On May 21st, Water Resource Team of the Earthquake Relief Headquarters (Water Resource Group) identified Tangjiashan barrier lake as Extremely-High-Risk one. With limited time, decision was made to conduct engineering measures and

Q. Yang and W. Yang, *Emergency Handling of the "5.12" Tangjiashan Barrier Lake*,
https://doi.org/10.1007/978-981-99-1706-8

non-engineering measures (evacuation of impacted residents) in parallel. CISPDR was assigned as the leading party to prepare disaster relief plan jointly with POWERCHINA Chengdu and CAPF Hydropower Force.

On May 22nd, head of MWR arrived in Mianyang leading the disaster relief work of Tangjiashan barrier lake. The Water Resources Team made a presentation on Engineering Technical Scheme on Drainage of Tangjiashan Barrier Lake. In the Evening, the then Premier Wen Jiabao made a site visit over Tangjiashan barrier body by helicopter, after which a special meeting session was held in Mianyang on emergency treatment of barrier lakes formed after earthquake, especially the Tangjiashan barrier lake. A general strategy was made on the meeting, suggesting taking active and fast actions and integrating relief with rescue. Tangjiashan Barrier Lake Emergency handling Command was established headed by provincial governor of Sichuan, with a vice minister of MWR as executive commander.

At 24:00 on May 23rd, the Water Resource Team submitted the *Technical Scheme for Design and Construction of Tangjiashan Emergency Handling Project.* A disaster relief scheme was recommended in their submission in consideration of actual construction conditions and capacities, construction period identified based on hydrological and meteorological forecast and geological composition of the barrier dam. Support from Air Force and local government on transportation of goods was also indispensable for implementation of the scheme. The recommended construction scheme suggested machinery excavation of discharge channel, which would be widened and deepened through scouring, thus enlarging discharge capacity of the channel and preventing dam burst. Taking no account of any possible rainstorm and landslides, the proposed construction period was 10 days. On May 24th, the site staff and equipment were scheduled to be mobilized to the site. The construction was planned to be commenced on May 25th and finished by June 3rd. The report analyzed three scenarios: entirely collapse, half collapse and 1/3 collapse of the barrier dam. It was demonstrated that the barrier would gradually collapse within a time duration of 1–6 h. Although the third scenario, i.e., 1/3 collapse of the barrier, was the most likely to occur, all these three scenarios should be taken into consideration in making emergency response plan for evacuation of residents downstream by three levels. Such emergency response plans would be launched immediately in accordance with actual circumstances.

On May 23rd, a 100 mm heavy rainfall during May 24th–May 26th was forecasted by relevant authorities, which would possibly leave no time for Tangjiashan barrier lake emergency handling. Evacuation of residents on downstream began.

During May 23rd–May 25th, the preparation for construction started. Personnel, equipment and goods were all gathered in Mianyang city and Leigu Town, Beichuan. Due to unfavorable weather conditions, helicopters could not take off and none of them were transported to Tangjiashan barrier dam.

In the morning of May 25th, the Earthquake Relief Headquarters held discussion on the method of blasting and discharge. The blasting work plan was assessed by the blasting expert team and got approved by the Headquarters. Late that night, 1200 officers and soldiers from PLA and CAPF Hydropower Force were mobilized to the site carrying 10t of explosives and blasting tools into Tangjiashan mountain.

On May 26th, it was sunny. Helicopters had smooth access in and out of the barrier area. The equipment were lifted and conveyed onto the barrier dam crest. Excavation with equipment was started. An on-site emergency rescue command was set up for disaster relief of Tangjiashan barrier lake, headed by a deputy director of CAPF Hydropower Force. Chief Engineer of MWR was appointed as the deputy commander and chief engineer for the on-site emergency rescue command.

During May 26th–May 31st, the water expert team prepared three documents on site, i.e., *Work Plan for Blasting of Discharge Channel, Emergency Response Plan on Barrier Crest,* and *Work Plan for Evacuation of Construction Staff, Equipment and Goods*, among which the first document was not implemented.

In the morning of June 1st, with endeavor of CAPF Hydropower Force, the construction of the discharge channel was finished ahead of schedule. The channel was 475 m long with a bottom width of 7–10 m. The bottom elevation at inlet of the channel was 740 m and that at the outlet of the channel was 739 m. Total earth rock excavation quantity was 135,500 m^3.

On June 5th, the then Premier Wen Jiabao made the second visit to Tangjiashan barrier lake. During a special meeting session, he convened to discuss disaster relief of Tangjiashan barrier lake, Premier Wen required the technical team deal with the emergency in a safe, scientific and fast manner.

On June 6th, Premier Wen made his third visit to Tangjiashan. In the afternoon, engineering measures were taken by CAPF Hydropower Force and PLA officers and soldiers, led by Deputy Chief of Joint Staff PLA, to mitigate resistance in the channel and expand water carrying capacity.

On June 7th, the on-site command finished the *Implementation Plan of Branch Discharge Channel of Tangjiashan Barrier Lake.* Water discharged from the barrier lake had flowed through the channel since 07:08.

On June 9th, the discharge channel underwent obvious retrogressively scouring. Construction of branch channel was halted at the outlet of the discharge channel.

At 11:30 a.m. on June 10th, the discharge in the channel reached its peak flow of 6500 m^3/s.

On June 11th, the temporarily evacuated 277,600 residents on downstream of the barrier dam returned home. At 14:00, the water level in the barrier lake had dropped from 743.23 m to 714.13 m, with reservoir volume reducing from 248 million m^3 to 86 million m^3 correspondingly. Through continuous scouring, the discharge channel had turned into a V-shaped valley with a length of 800 m, bottom width of 80–100 m and top width of 145–235 m. The bottom elevation of the channel was 714 m at inlet and less than 700 m at the outlet. This newly shaped river channel is capable of carrying flood with return period of 200 years.

On June 12th, CISPDR worked with POWERCHINA Chengdu in Mianyang to prepare a draft of the *Technical Assessment Report on Emergency Handling of Tangjiashan Barrier Lake.*

On June 13th, the experts team of MWR submitted to Earthquake Relief Headquarters *The Assessment Report on Emergency Handling of Tangjiashan Barrier Lake with Suggestions for Further Work.*

On June 14th, MWR convened a Wrap-up Meeting in Chengdu on emergency treatment and disaster relief of Tangjiashan barrier lake.

The success implementation of Tangjiashan barrier lake disaster relief wins praise and recognition from state leaders, under the instruction of whom the Earthquake Relief Headquarters under State Council issued a congratulatory letter praising the relief action as "a world wonder in handling large-scale barrier lakes".

The manufacturer's authorised representative in the EU is Springer Nature Customer Service Centre GmbH, Europaplatz 3, 69115 Heidelberg, Germany. If you have any concerns regarding our products, please contact ProductSafety@springernature.com

Printed and bound by CPI Group (UK) Ltd, Croydon, CR0 4YY

01/10/2026

02225855-0011